Also in the Variorum Collected Studies Series:

MAURICE CROSLAND
Scientific Institutions and Practice in France and Britain, c.1700–c.1870

MARTIN J.S. RUDWICK
Lyell and Darwin, Geologists
Studies in the Earth Sciences in the Age of Reform

MARTIN J.S. RUDWICK
The New Science of Geology
Studies in the Earth Sciences in the Age of Revolution

HUGH TORRENS
The Practice of British Geology, 1750–1850

RICHARD YEO
Science in the Public Sphere
Natural Knowledge in British Culture 1800–1860

DAVID ELLISTON ALLEN
Naturalists and Society
The Culture of Natural History in Britain, 1700–1900

ROY M. MACLEOD
The 'Creed of Science' in Victorian England

DAVID OLDROYD
Sciences of the Earth
Studies in the History of Mineralogy and Geology

WILLIAM H. BROCK
Science for All
Studies in the History of Victorian Science and Education

MAURICE CROSLAND
Studies in the Culture of Science in France and Britain since the Enlightenment

ROBERT FOX
Science, Industry and the Social Order in Post-Revolutionary France

ROBERT FOX
The Culture of Science in France, 1700–1900

VARIORUM COLLECTED STUDIES SERIES

The Earth Sciences
in the Enlightenment

Kenneth L. Taylor (The *Archives Nationales* user card dates from his doctoral-research year in Paris, 1965–6.)

Kenneth L. Taylor

The Earth Sciences in the Enlightenment

Studies on the Early Development of Geology

This edition © 2008 by Kenneth L. Taylor

Kenneth L. Taylor has asserted his moral right under the Copyright, Designs and Patents Act, 1988, to be identified as the author of this work.

Published in the Variorum Collected Studies Series by

Ashgate Publishing Limited
Gower House, Croft Road,
Aldershot, Hampshire
GU11 3HR
Great Britain

Ashgate Publishing Company
Suite 420
101 Cherry Street
Burlington, VT 05401–4405
USA

Ashgate website: http://www.ashgate.com

ISBN 978–0–7546–5930–3

British Library Cataloguing in Publication Data
Taylor, Kenneth L.
  The earth sciences in the Enlightenment: studies on the
  early development of geology. – (Variorum collected studies series)
  1. Earth sciences – History – 18th century 2. Geology – France – History –
  18th century 3. Geologists – France – Biography 4. Geologists – France –
  History – 18th century
  I. Title
  550.9'033
  ISBN 978–0–7546–5930–3

Library of Congress Cataloging-in-Publication Data
Taylor, Kenneth L.
  The earth sciences in the Enlightenment: studies on the early development of
  geology / Kenneth L. Taylor.
        p. cm. – (Variorum collected studies series)
    Includes bibliographical references.
    ISBN 978–0–7546–5930–3 (alk. paper)
    1. Earth sciences – History – 18th century. 2. Geology – France – History – 18th
  century. 3. Geologists – France – Biography. 4. Geologists – France – History
  – 18th century.
    I. Title.

QE13.F8T39 2007
550.9'033–dc22   2007022385

The paper used in this publication meets the minimum requirements of the American National Standard for Information Sciences – Permanence of Paper for Printed Library Materials, ANSI Z39.48–1984. ∞ ™

Printed by TJ International Ltd, Padstow, Cornwall

VARIORUM COLLECTED STUDIES SERIES CS883

For Mike

CONTENTS

Preface   xi–xv

Notes on the articles   xvii–xxi

Bibliography   xxiii–xxiv

Acknowledgements   xxv–xxvi

A KEY FIGURE IN ENLIGHTENMENT GEOLOGY: NICOLAS DESMAREST

I   Nicolas Desmarest and geology in the eighteenth century   339–356
Toward a History of Geology, ed. C.J. Schneer. Cambridge, MA: The M.I.T. Press, 1969

II   The beginnings of a geological naturalist: Desmarest, the printed word, and nature   1–24
Earth Sciences History 20. Morgantown, WV, 2001, pp. 44–61. Originally published in French as "La Genèse d'un Naturaliste: Desmarest, la Lecture et la Nature", De la Géologie à son Histoire, Ouvrage Édité en Hommage à François Ellenberger, eds. G. Gohau and J. Gaudant. Paris: Comité des Travaux Historiques et Scientifiques, 1997, pp. 61–74

III   Nicolas Desmarest and Italian geology   1–18
Rocks, Fossils and History, eds. G. Giglia, C. Maccagni and N. Morello. Florence: Edizioni Festina Lente, 1995, pp. 95–109

IV   New light on geological mapping in Auvergne during the eighteenth century: the Pasumot-Desmarest collaboration   129–136
Revue d'Histoire des Sciences 47. Paris, 1994

V	Buffon, Desmarest and the ordering of geological events in *Époques* *The Age of the Earth: From 4004 BC to AD 2002*, eds. C.L.E. Lewis and S.J. Knell. London: Geological Society, 2001, pp. 39–49	1–20

INVESTIGATING AND UNDERSTANDING TERRESTRIAL PHENOMENA

VI	Natural law in eighteenth-century geology: the case of Louis Bourguet *XIIIth International Congress of the History of Science, Proceedings VIII.* Moscow: Editions "Nauka", 1974, pp. 72–80	1–7
VII	Reflections on natural laws in eighteenth-century geology *Originally published in French as "Les Lois naturelles dans la géologie du XVIIIème siècle: Recherches préliminaires," Travaux du Comité Français d'Histoire de la Géologie, 3ème série, vol. II, no. 1.* Paris, 1988, pp. 1–28	1–36
VIII	The historical rehabilitation of theories of the earth *The Compass 69.* Norman, OK, 1992, pp. 334–345	1–17
IX	Volcanoes as accidents: how 'natural' were volcanoes to 18th-century naturalists? *Volcanoes and History*, ed. N. Morello. Genoa: Brigati, 1998	595–618
X	Two ways of imagining the earth at the close of the 18th century: descriptive and theoretical traditions in early geology *Abraham Gottlob Werner and the Foundation of the Geological Sciences: Selected Papers of the International Werner Symposium in Freiberg 19th to 24th September 1999*, eds. H. Albrecht and R. Ladwig. Freiberg: Technische Universität Bergakademie Freiberg, 2002	369–378

THE EMERGING NEW GEOLOGICAL SCIENCE

XI	Early geoscience mapping, 1700–1830 *Proceedings of the Geoscience Information Society 15.* Alexandria, VA, 1985, pp. 15–49	1–44

	CONTENTS	ix

XII Geology in 1776: some notes on the character of an incipient science 75–90
Two Hundred Years of Geology in America: Proceedings of the New Hampshire Bicentennial Conference on the History of Geology, ed. C.J. Schneer. Hanover, NH: The University Press of New England, 1979

XIII The beginnings of a French geological identity 65–82
Histoire et Nature 19–20. Paris, 1981–1982

XIV The *Époques de la Nature* and geology during Buffon's later years 371–385
Buffon 88: Actes du Colloque International pour le Bicentenaire de la Mort de Buffon, ed. J. Gayon. Paris: Librairie Philosophique Vrin, 1992

XV Earth and heaven, 1750–1800: Enlightenment ideas about the relevance to geology of extraterrestrial operations and events 1–15
Earth Sciences History 17. Morgantown, WV, 1998, pp. 84–91

Index 1–6

This volume contains xxvi + 300 pages

PUBLISHER'S NOTE

Eight of the fifteen essays published here have been reset for this volume. In most instances this is for reasons of legibility although in the case of essay VII because it is published here in English for the first time.

The articles in this volume, as in all others in the Variorum Collected Studies Series, have not been given a new, continuous pagination. In order to avoid confusion, and to facilitate their use where these same studies have been referred to elsewhere, the original pagination has been maintained wherever possible.

Each article has been given a Roman number in order of appearance, as listed in the Contents. This number is repeated on each page and is quoted in the index entries.

PREFACE

The fifteen essays in this volume represent more than three decades of research on the formative stages of geological science, with emphasis on some of the French participants in that historical process. The chronological scope of the topics treated in these papers occasionally reaches back into the seventeenth century, and ranges forward into the first few decades of the nineteenth. The main focus, however, is the eighteenth century, and especially the period 1750–1800. During this half-century, geological problems began to be addressed with unprecedented intensity and originality, and a distinct science of geology started to take shape and acquire recognition within the scientific world. I hope these papers contribute a little to our historical understanding of these developments.

I make no apology for having concentrated especially on the French scene. Geology's historical fashioning was a conspicuously international venture, with a wide and often distinguished cast of characters identifiable as British, German, Italian, Swedish, Swiss, Russian, Belgian, Dutch, Transylvanian, and so on. If a case needs to be made that particular attention should be given to the French-speaking characters, it should suffice to recall the historical reality of *lingua franca*: the French language had no rival then as the international medium of intellectual exchange. Furthermore, during this period France's general supremacy in the natural sciences was widely acknowledged. That some of early geology's key episodes had French connections can hardly be disputed; but I trust that these papers display no agenda to detract from the significance of work done in other lands and expressed in other languages. In any case I see no profit in contests over national leadership in geology's founding and early growth (nor do I think that geology is a science with any single identifiable 'father' or 'founder'; there must be few sciences whose multiple parentage is more evident). If the studies reproduced here serve to elevate the scientific reputation of one historical figure or another – or possibly, in certain cases, to diminish it – I would say this is not necessarily a bad thing; but rectification of scientific reputations has not been basic to my purpose.

My research on early French geology began with my doctoral thesis, completed in 1968, on Nicolas Desmarest (1725–1815). The outcome of this research was, it seemed to me, more illuminating around the edges than in central parts of the existing historical picture of this fascinating character, as drawn by such authors as Archibald Geikie and Alfred Lacroix. What I found out about Desmarest did nothing to call into question his significance as a geological innovator. His reputation as a perceptive observer in the field was, and is, richly deserved. His proposals for chronological interpretation of the Auvergne landscapes were indeed novel and influential. However, I did learn some unexpected things, and drew some modestly revisionist conclusions, some of which had historically interesting consequences. The breadth of Desmarest's geological interests surprised me a little. Perhaps my most intriguing discovery was that Desmarest's volcanological investigations never brought him to embrace a comprehensively 'vulcanist' doctrine, indeed he was in certain ways of a resolutely 'neptunist' persuasion. This observation contradicted parts of the standard story-line, in which advocates of those two views were supposed to be clearly differentiated. It also emerged that the seeds of Desmarest's geological career had an even more distant resemblance to modern sorts of geological experience than I had been led to suppose, whereas he had unexpectedly strong affinities with antiquarian scholars. It became apparent that I would need to learn much more about the broader scientific culture in which Desmarest lived, and the community of savants with whom he interacted. My effort to understand the world and work of Desmarest – who, incidentally, was also an important figure in technological and industrial developments of his period – widened into an inquiry into the ideas and circumstances guiding the early growth of geological thought and investigation during his lifetime.

This volume's organizational scheme may be seen as a loose recapitulation of my own experience, or as following a sort of inductive logic. It starts with papers on Desmarest's individual case, then moves to a set of studies that try to recapture some of the ways writers in eighteenth-century geology thought geological inquiry should be carried out. The essays in the last section treat aspects of geology's maturation process, and some features characterizing the emergent science near the close of the eighteenth century.

In the first of the three sections, three essays focus on Desmarest alone, while two others observe him in his historical partnership with a cartographic collaborator (Pasumot, in article IV) and with an iconic senior colleague who was both an intellectual model and a rival (Buffon, in article V). These studies describe (and, I would like to think, help to explain) Desmarest's pathway into scientific study of the earth, and his accomplishments as a geological investigator and thinker. I believe they are also suggestive of the difficulties confronted by those who, in common with Desmarest, were then attempting both to define the problems that a science of the earth should address, and to fashion the best ways for finding proper answers to those problems.

In the second part, the middle five essays are explorations in the thought-world of eighteenth-century savants who were trying to devise satisfactory ways to 'do' geology. Actually, a point of some consequence is to realize that their intentions were often far from congruent with those of later geologists, working when the science had become firmly established. However, a feature of fledgling geology that has endured across time, I think, was the challenge of reconciling steps taken to reach two legitimate but distinct goals: to acquire detailed knowledge of specific local circumstances, and to gain satisfactory generalized knowledge through universal explanatory principles. In its youthful stages just as later on, the nascent geological science exhibited tendencies both to address the complexities of individual cases and to strive for the simplicity of the general case.

Two intertwined themes are prominent in this middle cluster of studies. One concerns how far prevalent traditions of Theories of the Earth – and reactions against those traditions – governed geological opinion during the eighteenth century. The second pertains to what I perceive as a struggle, on the part of early geological thinkers, to adapt elements of existing explanatory models to their needs. These papers are largely about how the two classical ways of formulating scientific knowledge – natural philosophy with its habitual emphasis on general laws, and natural history with its emphasis on description and classification – provided many, but not all, of the theoretical ingredients that were sought. For the historical or developmental dimension that was manifestly required (at least, so it seems in retrospect), but was not sanctioned by the essentially static world vision represented by either natural philosophy or natural history, early geologists also borrowed from a third form of understanding, historical and antiquarian scholarship. These papers offer support for the idea that as they grappled with the problem of finding a suitable methodology for their scientific work, early geologists sought answers simultaneously for questions of definition, investigative procedure, and explanatory form: that is, about how to identify the objects belonging rightfully to their study, how to assemble information about them, and how to account for them.

My studies emphasize in particular a sustained influence on at least some eighteenth-century geological inquirers, deriving from the prestigious status of general laws in natural philosophy, of the ambition to establish natural laws specific to geological science through identification of dispositional regularities (in articles VI, VII, VIII, and X). Article IX discusses some of the historical consequences of a contemporary view, concomitant with privileging the general case, that only that which is 'normal' or 'usual' in nature counts as 'natural,' so that 'accidental' phenomena were often seen as marginal in importance.

In the third section, the last five studies are concerned with geology's formation as a new science, and with a growing sensibility among geological figures regarding a commonality of scientific purpose and method. This transformation is approached in one paper through an overview of mapping's development as a conceptual and communicative tool in the geosciences (article XI). A case is made in article XII for seeing geology's historical emergence as a hybridizing process, a reconfiguration in which elements from existing sciences were reshuffled. Article XIII is an attempt to show how a shared sense of disciplinary identity arose among the French in the later part of the eighteenth century. Through a discussion of how Buffon's *Époques de la Nature* (1778) was viewed by active geological investigators in the closing decade of his life, some of the maturing science's comparatively restricted objectives and investigatory conventions are outlined (article XIV). Finally, an argument is made that, in geology's founding stages, the exigencies of disciplinary cohesiveness played a role in defining which sorts of natural processes properly belong to the science: some early geologists saw advantages to their science's standing as a distinct discipline, in marking out the boundaries of their emerging science so as to minimize or even exclude extraterrestrial operations and events (article XV).

If there is some specific orientation or predisposition underlying these fifteen essays, it may be my conviction that it is worthwhile trying to examine basic elements of the conceptual worlds of my subjects, the historical characters whose ideas and deeds helped establish the new geological science. This conviction is expressed beautifully in the opening lines of L.P. Hartley's novel *The Go-Between*: "The past is a foreign country: they do things differently there". As a teacher for almost four decades, I have urged this idea on my students. As a student myself, I have made this idea one of my main historical axioms. Thus, without advocating neglect of historical continuities, I admit to seeing historical differences and discontinuities as particularly important. The studies reproduced here exemplify that emphasis.

It has never been difficult to see my subjects as foreign. They are, both geographically and chronologically. To figure out how these foreigners operate it is important (and, usually, fun!) to try to identify the beliefs and rules underlying their conduct. That these foreign friends – well, they may not always seem entirely friendly, but it is good policy to approach them sympathetically – are often silent about such beliefs and rules is to be expected. The relevant precepts frequently lurk partly or perhaps wholly beneath the horizon of their consciousness. It is the historian's job to bring them to light. This involves both detective work and exercise of the imagination, and when successful it yields one of the historian's greatest pleasures: getting partway into the mind of a distant figure, making the past a little bit more intelligible. Thus, one of my own rules: *Cherchez l'opinion préconçue* – seek out the hidden preconceptions standing behind what is puzzling or obscure about the ways your subjects think. In many of the studies reproduced here, the reader should be able to recognize my efforts to apply this rule.

Many of these papers were written for symposia or conferences where historians of science and scientist-historians came together to address common historical interests and problems. I trust that the research reported upon in this volume reflects, first and foremost, my own academic background and identity in the history of science (I had no advanced training in science itself). But I would be surprised if readers see no signs whatever of my associations with geologist-historians. Interaction between historically-trained scholars and scientists with active interests in geological science has been a common occurrence during recent decades, probably more so than in most subfields of the history of science. One reason this has happened may be that without the participation of both groups, a community of historians of geology would fall short of critical mass. The existence for more than twenty years of the History of the Earth Sciences Society – with its diversified membership – and of its journal *Earth Sciences History*, exemplifies the phenomenon. Another reason is the healthy perception of mutual advantage: each group within the community offers something of value to the other. One of the ways my own understanding has profited has been in seeing historically-significant geological localities in the company of trained geologists. It does pay, I believe, to have acquaintance with the ground of the action, the places where one's historical characters worked. The International Commission on the History of Geological Sciences (INHIGEO) currently follows a policy of scheduling its annual meetings in various countries, in part to create opportunities for seeing geological sites of historical interest. Similar goals are often in play at meetings organized by the historical branches within organizations like the Geological Society of America (History of Geology Division) and the Geological Society of London (History of Geology Group). The Comité Français d'Histoire de la Géologie (COFRHIGÉO), while it does not usually arrange field trips, also maintains a resolutely multi-disciplinary posture. The fact that I am a member of all five of the above-mentioned organizations, and have served in four of them as an officer, illustrates my support for this sort of cross-disciplinary cooperation and learning.

KENNETH L. TAYLOR

Norman, Oklahoma
March 2007

NOTES ON THE ARTICLES

Just over half of these papers (eight of fifteen) have been re-formatted for this volume. In the majority of cases this is for reasons of legibility: re-formatting of articles that were produced originally in large or two-column format makes them easier to read. One article that existed only in French (VII) is presented here for the first time in English. A major factor in deciding to re-format two of the articles was the inordinately large number of typographic errors in their original printings (III, VI). In accord with custom for volumes in this series, most abstracts and summaries are omitted. Studies that included abstracts or summaries in their original published form are: II, V, VIII, IX, X, XI, and XV.

Illustrations appear in four of the articles (V, VII, VIII, and XI). The two figures in article VII are new (the *Travaux* of COFRHIGÉO, in which this essay originally appeared, became equipped to carry illustrations only at a later time). All the previously-published illustrations (those in articles V, VIII, and XI) have been redone in digital format for this volume. This appreciably improves the image quality for article XI in particular. In that article on early geoscience mapping, in several illustrations the portions of maps chosen for reproduction are altered. Four figures are in now color. Some of the captions are expanded to supply additional pertinent information. Since the same portrait of Desmarest in academic garb was a component of two of the articles (V, VIII), it has been omitted from the second of these.

***

"Nicolas Desmarest and geology in the eighteenth century" (I), the first article I ever wrote up for publication, was prepared as I was nearing completion of my doctoral thesis. The occasion was the first New Hampshire conference on the history of geology, held in September 1967, and the results appeared in the volume edited by the conference organizer, Cecil Schneer: *Toward a History of Geology* (1969). The conference itself was unforgettable for its setting, an old seaside inn at Rye Beach, and unusual in its organizational format, through which participants were able to read one another's papers in advance and spend session time discussing them. "The Beginnings of a geological naturalist: Desmarest, the printed word, and nature" (II) was originally prepared for the *livre jubilaire* to celebrate the 80th birthday of François Ellenberger, the distinguished French geologist and irrepressible enthusiast for his science's history. François was the founder and leader of the Comité Français d'Histoire de la Géologie (COFRHIGÉO). I composed this essay initially in English, and my colleague Jean-Marc Kehres endured numerous discussions with me as he translated it into French for publication in *De la géologie à son histoire* (edited by Gabriel Gohau and Jean Gaudant) in 1997. The English version appeared in *Earth Sciences History* in 2001. "Nicolas Desmarest and Italian geology" (III) was presented in Pisa, as part of a splendid 'traveling' meeting of the International Commission on the History of Geological Sciences (INHIGEO) held in autumn 1987. Sessions were held in Pisa and then Padua, with most participants enjoying the prolonged mid-conference historical trek from Pisa to Padua by bus, with stops at (among other places) Carrara, the Apuane Alps, Scandiano, and the Euganean Hills.

"New light on geological mapping in Auvergne in the eighteenth century: the Pasumot-Desmarest collaboration" (IV) originated as a plenary lecture for the 12th International Conference on the History of Cartography, held in 1987 in Paris. This invitation arose out of François Ellenberger's recommendation to one of the conference organizers, Monique Pelletier, of the *Département des cartes et plans* of the French National Library. At the conference, I was fascinated to see that historians of cartography were experiencing their own version of the 'internalist-externalist' debates that historians of science were then finally putting behind them. It was instructive, for example, to hear some of the field's leaders insist on knowledge of the cartographic objects themselves, and of their construction, as the true and essential purposes of scholarship in the area; whereas a group of (mainly younger) reforming participants tried to gain a hearing for the idea that historians of cartography should also investigate the ways maps have been *used*.

"Buffon, Desmarest, and the ordering of geological events in *Époques*" (V) resulted from an invitation to participate in the 2000 Millennium Meeting of the Geological Society of London. The program for this meeting, called *Celebrating the Age of the Earth*, featured a blend of historical and contemporary scientific presentations. Cherry Lewis organized the meeting, in collaboration with her colleagues in the Society's History of Geology Group. Cherry and Simon Knell edited the volume *The Age of the Earth*, a Geological Society Special Publication.

"Natural Law in eighteenth-century geology" (VI) was presented at the International Congress of the History of Science, held in Moscow in summer 1971. (If any hyper-attentive reader should notice that the paper as reproduced in this volume closes with a sentence not found in the original publication, I can only speculate that a possible reason the Russian editors omitted that sentence from the Congress Proceedings volume was an economic one: it kept the article from requiring one additional page.) I later expanded on this paper's fundamental idea in article VII, a presentation originally delivered in French at a meeting of COFRHIGÉO in Paris, in February 1988. It is published here for the first time in English. Two illustrations not included in the 1988 *Travaux* are added. In this English version, I have not attempted to eliminate features that reflect its oral delivery. (These include personal mention of COFRHIGÉO members present, notably François Ellenberger, Jacques Roger, and Gabriel Gohau, as well as conspicuous mention of the book that Rachel Laudan had published recently. At the Comité's invitation I presented a review-discussion of this important book at one of its later meetings.) Of the eight re-formatted articles in this volume, this is the only one in which the notes are presented as end-notes, rather than footnotes. This seemed advisable in view of the unusual size of some of the notes.

"The historical rehabilitation of theories of the earth" (VIII) was written at the invitation of Dan Merriam, who edited a special historical issue of *The Compass*, the journal of the American national honorary society for the earth sciences, Sigma Gamma Epsilon. This was intended as one of a cluster of articles written to demonstrate to geology students that historical perspectives on science can be valuable and interesting.

"Volcanoes as accidents: how 'natural' were volcanoes to 18th-century naturalists ?" (IX) is the expanded version of a paper presented in 1995 at another memorable INHIGEO meeting. This one began in Naples, travelled to Vulcano in the Aeolian Islands, and ended at Catania in Sicily. The setting for my session was the Osservatorio Vesuviano, situated spectacularly on the flank of the great volcano. "Two ways of imagining the earth at the close of the 18th century: descriptive and theoretical traditions in early geology" (X) was presented at the international meeting held at Freiberg in 1999, to commemorate the 250th birthday of A.G. Werner. This was also an INHIGEO symposium.

"Early geoscience mapping, 1700–1830" (XI) was first presented to an audience of geoscience information professionals. It was written for a symposium on "Maps in the Geoscience Community", organized by the Geoscience Information Society at the 1984 annual meeting of the Geological Society of America in Reno. My Oklahoma colleague Claren M. Kidd, the Geology Librarian, was President of the GIS, and thus had responsibility for organizing the symposium. Preparing this talk, I was motivated to look more broadly and with fresh eyes at the cartographic components of our History of Science Collections. (On one opinion expressed in this article, to the effect that one finds few significant connections between theories of the earth and local geoscience observation, I would soon change my mind: see in particular articles VII and VIII.)

"Geology in 1776: some notes on the character of an incipient science" (XII) was presented at the second major conference on the history of geology organized at the University of New Hampshire by Cecil Schneer, in 1976. This was an American bicentennial event, and thus had a deliberately American focus. Cecil and most of the other participants kindly tolerated my paper notwithstanding its questionable relevance. As at the 1967 New Hampshire conference, interchange among participants was encouraged by emphasis on discussion of pre-circulated papers.

"The beginnings of a French geological identity" (XIII) is a paper prepared for delivery in a symposium organized by François Ellenberger, with the cooperation of Reijer Hooykaas and Jacques Roger, on the development of francophone geology in its international relations to the time of Cuvier. This was an INHIGEO event, within the framework of the 26th International Geological Congress held in Paris in 1980. (To the best of my recollection, this was the first time I participated in an INHIGEO meeting.) Before the Congress, François Ellenberger (in collaboration with Georgette Legée) led a memorable historical field trip of some ten days duration through France. The field guide-book published for that trip can still be recommended, for those wishing to gain historical insight by journeying to key French localities.

"The *Époques de la Nature* and geology during Buffon's later years" (XIV) was written for presentation at a conference organized to observe the 200th anniversary of the death of Buffon, held in June 1988 in Paris and then at Montbard and Dijon. The volume of conference papers, edited by Jean Gayon, remains an important contribution to research on Buffon. "Earth and Heaven, 1750–1800: Enlightenment ideas about the relevance to geology of extraterrestrial operations and events" (XV) was presented at the annual meeting of the Geological Society of America held in Seattle in 1994. Joanne Bourgeois and Mott Greene organized a session with historical papers on ideas about the influence of extraterrestrial forces in Earth history.

***

I should like to mention, in closing, a few articles not included in this volume, where I have recently addressed closely related topics. A recently-published encyclopedia of early modern Europe includes two synoptic articles, on Geology and on Theories of the Earth, which I wrote in collaboration with my former student and present colleague, Kerry V. Magruder. An examination and transcription of an unpublished critique, by Desmarest, of William Hamilton's *Campi Phlegraei* (1776) was published (in French) in the *Travaux* of COFRHIGÉO. My inquiry into the ideas that motivated É-C. Marivetz and L-J. Goussier, in claiming that their *Physique du Monde* (1780–1787) would derive a comprehensive theory of physical geography from general physical principles, appeared recently in the journal *Centaurus*. An investigation of geological travel in Auvergne during the second half of the eighteenth century is scheduled to appear at about the same time as publication of this volume.

BIBLIOGRAPHY

Albrecht, Helmuth, and Roland Ladwig (eds.). 2002. *Abraham Gottlob Werner and the Foundation of the Geological Sciences: Selected Papers of the International Werner Symposium in Freiberg, 19th to 24th September 1999.* Freiberger Forschungshefte, vol. D207. Freiberg: Technische Universität Bergakademie Freiberg. (2nd ed., 2003.)

Ellenberger, François (ed.). 1981–1982. *Le développement de la géologie de langue française dans ses relations internationales des origines à la mort de Cuvier (1832).* [Actes, Symposium in history of geology, 26th International Geological Congress, Paris, 1980.] *Histoire et Nature: Cahiers de l'Association pour l'histoire des Sciences de la Nature*, nos. 19–20.

Ellenberger, François, and Georgette Legée. 1980. *Histoire de la géologie française. "Aux sources de la géologie française" – Guide de voyage à l'usage de l'historien des Sciences de la Terre.* 26e Congrès Géologique International, Paris. Livret-Guide G–30, Excursion 138–A.

Gayon, Jean (ed.). 1992. *Buffon 88. Actes du Colloque international pour le bicentenaire de la mort de Buffon (Paris, Montbard, Dijon, 14–22 juin 1988).* Paris: Librairie Philosophique Vrin.

Giglia, Gaetano, Carlo Maccagni and Nicoletta Morello (eds.). 1995. *Rocks, Fossils and History. Proceedings of the 13th INHIGEO Symposium, Pisa – Padova (Italy), 24 September – 1 October 1987.* Florence: Edizioni Festina Lente.

Gohau, Gabriel, and Jean Gaudant. 1997. *De la géologie à son histoire. Ouvrage édité en hommage à François Ellenberger.* Paris: Comité des Travaux Historiques et Scientifiques.

Laudan, Rachel. 1987. *From Mineralogy to Geology: The Foundations of a Science, 1650–1830.* Chicago and London: University of Chicago Press.

Lewis, C. L. E., and S. J. Knell (eds.). 2001. *The Age of the Earth: From 4004 BC to AD 2002.* London: The Geological Society [Special Publication 190].

Magruder, Kerry V., and Kenneth L. Taylor. 2004. "Earth, Theories of the," in *Europe 1450 to 1789: Encyclopedia of the Early Modern World*, ed. by Jonathan Dewald. New York: Charles Scribner's Sons, vol. 2, pp. 222–226.

Morello, Nicoletta (ed.). 1998. *Volcanoes and History. Proceedings of the 20th INHIGEO Symposium, Napoli – Eolie – Catania, 19–25 September 1995.* Genoa: Brigati.

Schneer, Cecil J. (ed.). 1969. *Toward a History of Geology. Proceedings of the New Hampshire Inter-Disciplinary Conference on the History of Geology, September 7–12, 1967.* Cambridge, Massachusetts, and London: M.I.T. Press.

Schneer, Cecil J. (ed.). 1979. *Two Hundred Years of Geology in America: Proceedings of the New Hampshire Bicentennial Conference on the History of Geology.* Hanover, New Hampshire: The University Press of New England.

Taylor, Kenneth L. 1968. Nicolas Desmarest (1725–1815): Scientist and Industrial Technologist. Ph.D. thesis, Harvard University, Cambridge, Massachusetts.

—. 1988. "Sur les origines minéralogiques de la géologie: une nouvelle analyse, par Rachel Laudan 1987, des fondements de la science géologique," *Travaux du Comité Français d'Histoire de la Géologie*, ser. 3, vol. 2, pp. 61–65.

—. 2001. "Un commentaire anonyme inédit sur les observations et les idées de William Hamilton (1730–1803) relatives aux phénomènes volcaniques de la région de Naples," *Travaux du Comité Français d'Histoire de la Géologie*, ser. 3, vol. 15, pp. 1–35. (An article based on this study, 'La volcanologie au XVIIIe siècle', appeared in *Pour la science*, no. 286, August 2001, pp. 8–10.)

—. 2006. "Marivetz, Goussier, and Planet Earth: A Late Enlightenment Geo-Physical Project," *Centaurus*, vol. 48, pp. 258–283.

—. [forthcoming in 2007]. "Geological Travellers in Auvergne, 1751–1800," in *Four Centuries of Geological Travel*, edited by Patrick Wyse Jackson. London: The Geological Society [Special Publication 287], pp. 73–96.

Taylor, Kenneth L., and Kerry V. Magruder. 2004. "Geology," in *Europe 1450 to 1789: Encyclopedia of the Early Modern World*, ed. by Jonathan Dewald. New York: Charles Scribner's Sons, vol. 3, pp. 39–42.

ACKNOWLEDGEMENTS

Grateful acknowledgement is made to the following persons, journals, institutions and publishers for their kind permission to reproduce the essays included in this volume: The M.I.T. Press, Cambridge, MA (for essay I); The Comité des Travaux Historiques et Scientifiques, Paris and Patrick Wyse Jackson, Editor, *Earth Sciences History*, Dublin (II); Paolo Gori Savellini, Edizioni Festina Lente, Florence (III); Michel Blay, Editor, *Revue d'Histoire des Sciences*, Paris (IV); The Geological Society [London] Publishing House, Bath (V); A. Mozdakov, Nauka Publishers, Moscow (VI); Gabriel Gohau on behalf of the Comité Français d'Histoire de la Géologie, Paris (VII); Donald W. Neal, Editor, *The Compass*, on behalf of Sigma Gamma Epsilon (VIII); Brigati Glauco, Genoa (IX); Technische Universität Bergakademie, Freiberg (X); Patricia Yocum, President of the Geocience Information Society, on behalf of the *GSIS Proceedings* and The Geological Society of America, Boulder, CO (XI); The University Press of New England, Hanover, NH (XII); Librairie Philosophique Vrin, Paris (XIV); Patrick Wyse Jackson, Editor, *Earth Sciences History*, Dublin (XV).

I owe thanks to Trevor Levere for the initial suggestion resulting in this volume; to Martin Rudwick for pressing the idea forward; and to John Smedley for seeing it done.

For help in preparing the re-formatted texts for this volume, I am grateful to Linda Baker, Carol Clayton, and Colleen Wilson. For valuable assistance with the illustrations I thank Melissa Rickman, Mark Hopkins, and Julia Daine. Most especially I am grateful to Kerry Magruder, for his masterful help in reconstituting the illustrations.

I appreciate the generous support provided by the College of Arts and Sciences and the Vice President for Research, University of Oklahoma, which makes possible the set of four color plates in article XI.

My studies would not have been possible without opportunities to pursue research at a number of great libraries and archives. Chief among these in my own country are the libraries of Harvard University, and of the University of Oklahoma, where the History of Science Collections have been my treasured resource for four decades. In Paris, I have depended especially on the resources of the Bibliothèque nationale de France, the Bibliothèque centrale du Muséum national d'histoire naturelle, the Archives nationales, and the Archives de l'Académie des Sciences. I am grateful to the staffs of these and the many other institutions whose holdings I have been privileged to use.

In several of this volume's papers, the reader will see acknowledgements representing only a small fraction of the debts of gratitude I have incurred in my scholarly work. Over a long period, a great many teachers, colleagues, and friends have given me the benefit of their instruction, advice, criticism, and encouragement, in addition to the example of their own work. It is of course impossible to thank them all by name in this space. Invidious though it is to select only a few for explicit thanks, thus committing the injustice of omitting a great many others, still I wish to make particular mention of the following: Ken Bork, Claudine Cohen, François Ellenberger†, Jean Gaudant, Gabriel Gohau, David Kitts, Kerry Magruder, Alex Ospovat, Rhoda Rappaport, Jacques Roger†, Martin Rudwick, Cecil Schneer, René Taton†, and Hugh Torrens.

I

Nicolas Desmarest and Geology in the Eighteenth Century

Nicolas Desmarest is often remembered as the scientist who first stated the volcanic origin of basalt, and as the worthy successor of Guettard in studying the volcanoes of Auvergne. In connection with his volcanic researches, he has also been credited with an important discussion of the development of landforms through systematic degradation of geological formations, and he has even been named as an early exponent of the uniformitarian principle in geology. Moreover, Desmarest's geological thought has sometimes been regarded as one of the origins of a vulcanist school that in time clashed with Wernerian neptunism in one of the more important controversies in the history of geology. This paper offers a brief survey of Desmarest's geological work, with a view toward determining whether this summary accurately describes his significance in the history of geology.[1]

---

[1] The most readily available account of Desmarest's life and work is in Sir Archibald Geikie, *The Founders of Geology* (2nd ed.; London and New York: Macmillan, 1905; reprint, New York: Dover, 1962), pp. 140–175. It is based to a great extent on Georges Cuvier's eulogy, delivered before the Academy of Sciences in 1818: "Eloge historique de Nicolas Desmarets," *Recueil des éloges historiques lus dans les séances publiques de l'Institut Royal de France* (3 vols.; Strasbourg and Paris: F. G. Levrault, 1819–1827), Vol. II, pp. 339–374. Cuvier's biographical sketch is, in turn, highly dependent on a manuscript outline prepared by Desmarest's son, Anselme-Gaëtan (1784–1838), "Notes et renseignements sur la vie et les ouvrages de mon père," now in the Bibliothèque de l'Institut de France, Fonds Cuvier, MS 3199. Another *éloge*, given by Augustin-François Silvestre to the National Society of Agriculture (of which Desmarest was a member), deals less with geology than with Desmarest's career in industrial technology: "Notices biographiques sur MM. Journu-Auber, Cotte, Allaire, Desmarets et Tenon, membres de la Société Royale et Centrale d'Agriculture de Paris: Lues à la séance publique de la Société, le 28 avril 1816," *Mémoires d'agriculture, d'économie rurale et domestique, publiés par la Société Royale et Centrale d'Agriculture* (1816), pp. 80–123. The many biographical notes and articles that have appeared in various dictionaries and other compilations add little to Cuvier's and Silvestre's versions of Desmarest's life. The only significant addition to the biographical literature, aside from Geikie's, is by Alfred Lacroix, "Nicolas Desmarest," *Figures de savants* (4 vols.; Paris: Gauthier-Villars, 1932–1938), Vol. I, pp. 7–18.

The study of volcanoes in France began with the dramatic realization by Guettard and Malesherbes that extinct volcanoes existed in Auvergne. According to Desmarest's account, his own discovery of basalt among the volcanic rocks of Auvergne had an impact on him just as decisive as the first recognition of dark Volvic rock had had on his predecessors. Desmarest was first in Auvergne in 1763 on one of his tours to gather industrial information for Daniel-Charles Trudaine, the Director of Commerce; he made his observations during a natural history excursion. This discovery so intrigued him, he relates, that he was determined to investigate the matter thoroughly.[2] It is a fact that his scientific career was significantly altered by this episode. His belief that the Auvergne basalt was a volcanic rock gave him a taste of a kind of scientific achievement he had not known before, and it no doubt played a role in his election to the Royal Academy of Sciences in 1771, after more than a decade of frustration.[3] Behind Desmarest's insistence that the sudden discovery of basalt in an unexpected context was decisive for his career is the suggestion that before this incident he had had no special interest in or knowledge of volcanic geology.

Desmarest had done scientific work that may be called geological since his mid-twenties, yet none of it bears any marks of a genuine interest in volcanic phenomena. His first scientific paper, an essay on the supposed former land connection between England and France

[2] Desmarest, "Mémoire sur l'origine & la nature du basalte à grandes colonnes polygones, déterminées par l'histoire naturelle de cette pierre, observée en Auvergne," *Mémoires de mathématique et de physique, tirés des registres de l'Académie Royale des Sciences*, in *Histoire de l'Académie Royale des Sciences. Avec les mémoires de mathématique & de physique, pour la même année. Tirés des registres de cette académie, 1771* (1774), 706–708. Hereafter, this journal is cited as *Mémoires de l'Académie Royale des Sciences*. Much later than this, Desmarest reflected on the importance of his earliest studies in Auvergne in "Mémoire sur les prismes qui se trouvent dans les couches horizontales de plâtre et de marnes des environs de Paris, et sur leur analogie avec les prismes du basalte," *Mémoires de la classe des sciences mathématiques et physiques*, in *Mémoires de l'Institut National des Sciences et Arts. Sciences mathématiques et physiques*, IV (an XI [1803]), 219–220. Hereafter, this journal is cited as *Mémoires de la classe des sciences mathématiques et physiques*.

[3] Desmarest's attempts to gain entrance to the Academy date back to at least as early as 1757, when one of his friends and mentors in his native Troyes, Jean-Baptiste Ludot (1703–1771), an eccentric savant with some influential friends, wrote to Pierre Bouguer on his behalf in hopes of having the young aspirant placed in a vacant slot (copy of letter from Ludot to Bouguer, December 20, 1757, Bibliothèque Municipale de Troyes, MS 2584.3). There followed many years of disappointment, when Desmarest often thought that election was close, but success finally came when he was backed by a winning combination of academicians, including d'Alembert and Condorcet, as well as influential outsiders, such as Turgot, Archbishop Loménie de Brienne, and the Duchesse d'Enville. For a glimpse of some of the politics involved in Desmarest's election, see Charles Henry (Ed.), *Correspondance inédite de Condorcet et de Turgot, 1770–1779* (Paris: Charavay Frères, 1883).

that won first prize in the competition of the Academy of Amiens in 1751,[4] referred to volcanic phenomena only incidentally, in a discussion of the causes that might possibly have ruptured the isthmus. In this essay, Desmarest was concerned principally with the influence of ocean currents and the destructive and constructive power of waves beating on the shore, as means of upholding the thesis that a land bridge joining France and England was destroyed naturally. Considering the possibility that it was volcanic action that had removed the isthmus, Desmarest dismissed this as a serious likelihood on the grounds, first, that volcanoes are irregular phenomena that occur by accident, rather than through ordinary processes; second, that volcanoes, when they do occur, leave chaotic traces of their action, which are not to be observed on the bottom of the English Channel (the *Dissertation* was accompanied by two maps and a cross-section of the English Channel and a part of the North Sea, prepared by the geographer Philippe Buache, which gave the ocean depths at intervals of ten fathoms, showing smooth contours for the bottom); and third, that the source (*foyer*) of a volcano required to create the English Channel would have been, of necessity, quite deep, which was "contrary to experience."[5] As authorities for the alleged shallowness of volcanic sources he cited Johann Anderson's natural history of Iceland and Pierre Bouguer's account of his travels in Peru.[6]

In his second paper on a geological subject, Desmarest addressed himself to the mechanism of earthquake propagation, a timely subject because of the recent occurrence of the destructive Lisbon earthquake of November 1, 1755.[7] In this pamphlet, which was intended as an

---

[4] *Dissertation sur l'ancienne jonction de l'Angleterre à la France, qui a remporté le prix, au jugement de l'Académie des Sciences, Belles-Lettres & Arts d'Amiens, en l'année 1751* (Amiens: Chez la Vve Godart, 1753). This essay was republished in the late nineteenth century, under the auspices of a company that hoped to stir up interest in the construction of a tunnel beneath the English Channel: *L'ancienne jonction de l'Angleterre à la France, ou le détroit de Calais, sa formation par la rupture de l'isthme, sa topographie et sa constitution géologique* (Paris: Isidore Liseux, 1875).

[5] *Dissertation*, pp. 128–129.

[6] Johann Anderson, *Histoire naturelle de l'Islande, du Groenland, du détroit de Davis*, trans. Gottfried Sellius from the German edition of 1746 (2 vols.; Paris: S. Jorry, 1750). Bouguer's travels are recounted in several papers in the Academy's *Mémoires* and in *La figure de la terre, déterminée par les osbervations de MM. Bouguer et de La Condamine envoyés par ordre du roy au Pérou pour observer aux environs de l'équateur, avec une relation abrégée de ce voyage qui contient la description du pays dans lequel les opérations ont été faites* (Paris: Chez Charles-Antoine Jombert, 1749).

[7] *Conjectures physico-mechaniques sur la propagation des secousses dans les tremblemens de terre, et sur la disposition des lieux qui en ont ressenti les effets* (n.p., no pub., 1756). It was published anonymously, but evidence from numerous independent sources shows beyond doubt that Desmarest was the author.

explanation of earthquakes for a certain gentleman of eminence (probably the Attorney-General Guillaume-François Joly de Fleury),[8] Desmarest did not assign to volcanic phenomena any more importance than he had in his earliest publication. Here he was primarily interested in the distribution of mountain ranges throughout the world, and in the probability that an interlocking network of mountain ranges constitutes the medium through which earthquake disturbances travel. For his geographical ideas, some of them highly speculative, Desmarest was indebted to Buache, the geographer who is perhaps best remembered for his insistence on the importance of drainage basins as units is geographical inquiry. Buache emphasized the distinct nature of different drainage areas, but it was the high ground separating one region from another that particularly interested Desmarest. Between these natural basins stand solid mountains, which Buache believed were composed of hard material unlike that of the intervening areas, forming a closed system of interconnecting chains extending beneath the sea and through low-lying areas where their presence may be difficult to detect. Buache extended his concept far enough so that Desmarest had only to adopt it for his use. To Buache, the interest and value of this interlocking mountain system, apparently borrowed from Athanasius Kircher, lay in its function as the structure that holds the earth together and defines the main features of the continents and oceans.[9] However, Desmarest, adopting Buache's idea, turned it to a different use, as a propagator of seismic shock. Most of his essay is devoted to establishing the plausibility of this means of earthquake propagation, which he compared to the delivery of shock through a row of billiard balls.[10]

[8] Aside from its deferent tone, the text contains no hint of the gentleman's identity, but a letter from Desmarest to his friend Pierre-Jean Grosley, in Troyes, indicates that Joly de Fleury had solicited the essay (letter dated March 27, 1756, Bibliothèque Nationale, Fonds Français, Nouvelles Acquisitions, MS 803, fol. 91–92).

[9] Buache's geographical ideas can be examined in his "Essai de géographie physique où l'on propose des vues générales sur l'espèce de charpente du globe composée de chaînes de montagnes qui traversent les mers comme les terres, avec quelques considérations particulières sur les différents bassins de la mer et sur sa configuration intérieure," *Mémoires de l'Académie Royale des Sciences, 1752* (1756), 399–416; reprinted (without the maps in the original) in *Revue de géographie, XXII* (1888), 293–303. Buache's geography has been treated by Ludovic Drapeyron, "Les deux Buache, ou l'origine de l'enseignement géographique par versants et par bassins," *Ibid., XXI* (1887), 6–16, and by J. Thoulet, "L'étude de la mer au XVIIIe siècle: De Maillet, Buache et Buffon," *Mémoires de l'Académie de Stanislas, VI* (1908–1909), 214–256, as well as in the "Eloge de M. Buache" by Grandjean de Fouchy, *Histoire de l'Académie Royale des Sciences*, in *Histoire de l'Académie Royale des Sciences. Avec les mémoires de mathématique & de physique, pour la même année. Tirés des registres de cette académie, 1772*, pt. 2 (1776), 135–150.

[10] *Conjectures*, p. 18ff.

What is of interest to us for the moment, however, is the later section of Desmarest's tract, in which he discussed the possible sources of earth tremors, a problem he regarded as more difficult than that of the disturbance's propagation. He did not commit himself to any hypothesis of earthquake source, but only mentioned briefly the widespread view that earthquakes result from the explosive expansion of the air produced by underground fires. He believed this opinion to have its difficulties, but these did not touch on the existence of burning materials underground, which Desmarest accepted without question. He did not relate the incidence of earthquakes to regions of volcanicity, although, of course, his main thesis related them to mountainous areas. He rejected the often-repeated hypothesis of the underground connection between separate regions of volcanic action — Vesuvius, Etna and the Lipari Islands had frequently been associated in this way in literature since antiquity.[11] But these do not represent extensive remarks about volcanoes, and all in all Desmarest seemed impressed neither with their inherent interest nor with their potential power to cause the earthquake shocks whose dissemination concerned him. He included a passing reference to the volcanoes of Auvergne (which he had not yet seen), but only to remark that they were burned out and incapable of causing damage: "... ainsi la nature est tranquille dans le beau pays que nous habitons. Heureux!"[12]

1757 witnessed the publication of the seventh volume of the *Encyclopédie*, including two geological articles by Desmarest: "Fontaine" and "Géographie physique."[13] The former article, dealing with springs and related phenomena, did not touch on volcanoes at all. The second article, however, briefly approached several questions relating to volcanism and heat in the earth, but did not deal with them at any great length, even though Desmarest's writing was here characterized by a degree of generality and scientific temperateness that had been less in evidence in his earlier efforts, which, it may be noted, constituted arguments for particular theories. In the *Encyclopédie*, Desmarest was concerned to deal with his appointed subjects in a comprehensive fashion, and here, for the first time, he formulated for publication his opinions on

[11] This idea was expressed, in fact, in a book edited by Desmarest, a posthumous collection of the wisdom of the cleric Longuerue (1652-1733): *Longueruana, ou recueil de pensées, de discours et de conversations, de feu M. Louis du Four de Longuerue* (2 vols. in 1; Berlin [i.e., Paris]: no pub., 1754), Vol. I, p. 182.
[12] *Conjectures*, p. 60.
[13] "Fontaine," *Encyclopédie, ou dictionnaire raisonné des sciences, des arts et des métiers*, Vol. VII (Paris: Briasson, David, Le Breton, Durand, 1757), pp. 80-101; "Géographie physique," *Ibid.*, pp. 613-626.

the methods and principles of the science of the earth. He distinguished between the primary and secondary segments of the earth's crust, and followed the general opinion of his time in declining to link the earliest strata with any volcanic action.[14] The existence of modern volcanoes was noted, as was the presence of extinct volcanoes, which could be recognized by their shape and the burned nature of their ejecta.[15] Again he declared the source of volcanic heat to be at a shallow depth. In general, though, Desmarest again gave no sign of special concern with volcanoes.

Between the publication of the *Encyclopédie* articles in 1757 and Desmarest's first visit to Auvergne six years later, he produced three more geological writings that have survived. The first of these, some geological observations made in Champagne in 1760, consists mainly of descriptions of local soils and includes considerable discussion of the destructive effects of running water upon land.[16] Here he viewed the cumulative effect of small-scale changes, such as the rounding of pieces of gravel in running water, as amounting in time to alterations of great magnitude — the destruction and removal of entire sets of thick beds, for example.[17] The other two writings, which have never been published, show that Desmarest's attention continued to focus on the alteration of geological features. They are accounts of Desmarest's travels in the southwest part of France in 1761 and 1762, when he was employed to help carry out a statistical survey of agricultural resources for Charles-Robert Boutin, the new Intendant of the *généralité* of Bordeaux.[18] As might be expected from geological excursions in the Gironde and neighboring districts, Desmarest nowhere discussed any aspects of volcanic

---

[14] "Géographie physique," p. 622.
[15] *Ibid.*, p. 624.
[16] "Observations géologiques," in Pierre Samuel Du Pont de Nemours (Ed.), *Œuvres de Mr. Turgot* (9 vols.; Paris: A. Belin and Delance, 1808–1811), Vol. III, pp. 376–447. These notes were erroneously attributed to Turgot, who was Desmarest's benefactor and superior as Intendant in the *généralité* of Limoges, where Desmarest was Inspector of Manufactures from 1762 to 1771. Part of the notes were reprinted as "Observations géologiques, faites par Desmarets, en 1760. dans la partie de la Champagne qui forme actuellement le département de l'Aube," *Annuaire administratif et statistique du département de l'Aube* (1841), pt. 2, pp. 29–48.
[17] "Observations géologiques," pp. 395, 402, 404.
[18] "Voyage dans une partie du Bordelois et du Périgord," Bibliothèque Municipale de Bordeaux, MS 721. "Remarques de Mr. Desmarest (de l'Académie des Sciences) sur la géographie physique, les productions & les manufactures de la généralité de Bordeaux, lors de ses tournées depuis 1761 jusqu'en 1764," Archives départementales de la Dordogne, MS 26. The first of these manuscripts is anonymous, but circumstantial evidence, including proof that he purchased a horse and chaise in Bordeaux the day before the journey's beginning (Archives départementales de la Gironde, C 2891, no. 4), indicates that Desmarest is the author. The title of the second is in error, as Desmarest's travels in Gironde ended in 1762.

geology in these documents. What attracted most of his attention, aside from the nature and quality of the soils, was the destruction of rock and soil and the configuration of rivers in beds of their own creation. In both of these manuscripts, Desmarest frequently remarked on the influence of water on terrain, employing the terms *degradation* and *transport*, and relating changes in type of deposit to "revolutions" in the past. He was quite specific about the action of rivers in removing material from one place to another in creating the terrain's main features. Noticing torrential deposits, for example, high above the present level of a river bed, he wrote:

> At first glance it seems rather peculiar that these torrential deposits should be placed on the highest summits on the banks along our river-channel, and at over 200 or 300 feet above the present level of running water: the only reasonable explanation for this circumstance is in supposing that since [these deposits were laid down] the water has excavated its channel and that before the deepening of the valleys the products of torrential floods . . . occupied the heights in question, which served as beds. . . .[19]

He also stated clearly that this type of process was continuing in the present, although perhaps not at the same rate. Thus, Desmarest conceived of a kind of subaerial denudation, and saw valleys as carved by their own streams, ideas that he used to advantage in his studies in Auvergne, as shall be seen momentarily.

The overwhelming impression gained of Desmarest's geological work before 1763 is that up to that time he had no particular interest in volcanic geology, although he had shown a solid interest in the natural history of the earth in general for over a decade. On those occasions when he had referred to volcanoes or to subterranean heat, it was without much detail and often in a way not directly related to the subject at hand. What ideas he had formed about volcanoes he had probably derived from the texts of other naturalists and travelers. This general impression is sustained from what is known of Desmarest's correspondence.[20] It is certain that his opportunity for direct observation of volcanoes was very limited. He traveled little until 1757, when he began to carry out industrial investigations for Trudaine in various parts of France, but he did not go abroad until 1765, except for a brief excursion to San Sebastián. That his travel was restricted in these early decades is

---

[19] "Remarques de Mr. Desmarest," fol. 4.
[20] What I have been able to find of Desmarest's correspondence, numbering somewhat over 100 letters, dates mainly from the period 1756 to 1772. His principal correspondents were Pierre-Jean Grosley, Turgot, François Pasumot, and a certain Gonthier, of Troyes. The extant correspondence consists rather largely of letters received by Desmarest.

borne out by the fact that it was only in 1761, when he was 36 years of age, that he saw the ocean for the first time in his life.[21]

What captured Desmarest's attention in Auvergne, and converted him almost instantly into a student of volcanic geology, was basalt — not mere ordinary basalt, but "prismatic" basalt, formed into columns and other regular shapes. Desmarest had been familiar with this form of basalt, not from immediate experience, but from second-hand acquaintance with the most famous example of the phenomenon, the Giant's Causeway in County Antrim, Northern Ireland. The Giant's Causeway had been well known since it had been described in scientific fashion in the late seventeenth century, and its fame on the Continent grew enormously after 1740, when two engravings made from Mrs. Susannah Drury's paintings of the location were widely circulated.[22] But these curious columns, regarded as unique in the world (much significance was placed in the fact that Giant's Causeway columns are articulated, or divided into sections at regular intervals, whereas the basalt columns of Saxony are monolithic), were not visibly associated with volcanoes, active or extinct, and it occurred to no one that they might be of anything other than aqueous origin, except for those who assigned to them a human or supernatural cause (a cause implied by their name). The Giant's Causeway was an ornament of some consequence, and the discovery of similar forms of basalt in the heart of France was, in itself, sufficient reason for Desmarest to be excited by his find. But he was not content to claim a similar formation for France; in addition he sought to elucidate the origin of prismatic basalt, which he immediately presumed to be of the same origin everywhere, and therefore volcanic in Ireland if volcanic in France.

Desmarest's excursion of 1763 permitted him little more than a strong suspicion of his conclusion. But the next year he returned to Auvergne, this time with a cartographer, François Pasumot, and an artist, with a plan to map the volcanic regions, aided and protected by the Intendant, Simon-Charles-Sébastien Bernard de Ballainvilliers.[23] In 1765, Desmarest delivered a preliminary report of his findings to the Academy of Sciences, before departing for a year-long tour of Switzerland and

---

[21] Evidence of this astonishing fact is found in a letter to Grosley of September 7, 1761, wherein Desmarest stated that he had seen the sea the previous day "for the first time." He marveled at the "sublime" spectacle, and was baptized by his companions "according to the custom." (Bibliothèque Nationale, F.F., N.A., MS 803, fol. 133–134.) It is interesting that Desmarest devoted a substantial section of his *Dissertation* to the power of ocean currents and shore waves without ever having seen the sea.
[22] Sergei Ivanovich Tomkeieff, "The Basalt Lavas of the Giant's Causeway District of Northern Ireland," *Bulletin volcanologique*, VI (1940), 89–143.
[23] [L. Welter], "Carte des volcans d'Auvergne préparée par Pasumot," *Bulletin historique et scientifique de l'Auvergne*, LXIV (1944), 215–216.

Italy in the company of the young Duke Louis-Alexandre de La Rochefoucauld. Pasumot continued his mapping while Desmarest and La Rochefoucauld inspected the volcanic phenomena of Tuscany, Umbria, Latium, Campania, and the Euganean Hills in the district of Padua and Vicenza. Desmarest rejoined Pasumot in 1766, and they completed a geological map of the southern, Mont-Dore district that was published with Desmarest's report to the Academy of 1771. This article is generally taken as the first published announcement that basalt was volcanic. However, the discovery came to be known a good deal in advance of 1771, in part through Desmarest's unpublished announcement of 1765, which was probably discussed by members of the Academy, but also through the subsequent publication of the substance of this report with the plates of the *Encyclopédie* in 1768.[24] The existence of this report, published six years before the fuller memoir of the Academy, appears not to have been known to any of Desmarest's biographers, but it is not clear whether or not it was equally ignored by his contemporaries. It may well have been, for it is found as an explanation for one of the plates, unsigned; but at the very least it was read by Rudolf Erich Raspe, who acknowledged it in his letter establishing the igneous nature of German basalts.[25] Proof that word of Desmarest's 1765 report was spread in France is found in a 1766 memoir of Jacques Montet, who referred to the discovery in his account of volcanic mountains near Montpellier.[26]

The brief article in the *Encyclopédie* was long enough to do little more than describe the types of prismatic basalt found and to declare their certain association with volcanic flows. Desmarest was concerned there only with the configuration of basalt columns as he found them at the extremities of flows. In his two long memoirs on the Auvergne basalts, however, he turned to a historical task, that of reconstructing the former configuration of volcanic flows from the evidence at hand. These memoirs were presented to the Academy of Sciences in 1771 and 1775, respectively, and represented his fullest and most important contribu-

[24] "Basalte d'Auvergne," *Recueil de planches, sur les sciences, les arts libéraux, et les arts méchaniques avec leur explication*, Vol. VI (Paris: Briasson, David, Le Breton, 1768), pp. 3-4 of section "Histoire naturelle, règne minéral." The two plates that this article accompanies depict a butte of articulated basalt columns upon which the château of La Tour d'Auvergne once stood (plate VII, section "Histoire naturelle, minéralogie, 6e collec., volcans"), and a mass of basalt in "boule" or ellipsoidal form, located by the château of Pereneire, near Saint-Sandoux (plate VIII).
[25] "A Letter from Mr. R. E. Raspe, F.R.S., to M. Maty, M.D. Sec. R.S., Containing a Short Account of Some Basalt Hills in Hessia," *Philosophical Transactions*, LXI (1771), 580-583.
[26] "Mémoire sur un grand nombre de volcans éteints, qu'on trouve dans le Bas-Languedoc," *Mémoires de l'Académie Royale des Sciences*, 1760 (1766), 466-476. Although Montet's memoir is published with the proceedings of 1760, it was read in 1766.

tions to volcanic geology.[27] It was in the historical approach that Desmarest brought his earlier geological ideas to bear on the study of volcanic phenomena, and the result was an analysis of the Auvergne terrain that was original and effective. Applying the same principles of degradation and river-valley carving that he had begun to develop in the Champagne and Gironde regions, Desmarest was able to envision the former unity of lava structures now separated into distinct entities. He observed certain columnar basalt groups having no apparent connection with lava flows, resting in isolated patches, often perched atop hills and peaks whose bases were manifestly nonvolcanic. But being convinced now of the volcanic origin of all such basalts, without exception, he concluded that except in cases where they were distinct eruptions themselves, they must have been originally joined with neighboring lava flows, a connection demonstrable by identity in number and mineralogical type of corresponding layers. He ascribed the separation of such masses to "obvious alterations in the primitive disposition of the flows, either by the destruction of subsequent eruptions or, chiefly, by the degradation of water. . . ."[28] He was distinctly conscious of the power in connecting a current destructive agent with the reconstruction of older conditions.

> Water from rain and melted snow appeared to me as agents quite capable of working a large part of these degradations. Prisms, detaching easily from one another, can collapse by the removal of unsteady and light materials that serve as their base; for burned earth, pumice, scoriae, and porous lavas yield with the greatest ease to the action of the least trickle of water; and this play of water, whose last traces and most recent marks I had examined, once recognized, helped me, in my thought, to replace all the removed materials into their primitive state, and to restore without effort the ancient configuration of the terrain, as it was after the volcanic eruptions.[29]

---

[27] The 1771 paper, read May 11, was published in expanded form in three parts, the first two together, under the title "Mémoire sur l'origine & la nature du basalte à grandes colonnes polygones, déterminées par l'histoire naturelle de cette pierre, observée en Auvergne," *Mémoires de l'Académie Royale des Sciences, 1771* (1774), 705-775, and the third, "Mémoire sur le basalte, troisième partie, où l'on traite du basalte des anciens, & où l'on expose l'histoire naturelle des différentes espèces de pierres auxquelles on a donné, en différens temps, le nom de basalte," in *Mémoires, 1773* (1777), 599-670. The memoir that he read on St. Martin's Day (November 11) 1775 was published in abbreviated form as "Extrait d'un mémoire sur la détermination de quelques époques de la nature par les produits des volcans, & sur l'usage de ces époques dans l'étude des volcans," *Observations sur la physique, sur l'histoire naturelle et sur les arts*, XIII (1779), 115-126. Much later, on 1 prairial an XII (May 20, 1804), he delivered essentially the same paper to the Academy: "Mémoire sur la détermination de trois époques de la nature par les produits des volcans, et sur l'usage qu'on peut faire de ces époques dans l'étude des volcans," *Mémoires de la classe des sciences mathématiques et physiques*, VI (1806), 219-289.
[28] "Mémoire sur l'origine & la nature du basalte," p. 720.
[29] *Ibid.*, p. 738.

NICOLAS DESMAREST 349

Desmarest's disposition to work from the present backward is illustrated in his comparison of different effects in different lava flows.

> I noticed that the deepening of these valleys was in proportion to the antiquity of these flows and to the abundance of the streams' waters; that the streams that separated the different parts of the old flows had achieved the excavation of very deep cuts between them, whereas they were only rough-hewn and studded with falls and cascades when they were found cutting between parts of the more modern flows.[30]

His observations on the relative degree of degradation of lava flows of varying age were generalized into a principle of uniform action, which he enunciated in direct connection with the history of Auvergne terrain.

> The results of the last operations are simpler and less modified by the changes which the primitive forms daily undergo, and that one recognizes there the agents more easily when the traces of their march are more manifest; moreover, this primitive state is a standard for comparison, to be kept continually before the eyes of the observer who would judge of the extent and progress of the successive alterations.[31]

Among the general conclusions Desmarest drew from his research in the Auvergne volcanoes, some of the most interesting include this link he saw between the process of degradation and uniform change in time. His proposition that "the same causes or parallel causes have equally altered, in the same place, different primitive conditions"[32] classed these causes among the "order of the most ordinary phenomena."[33] His method of research, he believed, showed that "Nature has been subject to the same order in the most remote ages as in the most recent times."[34]

Hooykaas rightly points out that although Desmarest claimed the uniformity of natural processes as a result of his investigation, he actually argued in a circle, as the assumption that nature proceeds uniformly was implicit in his method.[35] This circularity is illustrated in the passage already quoted (see note 30), in which Desmarest recognized a correlation between the age of a lava flow and its degree of degradation. For how did he determine the age of a lava flow? On this question he was not entirely explicit; but aside from cases in which he could assign relative

---

[30] Ibid.
[31] "Extrait d'un mémoire sur la détermination de quelques époques," pp. 117–118. This translation is taken from R. Hooykaas, *Natural Law and Divine Miracle: The Principle of Uniformity in Geology, Biology and Theology* (Leiden: E. J. Brill, 1963), p. 16.
[32] "Mémoire sur l'origine & la nature du basalte," p. 739.
[33] Ibid., p. 737.
[34] Extrait d'un mémoire sur la détermination de quelques époques," p. 117.
[35] Hooykaas, *Natural Law*, p. 17.

ages from the circumstance that one flow partially covers another, he seems simply to have presumed that the more thoroughly destroyed the flow appeared, the older it was. There is no need to criticize this presumption, which is of a kind often employed by geologists, for it worked well, and the phenomena described by Desmarest fit neatly and harmoniously into the scheme based on the assumption. Nevertheless, Desmarest was either unwilling or unable to declare the uniformitarian principle, such as he used it, a necessary part of his theoretical framework, rather than a result of a series of demonstrations.

Although Desmarest used uniformitarian arguments, it should be recognized that there were limits to his use of the concept. The idea that nature acts uniformly was generally confined to "ordinary" phenomena, a category that included subaerial denudation but not volcanic activity. He held to the "accidental" nature of volcanoes for most if not all of his years. "I comprehended from the beginning," he wrote in 1806, "that, volcanic eruptions being accidents among ordinary phenomena of nature, the return of their attacks had not been subject to any fixed period."[36] In Desmarest's view, even though volcanoes had been the cause of substantial change in the landscape, they could not be relied on to account for regular effects of great consequence in earth history. It is a mistake to attribute to him the conviction that the eruptivity of volcanoes is subject to a uniformitarian order.[37]

This limitation on the power of volcanoes as agents of geological change is clearly expressed in Desmarest's last and largest geological work, the compendium *Géographie physique* of the series *Encyclopédie méthodique*.[38] Here he offered his opinions on a very wide variety of geological subjects, amplifying his ideas on questions discussed in his earlier works, and providing an unusual record of his thought, not only on the subjects listed alphabetically in Volumes II through IV, but also on the 40 geological thinkers dealt with in the first volume.[39] His ideas

---

[36] "Mémoire sur la détermination de trois époques de la nature," p. 221. Desmarest made an almost identical statement in the 1779 version.

[37] Hooykaas errs slightly in emphasizing Desmarest's uniformitarianism in the study of volcanic history; he translates the passage quoted on page 349 *supra* (note 34) as follows: "Nature, in the eruptions of volcanoes, has been subject to the same order in the most remote ages as in the most recent times." However, the phrase "in the eruptions of volcanoes" is not present in the 1779 version, having been added only in 1806.

[38] *Encyclopédie méthodique, ou par ordre de matières: Par une société de gens de lettres, de savans et d'artistes: Géographie physique* (5 vols.; Paris: Chez H. Agasse, an III [1794/95]–1828).

[39] He refused to write on living geologists, with two exceptions: Hutton and Pallas. Of the remaining 38, most lived in the seventeenth or eighteenth centuries. The fifth volume, published long after Desmarest's death, was completed by Bory de Saint-Vincent, Doin, Ferry, and Huot, and the contents appear to be more their doing than Desmarest's.

NICOLAS DESMAREST 351

expressed here exclude the possibility that volcanoes or subterranean heat have operated throughout the earth's history. Following the accepted division of strata into three principal epochs, he explained his belief that the oldest, granitic formations (the *ancienne terre*) were not produced by fire or heat, as experiments showed that granite vitrifies when heated sufficiently.[40] Desmarest was following the lead of Guillaume-François Rouelle, whose courses he had taken and whose reflections on the history of the earth he greatly admired.[41] In Rouelle's opinion, the primitive core of the globe was crystallized out of an aqueous fluid.[42] Granite could not be formed from fusion, which results only in a homogeneous mass without distinct crystals of feldspar, quartz, or mica. Desmarest regarded granite as the primary stuff of lava, without supposing, however, that the *ancienne terre* necessarily represents the first and oldest stage of the earth's history — in fact, he speculated, without ever committing himself to an estimate of the earth's age, that a virtually endless succession of epochs and destructions may have preceded the three periods for which the evidence is clear.[43] There were then no signs of volcanic activity in the *ancienne terre*, but only in the secondary (*moyenne*) and tertiary (*nouvelle*) strata.[44]

This was consistent with Desmarest's conviction that the source of volcanic heat lay in subterranean masses of burning coal, which he preferred to the "fermentation" of pyrites favored by others. The massive and unstratified primary deposits did not contain significant collections of organic remains, and could not be presumed to have possessed coal beds adequate to produce volcanoes.[45] But later deposits had abundant coal, and with water this was enough to produce the required effects.

I admit that combustible materials alone are not capable of producing all the effects of volcanoes; but that should not cast any doubt on the necessity of these materials to serve as fuel for volcanic fire. Expanding water in

---

[40] *Géographie physique*, Vol. II, p. 502 ("Ancienne terre").
[41] That Desmarest attended Rouelle's public lectures in the early 1760's is inferred from two letters he wrote to Grosley, one [1761] in the Bibliothèque Nationale, F.F., N.A., MS 803, fol. 131-132, and the other [December 1762] in "Lettres inédites de Grosley et de quelques-uns de ses amis," ed. Truelle Saint-Evron and Albert Babeau, *Société Académique de l'Aube: Collection de documents inédits relatifs à la ville de Troyes et à la Champagne méridionale*, I (1878), 346-349.
[42] *Géographie physique*, Vol. I, pp. 409-431 ("Rouelle").
[43] *Ibid.*, Vol. II, p. 554 ("Anecdotes de la nature et de l'histoire de la terre").
[44] *Ibid.*, Vol. II, p. 308 ("Alpes"); Vol. II, p. 502 ("Ancienne terre"). At one point Desmarest dismissed underground fire even from the secondary strata; *Ibid.*, Vol. III, p. 508 ("Couches de la terre").
[45] *Ibid.*, Vol. I, p. 268 ("Lazzaro-Moro"); Vol. III, p. 366 ("Charbon de terre"). Desmarest set forth this opinion as early as 1754, in a note in his edition of Francis Hauksbee's *Expériences physico-méchaniques sur différens sujets, et principalement sur la lumière et l'électricité, produites par le frottement des corps* (2 vols.; Paris: Chez la Veuve Cavelier & Fils, 1754), Vol. II, p. 553.

the next instance seems to be the active force capable of carrying out the acts of eruption, and of producing the disturbance accompanying them; but it is no less true that the condition of all the products of subterranean fire shows that coal is the only material suitable not only to achieve their fusion, but also, by its immense quantity and its disposition in the bowels of the earth, to supply the subterranean fires....[46]

Furthermore, volcanic remains show remnants of the combustion of coal.

Moreover, if one examines the vicinity of the craters and funnels by which flame and smoke surge out in frightful clouds, it will be seen that these craters and funnels are covered with immense masses of slag perfectly similar to the residue of the combustion of coal....[47]

Coal was always Desmarest's preference for the cause of volcanic heat; he never accommodated himself to the opinion that related volcanic material to the earth's innate heat.[48]

Nor would Desmarest accept volcanic activity as a plausible cause of the uplift of mountains, a problem for which he was unable to offer a suitable mechanism. Recognizing that mountains and continents have incontestably risen, for whatever reason, he essayed an explanation only for their subsequent denudation. He rejected Anton Lazzaro Moro's suggestion that subterranean fire had caused mountains to rise early in the earth's history, by referring to the availability of coal only in the latter stages of the earth's development;[49] to the more ambitious system of Hutton, calling for both the consolidation of strata and mountain uplift by subterranean heat, he responded similarly.[50] As long as real and literal fires were the only source he could imagine for volcanic heat, and as long as he dismissed other sources of internal heat, it was difficult for him to think otherwise.

In short, Desmarest was no vulcanist. Instead, he was a neptunist, in the broad sense that he viewed water, rather than heat, as the main agent of geological change and the chief medium of creation of terrestrial features. This scientist, who had done much to extend the scope of volcanic action in geology, nevertheless thought volcanoes to be of far more limited importance in the earth's crust than water. Signs of his neptunist frame of mind are to be found in all of his geological works, dating back to his earliest publication, and it is unnecessary to document here the constancy of his attitudes. Suffice it to note that one of his later

---

[46] *Géographie physique,* Vol. IV, p. 164 ("Feu des volcans").
[47] *Ibid.,* Vol. IV, p. 163 ("Feu des volcans").
[48] *Ibid.,* Vol. I, pp. 382-409 ("Romé de l'Isle"), and Vol. III, pp. 344-349 ("Chaleur du globe").
[49] *Ibid.,* Vol. I, p. 268 ("Lazzaro-Moro").
[50] *Ibid.,* Vol. I, p. 749 ("Hutton").

papers deals with prismatic columns formed like those of basalt, but from aqueous clay.[51] Far from regarding lavas as the sole material capable of this behavior, he was pleased to be able to extend the process of formation of prisms by contraction to aqueous as well as igneous substances. Desmarest's neptunism is not altogether surprising, as others prominent in volcanic geology of the time held similar opinions, but it calls for a clarification of the so-called vulcanist-neptunist dispute and of the specific nature of the controversy over the origin of basalt. The task of tracing the basalt controversy through the last four decades of the eighteenth century cannot be done here, but a few remarks and tentative conclusions can perhaps be offered.[52]

First of all, Dr. Rappaport is quite correct in indicating that a geologist's position on the basalt controversy is not a reliable index for classification in the vulcanist-neptunist dispute.[53] Many geologists not ordinarily reckoned among the friends of vulcanism spoke in favor of basalt's igneous nature. The neptunist Guettard, long a defender of the creation of basalt "in the humid way," came over to Desmarest's point of view late in his career; his change of mind on this subject is implied, as Dr. Rappaport has suspected, in his *Mémoires sur la minéralogie du Dauphiné*.[54] Jean-André Deluc, chosen by Professor Gillispie as a paragon of neptunism, nevertheless described basalts as volcanic as early as 1778.[55] Jean-Claude de Lamétherie, whose theory of geology owed much to Werner, attributed basalt to an igneous cause no later than 1795.[56] Furthermore, some geologists aside from Desmarest who devoted

[51] "Mémoire sur les prismes qui se trouvent dans les couches horizontales de plâtre et de marnes des environs de Paris, et sur leur analogie avec prismes du basalte," *Mémoires de la classe des sciences mathématiques et physiques*, IV (an XI [1803]), 219–231; "Seconde mémoire sur la constitution physique des couches de la colline de Montmartre et des autres collines correspondantes," *Ibid.*, V (an XII [1804]), 16–54.
[52] On the basalt controversy, see Otfried Wagenbreth, "Abraham Gottlob Werner und der Höhepunkt des Neptunistenstreites um 1790," *Freiberger Forschungshefte* (D 11), (Berlin: Akademie-Verlag, 1955), pp. 183–241; and Arnold von Lasaulx, "Der Streit über die Entstehung des Basaltes," *Sammlung gemeinverständlicher wissenschaftlicher Vorträge*, ser. 4, LXXVI (1869).
[53] Rhoda Rappaport, "Problems and Sources in the History of Geology, 1749–1810," *History of Science*, III (1964), 66.
[54] (2 vols.; Paris: De l'Imprimerie de Clousier, 1779). See Vol. I, pp. 128–154.
[55] *Lettres physiques et morales sur l'histoire de la terre et de l'homme: Adressées à la reine de la Grande Bretagne* (5 vols. in 6; La Haye and Paris: De Tune and Veuve Duchesne, 1779–1780), Vol. II, 478–481; Vol. IV, pp. 146–161; *Geological Travels in Some Parts of France, Switzerland, and Germany* (2 vols.; London: Printed for F. C. and J. Rivington, 1813), esp. Vol. I, p. 232; Vol. II, pp. 264, 284. Gillispie deals at length with Deluc in *Genesis and Geology: A Study in the Relations of Scientific Thought, Natural Theology, and Social Opinion in Great Britain, 1790–1850* (Cambridge: Harvard University Press, 1951; reprinted Harper Torchbook, 1959), pp. 56–66.
[56] *Théorie de la terre* (3 vols.; Paris: Chez Maradan, 1795), Vol. II, pp. 47, 55–61.

substantial parts of their attention to volcanic activity nevertheless refused to behave as though a final choice had to be made between fire and water. Dolomieu and de Saussure fall into this class. It is not surprising, though, that there is confusion in the relation between the basalt controversy and the vulcanist-neptunist debate, as the participants themselves, and interested bystanders, often failed to distinguish between them. There are signs that the term *vulcanist* was first applied as an indication of belief only in the volcanic origin of basalt, and if this is so, then its quite natural subsequent extension to a more general geological doctrine has exacted its cost in historical clarity.[57] A few historians have since attempted to remedy the situation by employing the term *plutonism* for vulcanism of the variety requiring for heat a prime place in geological change, but this is difficult to carry through, and perhaps of doubtful desirability, since the principals in question frequently called themselves vulcanists.[58]

It is not unlikely that the stances of individual geologists regarding the basalt controversy were substantially influenced by their methodological predispositions. Perhaps the historian would find it useful to consider the opinions of the antagonists in the light of the various modes of research they most favored. Opposition to the igneous origin of basalt seems to have come particularly from mineralogists — aside from Werner, Bergman, Wallerius, Klaproth, and Romé de l'Isle were prominent mineralogists opposed to Desmarest on this count. On the other hand, support for the igneous view came especially from geological travelers who observed columnar basalt in its natural location and were less apt to study rocks from the standpoint of chemistry and mineralogy than from a dynamic and physiographical framework. The Auvergne basalts, in particular, were capable of convincing visitors, and Desmarest himself said that he might not have made his discovery anywhere else.

[57] *Vulcanist* and *neptunist* are used in a somewhat narrow sense by Saussure, "Observations sur les collines volcaniques du Brisgaw," *Journal de physique, de chimie et d'histoire naturelle,* XLIV (*an* II [1794]), 356, and by Faujas de Saint-Fond, "Essai de classification des produits volcaniques, ou prodrome de leur arrangement méthodique," *Annales du Muséum National d'Histoire Naturelle,* III (1804), 85–86. In the second edition of his *Théorie de la terre* (5 vols.; Paris: Chez Maradan, 1797), Lamétherie used the term *vulcanistes* in specific reference to believers in the igneous origin of basalt (Vol. II, p. 57).
[58] Geikie is perhaps the most notable example of a historian distinguishing vulcanists from plutonists. Frank Dawson Adams, *The Birth and Development of the Geological Sciences* (Baltimore: Williams and Wilkins, 1938; reprinted Dover, 1954), used the term *plutonist,* but then incorrectly summoned Desmarest to the anti-neptunist cause (p. 245).

Clarification of the role played by the basalt controversy in the development of plutonism will have to await research in the geological thought of a number of eighteenth-century scientists whose work has not been examined adequately. Too many geologists of this period have been understood only through the distorted pictures given us by other scientists in funerary orations; the alleged antagonism between Desmarest and Werner, for example, derives mostly from Cuvier's artistic sense, for in their rivalry he found a convenient common theme for their eulogies, which he delivered on the same day. Modern scholarship will be able to eliminate misrepresentations such as this, and perhaps will provide new categories through which the development of eighteenth-century geology can be better understood.

To make a clear distinction between the basalt controversy on the one hand and the vulcanist-neptunist dispute on the other is not to say that they were unrelated, but only that it is plain that many who believed in the volcanic origin of basalt would have found it necessary to change some of their basic presuppositions before they could have become vulcanists of the Huttonian mold. In Desmarest's case, at least, it was impossible to accept the action of volcanoes throughout all time or to treat them as geological agents of the highest importance because of his particular idea about their cause. A thoroughgoing vulcanist, on the contrary, required a vaster and more constant source of heat, and this could only come from the interior of the globe. Perhaps the radical vulcanists learned something from the adherents of "central heat" — Dortous de Mairan, Buffon, and Bailly. But it seems true, too, that vulcanists were capable eclectics, able to select useful ideas and information from their neptunist colleagues without being required to accept all neptunist explanations. Desmarest opened the way for later investigators in Auvergne without convincing them either that lava derives from granite or that volcanoes are burned-out vents from coal beds. And he was not the only neptunist who contributed to the development of vulcanism.

In conclusion, it can be said that Desmarest's geology is somewhat more difficult to characterize in the framework of eighteenth-century geology in general than has sometimes been thought, and that this perhaps reflects on the depth of our understanding of geological thought in this period. In particular, it may be suggested that vulcanism and neptunism should not be regarded as altogether hostile schools, but that valuable elements of one may have been derived from the other. The career of Nicolas Desmarest is an example of how some vulcanist concerns could play an important part in a geological outlook that always favored the primacy of neptunist agencies. In addition, the study of Desmarest's geology shows that actualistic ideas were present in his mind as early as the 1760's, even before his first visit to Auvergne, and that this actualism was strongly associated with that part of his geology that was not specifically volcanic; in fact, for him, volcanic activity lay beyond the limits of uniformitarianism. When one reflects that the fusion of plutonism with uniformitarian thinking helped geology take a new turn at the end of the eighteenth century, it seems pertinent to ask what alterations the two concepts needed to undergo in order to make this fruitful union possible.

II

The Beginnings of a Geological Naturalist: Desmarest, the Printed Word, and Nature

> Let us consult nature herself, who ordinarily leaves recognizable marks of her operations, even if she likes most often to cloak them in disguise, to hide them from minds unfamiliar with her tricks, or heedless in following their solution.
>
> Nicolas Desmarest, *Dissertation sur l'ancienne jonction de l'Angleterre à la France*, 1753[1]

"Let us Consult Nature Herself"

The scientific achievements of Nicolas Desmarest (1725–1815), which earned him a place of some distinction in the history of the natural sciences, are understood to have been founded in large part on his shrewdness in observing and interpreting geological phenomena. Commentators on geology's early development have frequently recognized Desmarest's empirical skills (and we should bear in mind that sound observation consists of much more than an acute sensory ability, involving choices in both selection of and assignment of signification to what is seen). If we were to search among Desmarest's own writings for a concise expression of his scientific credo, we would find few statements more aptly applicable to his best work than his injunction (in the epigraph above) to consult nature itself.[2]

---

[1] Nicolas Desmarest, *Dissertation sur l'ancienne jonction de l'Angleterre à la France, qui a remporté le prix, au jugement de l'Académie des Sciences, Belles-Lettres & Arts d'Amiens, en l'année 1751* (Amiens: Chez la Vve Godart; Et se vend à Paris: Chez Ganeau, Chaubert, Lambert, 1753), pp. 97–98. (*"Consultons la nature elle-même, qui laisse ordinairement des traces reconnoissables de ses opérations, quoiqu'elle se plaise le plus souvent à les couvrir d'un voile épais pour les cacher aux esprits peu familiarisés avec ses jeux, ou peu attentifs à en suivre le dénouement."*) Translations throughout this essay are mine unless otherwise indicated. This passage offers some typically interesting and problematic choices in translation into English, notably in deciding whether or not to personalize nature as female.

[2] Already in Cuvier's 1818 *Éloge* of Desmarest one sees developed the theme of Desmarest as virtuously restrained observer, in contrast to the vices of speculation and doctrine, implicitly personified in Buffon and Werner, respectively. (Also prominent in the moral burden of Cuvier's *éloge* is Desmarest's personification of the values of useful knowledge and of disinterested, unaffected personal simplicity.) See Georges Cuvier, Éloge historique de Nicolas Desmarets [sic], lu le 16 mars 1818, in *Recueil des éloges historiques lus dans les séances publiques de l'Institut Royal de*

The fact that we find such a statement in the *Dissertation*, which won Desmarest a prize in 1751 from the Académie des Sciences, Belles-Lettres & Arts d'Amiens, a prize which in a sense launched his scientific career, might easily mislead us into supposing that he had adopted an inductive approach to scientific problems practically from the start of his work. We might also think, from his empiricist declaration, that Desmarest's *Dissertation* was itself an early result of his own empirical research. But this would be an error. There is good reason to believe that at the time he composed his *Dissertation* on England's former connection to the European continent, Desmarest had very little if any experience of what we would now call scientific investigation, other than the experience of broad and intelligent reading. On close inspection of the *Dissertation*, it is apparent that Desmarest's exclusive resources for this essay were those of the printed word. This composition was not founded on empirical inquiry; it was a library research project. The direct experiences of nature which could have been most relevant in this prize competition, we might think, would have come from a journey to the Channel coast. It is edifying to realize, then, that Desmarest had never seen a seacoast when he composed the *Dissertation*, indeed that he was witness to a seashore for the first time in his life only in 1761, a decade following his success with the Amiens essay.[3]

France, 3 vols. (Strasbourg and Paris: F. G. Levrault, 1819–1827), 2:339–374. The informative account of Desmarest's life and work given by Geikie in the twentieth century's most influential anglophone history of geology follows Cuvier closely in several respects, not least in emphasis on Desmarest as a keen observer: Archibald Geikie, *The Founders of Geology*, 2nd ed. (London and New York: Macmillan, 1905). For more recent historical treatment of Desmarest, see Kenneth L. Taylor, Nicolas Desmarest and Geology in the Eighteenth Century, in *Toward a History of Geology*, ed. Cecil J. Schneer (Cambridge, Massachusetts: MIT Press, 1969), 339–356; Kenneth L. Taylor, Desmarest, in *Dictionary of Scientific Biography*, ed. Charles C. Gillispie, 16 vols. (New York: Charles Scribner's Sons, 1970–1980), 4:70–73; and François Ellenberger, *Histoire de la géologie*, 2 vols. (Paris: Technique et Documentation [Lavoisier], 1988–1994), 2:233–245.

[3] Desmarest's encounter with the sea occurred somewhere along the Atlantic coast of Guienne, presumably not greatly distant from Bordeaux, whence he wrote: "I arrived yesterday from a journey when I saw the sea; this for the first time. M. [Mathieu] Tillet was one of the party, as were the Intendant and his wife [Charles-Robert Boutin, Intendant of the Généralité of Bordeaux]. I was baptised in accord with the custom. The spectacle of the sea is sublime; besides there are the pleasures of fishing and collecting shells, etc." ("*Je suis arrivé hier d'un voiage ou jai vu la mer: cest pour la premiere fois. M. Tillet etoit de la partie ainsi que M. et M^{de} l'intendante: on m'a baptisé suivant la coutume: ce spectacle de la mer est sublime: sans compter les agremens de la peche et de la collection des coquillages &c.*") Letter from Desmarest to Grosley, 7 September [1761], Bibliothèque Nationale, MS 803, Fonds français, nouvelles acquisitions françaises, fol. 133–134. Hereafter this collection of letters is cited as BN/803.

In this and other quotations of manuscript materials throughout this essay, transcriptions in the French are rendered as exactly as possible consistent with intelligible presentation, retaining

To recognize the dependency of Desmarest's *Dissertation* on scholarship, rather than on his own direct empirical investigation, is not to dismiss it as inconsequential or even to mark it as totally derivative. Desmarest assimilated information and ideas, in this small volume, not without imagination and ingenuity. Nonetheless, this was a product of a style of work fundamentally different from Desmarest's later field investigations, beginning a dozen years after the Amiens competition, amidst the *puys* and *cheires* of Auvergne, that were to earn Desmarest a lasting place in the annals of geology. The investigative approaches and methods used by Desmarest in his mature scientific research were not fully developed in him from the start; they had to be acquired and cultivated. He had to learn as an adult to be a truly empirical naturalist.

In this essay I wish to present the case for seeing Desmarest's early scientific work as that of an essentially bookish scholar. This is not done with any intent of portraying Desmarest's intellectual trajectory as aberrational; to the contrary, it might be representative of the paths followed by a number of others who found their way into the developing field sciences during this period. I will argue that before Desmarest's transformation into an empirically investigative scientist, a change that occurred in the closing years of the 1750s and the beginning of the 1760s, his general orientation as a scientific scholar was more toward *physique*, or natural philosophy, than toward descriptive natural history. I will also try to indicate how Desmarest's early intellectual experiences, including his strong associations with antiquarian scholars, left indelible marks on his attitudes and mental habits. Finally, in a brief rehearsal of some of the circumstances of Desmarest's conversion into a genuinely empirical naturalist, I will suggest that his embrace of learning by direct interrogation of nature may have been influenced significantly by his initiation into the government bureaucracy responsible for analyzing and regulating industrial technology.

Empiricist in Principle

Desmarest lived in an age coming increasingly to believe in the supremacy of knowledge established on the basis of experience, and in the wide applicability of such knowledge. We may need to remind ourselves, however, of an elementary distinction between acceptance of this belief and the illusion that anyone's knowledge can rest wholly or even mainly on personal experience. Permit me to speak for a moment simply as a citizen of today's empirical-minded world. I cannot doubt that I am, as a child of my time, dedicated to a largely empiricist

---

contemporary spelling and punctuation, and without seeking to correct idiosyncrasies or errors of Desmarest's writing.

ideology. But that portion of my knowledge of the natural world which comes from direct experience is, I realize, extremely small. I must rely mainly on authorities to inform me, authorities who are viewed (in all reputable cases) as participants in an intricate network of critical judges of knowledge, each of whom is however the possessor of direct empirical warrant for only a very modest fraction of what is known. People who are essentially passive believers in the proposition that reliable knowledge originates in experience – as distinct from those who are actively and significantly engaged in the further practice of actually deriving natural knowledge from experience – might well be called *empiricists in principle*. In this respect I have much in common with Desmarest; as regards natural science, I am an empiricist in principle, just as was the young Desmarest. A difference between us, of course, is that Desmarest eventually became an empirical scientist, an examiner and analyst of natural phenomena (whereas my own active empiricism, in practicing the craft of history, is applied mainly to documents).

A scientist in the eighteenth century (not much differently from one at the turn of the twenty-first) might have direct experience of nature in several ways: by making unmediated observations of phenomena as they exist or occur in nature; by examining collected specimens; or by observing or experimenting in the controlled circumstances of a laboratory or *cabinet*. As far as we know, Desmarest's education seems to have given him little or no scientific experience of nature in the raw. However, while still a student, he may have had a modest exposure to experimental and observational science at the Oratorian Collège de Troyes, in the course of *physique* taught by Jean-Baptiste Canto, which concluded his curriculum.[4] Scholars who have studied the Oratorian curriculum have shown that in the *classe de physique* students at Troyes were required to discourse on propositions selected from mathematics, experimental physics, cosmography, mechanics, and anatomy. And the Collège at Troyes possessed a physical *cabinet*, with some scientific instruments or equipment, used (for a fee) by pupils in *logique* and *physique*.[5] However, a glance at the Latin thesis upheld by Desmarest for his graduation raises serious doubts about how deeply the curriculum could have been permeated by such experimental or observational instruction. The thesis is a traditional scholastic disputation.[6]

---

[4] Bibliothèque Municipale de Troyes, MS 357: *Catalogus Scholasticorum, Collegii Treoensis*, 3 vols., in vol. 3.

[5] Arsène Thévenot, Notice historique sur l'ancien collége et le lycée de Troyes, *Annuaire administratif et statistique du département de l'Aube* (1877), 79–132, on 102; Paul Lallemand, *Histoire de l'éducation dans l'ancien Oratoire de France* (Paris: Ernest Thorin, 1888), 259.

[6] Nicolas Desmarest, *Theses philosophicae. Has Theses, Deo duce & auspice Dei-parâ propugnabit Nicolaus Desmarets [sic] Trecensis, In Aula Collegii Treco-Pithoeani Sacerdotum Oratorii Domini Jesu, die*

Beginnings of a Geological Naturalist

Desmarest came to Paris at the beginning of 1747. According to biographical notes left by his son Anselme-Gaëtan, beginning soon after his arrival in the capital Desmarest began a pattern of enthusiastic and faithful attendance at public courses of instruction, especially in physics.[7] These courses may well have included the popular demonstration lectures offered by Nollet; but at this early date Desmarest probably was not yet going to Rouelle's chemistry lectures, which were to have an important influence on him when he did eventually attend them (Anselme-Gaëtan Desmarest wrote that this began in 1760).[8] Within a decade of settling in Paris, Desmarest had begun to establish a reputation as a scientific author and, indeed, as a writer with a certain concentration of interest in topics we recognize as geological in character. Such a concentration is apparent to us, at least, on the basis not only of the *Dissertation* on the former connection of England to the Continent, but also of his tract on

---

Lunae Decimâ Sextâ Anno Domini millesimo septingentesimo quadragesimo-sexto, horâ post meridiem tertiâ (Troyes: Bouillerot, 1746). Eight pages in length. Copy in Bibliothèque Municipale de Troyes.

[7] Anselme-Gaëtan Desmarest, Notes et renseignements sur la vie et les ouvrages de mon père, Bibliothèque de l'Institut de France, MS 3199, Fonds Cuvier. ("*Suivant les cours publics et notamment ceux de physique avec beaucoup de zèle et de constance – perfectionnant ainsi ses premieres etudes, et quoique destiné à suivre la carriere de l'Enseignement; il prefera de gagner peu et de vivre tres frugalement plutot que de prendre une place de precepteur qu'on lui offre à plusieurs reprises.*")

[8] A.-G. Desmarest, Notes et renseignements, entry for 1760: "*Commence a suivre les cours de Rouelle auquel il s'attache beaucoup.*" Desmarest's letters provide evidence that he was attending Rouelle's lectures in early 1761 (Desmarest to Grosley, 20 February 1761, BN/803, fol. 131–132) and in December 1762 (Desmarest to Grosley, December 1762, mentioning that on his return to Paris: "je suis arrivé à temps pour Rouelle"; in Pierre-Jean Grosley, Lettres inédites de Grosley et de quelques-uns de ses amis, Recueillies par M. Truelle Saint-Evron . . . et annotées par M. Albert Babeau, in *Collection de documents inédits relatifs à la ville de Troyes et à la Champagne méridionale, publiés par la Société Académique de l'Aube* (Troyes: Librairie Dufey-Robert, Léopold Lacroix successeur, 1878), 1:219–433, on 347). I have encountered no documentation for Desmarest's attendance at Rouelle's lectures before 1760, nor any references to Rouelle in Desmarest's publications before that date. Desmarest's familiarity, in the early 1750s, with some of Nollet's popular publications is clear; this is seen, for example, in his critical discussion of a book on electricity by Mangin: Desmarest, Lettre à l'auteur du journal historique, sur l'Histoire générale & particulière de l'électricité, &c., *Suite de la clef, ou journal historique sur les matières du tems*, (1753), 73:276–291. It is even more impressively shown in Desmarest's commentary for the French translation of Francis Hauksbee's *Physico-mechanical experiments*. This annotated translation appeared as *Expériences physico-méchaniques sur différens sujets, et principalement sur la lumière et l'éléctricité, produites par le frottement des corps. Traduites de l'anglois de M. Hauksbée, par feu M. de Brémond, de l'Académie Royale des Sciences. Revûes et mises au jour, avec un discours préliminaire, des remarques & des notes, par M. Desmarest*, 2 vols. (Paris: Chez la Veuve Cavelier, & Fils, 1754). In Desmarest's contributions to this edition I have counted references to Nollet on 61 pages (a number second only to 117 pages for C. F. Dufay, followed after Nollet by 49 for Boyle, 48 for Newton, and 47 for Musschenbroek).

the propagation of earthquakes, and two major articles published the following year in the *Encyclopédie*, treating "Fontaine" and "Géographie physique."[9]

It was hardly possible for Desmarest's contemporaries to identify him as a geologist. No doubt some of them recognized the kind of scientific preoccupations toward which he was moving. But they lacked our geological vocabulary for organizing such matters. This lack was more than a mere accident of language; at that time there was no recognized geological specialization, as there came to be by the beginning of the nineteenth century. One could do geography, to be sure, or mineralogy, both in their way accepted branches of knowledge and inquiry. (And one could *discuss* the Theory of the Earth, although this way of speaking designated more a form of multi-disciplinary discourse than a specific scientific *status*: it meant something to be designated a *géographe* or even a *minéralogiste*, but it is very seldom that one encounters contemporary reference to a *théoricien de la Terre*.)

There is little in Desmarest's early writings to suggest that he was especially concerned about what name to give to the collective of earth-centered interests he was cultivating. When applying a label to himself during this period, whether explicitly or only indirectly, he was most apt to say he was a *physicien*. This perhaps tells us little more than that he viewed study of nature as an arena in which thinkers seek to explain phenomena. But it is not a trivial point that the young Desmarest tended **not** to speak of himself very often as a *naturaliste*, nor to refer to his interests as belonging to *histoire naturelle*. Not too much should be made of this, however: compared with the rather well defined traditions of *physique*, which dealt centrally with nature's order and principles, there was an amorphousness to *histoire naturelle* that deprived it of parity in cognitive stature. Furthermore, there were reasons a respectable student of the world's natural objects (animal, plant, or mineral) might wish to avoid willing appropriation of the term *naturaliste*: the word suffered from taint by its possible application to persons professing atheism, spinozism, or materialism.[10] Nonetheless, merely on the point of emphasis on nature's order and principles rather than on the things of the world, by education and early inclination Desmarest's scientific interests were founded more along the lines of *physique* (or what the English called natural philosophy) than of *histoire naturelle*. This orientation is apparent in nearly all that Desmarest published during his first decade in Paris.

---

[9] [Nicolas Desmarest], *Conjectures physico-méchaniques sur la propagation des secousses dans les tremblemens de terre, et sur la disposition des lieux qui en ont ressenti les effets* (N.p., 1756) [published anonymously, 63 pp.]; Fontaine, in *Encyclopédie, ou dictionnaire raisonné des sciences, des arts et des métiers, par une société de gens de lettres* (1757), 7:81b-101a; Géographie physique, in the same volume of the *Encyclopédie*, 7:613b-626a.

[10] Naturaliste, in *Encyclopédie* (1765), 11:39b.

Beginnings of a Geological Naturalist

Close examination of the contents of these early publications strongly suggests, as well, that for all of them Desmarest's grounding in *physique* was basically literary: that is, his knowledge of natural philosophy was not mainly experimental or observational, but was built upon wide reading. Much of what he read, to be sure, was **about** experimental and observational physics; but to judge from his published writings (there is little unpublished material to go on before about 1756), he did little in the way of experiment or observation himself. The breadth of his reading is notable. In several articles on electricity that he prepared in 1752–53 in the course of general editorial work for the journal *Suite de la clef* – edited by Pierre-Nicolas Bonamy (1694–1770), and commonly known as the *Journal de Verdun* – he showed within a few years of his arrival in Paris his familiarity with much that had been written in experimental investigation in this area, from Gilbert and Boyle to Nollet and Franklin. Not insignificantly, this familiarity displays also Desmarest's confidence and judgment in evaluating the meaning of past and ongoing investigations.[11]

The most ample exposition of Desmarest's early critical-empirical temper may be found in his extensive additions to François de Brémond's French translation of *Physico-Mechanical Experiments*, by Francis Hauksbee (1666–1713). The introduction, notes, and commentary provided by Desmarest for this French edition, published in 1754, are more lengthy, even, than Hauksbee's original text (1709). Like the book to which his discussions are appended, Desmarest's remarks are about topics in experimental physics: electricity and magnetism, optical phenomena, capillary action, pneumatic devices and physical effects in a vacuum, and the like. Very little is said about natural history (the main exception is a short section on the parallelism and composition of crustal beds and the possible causes of their regularity). His command of literature treating experimental physics – including his evident critical comprehension of significant issues and their relations with reported evidence – is quite impressive. Signs of his own first-hand study of nature, however, are absent.

Or rather, to speak generally, such evidence as one can find of Desmarest's own active investigations of nature, in all of his work up through 1757, is slight. It is true that, by his own account, his readings concerning electricity evidently

---

[11] A series of articles by Desmarest in *Suite de la clef, ou journal historique sur les matières du tems*: Remarques sur les nouvelles expériences de l'électricité pour se préserver des effets du Tonnerre (1752), 72:50–61; Nouvelles expériences sur l'Electricité des nuages orageux (1752), 72:134–137; Lettre à l'auteur du journal historique, sur l'Histoire générale & particulière de l'électricité, &c. (1753), 73:276–291; Observations sur la réponse à la lettre du 10 décembre 1752, touchant l'histoire générale & particulière de l'électricité: Insérée dans le journal de septembre, pag. 196 (1753), 74:434–447. Desmarest's authorship of these articles is known from the journal's *Table générale* for 1697–1756 (Paris: Chez Ganeau, 1759), 3:374, 4:31.

were supplemented by some direct experience. At least once he referred to his repetition of experiments described by earlier authors.[12] Desmarest's attendance at later performances of electrical experiments is known from private correspondence.[13] For his article "Fontaine," he reported his own sustained observations of the varying rate of flow of the spring in his native village of Soulaines.[14] But these are infrequent exceptions to the rule. In most instances where Desmarest used the language of an observer or experimenter, these were to all appearances rhetorical appeals to everyday, universal experience, or to common-sense expectations in hypothetical circumstances. Thus, his brief description of a way to confirm experimentally his theory of earthquake shocks transmitted through interlocking mountain chains is clearly no more than a thought experiment.[15] Many other passages wrapped in phraseology of experience are transparently rhetorical. Some examples from his 1753 *Dissertation*: "I cast my eyes over all the coasts of Holland, I see that most of the lands . . . bear everywhere the distinctive marks of the sea's [previous] presence"; "If we examine carefully the French coasts, we will be convinced quite easily that the sea has worn away the shores and encroached on the land."[16] Or the sentence in "Géographie physique" which begins: "What strikes me first when I dig into the earth . . . ."[17] These can hardly have been intended even to

---

[12] Desmarest, Remarques sur les nouvelles expériences de l'électricité, 52.

[13] Desmarest to Grosley, Dec. 1756, in Lettres inédites de Grosley (ref. n. 8 above), 289.

[14] Fontaine (ref. n. 9 above), 91b-92a.

[15] *Conjectures* (ref. n. 9 above), 25.

[16] *Dissertation* (ref. n. 1 above), 100–101 and 117, respectively. (*"Je jette les yeux sur toutes les Côtes de la Hollande, je vois que la plûpart des terres ... portent partout des marques distinctives du séjour de la mer"; and "Si nous examinons avec soin les Côtes de France, nous nous persuaderons aussi aisément que la Mer en a miné les rivages & a gagné sur les terres."*)

[17] Géographie physique (ref. n. 9 above), 622a. (*"Ce qui me frappe d'abord en creusant dans la terre . . . ."*) This last formulation is reminiscent of Buffon's manner of expression, in appealing to common experience, seen in the opening pages of his *Théorie de la Terre* (1749): "Let us therefore begin by picturing to ourselves what general experience and what our own observations teach us on the subject of the earth. This immense globe shows us at its surface, . . . If we penetrate into its interior, there we find . . . ." (*"Commençons donc par nous représenter ce que l'expérience de tous les tems & ce que nos propres observations nous apprennent au sujet de la terre. Ce globe immense nous offre à la surface, . . . Si nous pénétrons dans son intérieur, nous y trouvons . . . ."*), in G.-L. Leclerc, Comte de Buffon, *Oeuvres philosophiques de Buffon*, Texte établi et présenté par Jean Piveteau (Paris: Presses Universitaires de France, 1954), 46. Buffon's theme, to the effect that our initial impression of disorder and confusion is overridden on closer inspection by recognition of regularities which are keys to understanding, is echoed strongly by Desmarest. The mildly disingenuous appeal to personal research is also similar in both cases. I call this disingenuousness mild, because it appears to me improbable that serious readers were expected actually to believe that the observation being

appear to be persuasive accounts of actual witnessing and examining, on the part of a deliberate observer.

A Phenomenalist Epistemological Vision

Desmarest's additions to Hauksbee's book of experiments reveal with special clarity another significant aspect of the empiricist attitudes he internalized during the time he became a vicarious student of natural experience: namely, his embrace of a **phenomenalist** conception of knowledge. Desmarest demonstrated enthusiasm for knowledge of nature's observed effects, coupled with great caution regarding the prospect of knowing the deeper causes of those effects. He linked this point of view with the *esprit systématique* that he extolled (in common with many enlightened figures of his time), as opposed to the pernicious *esprit de système* supposedly being driven out of progressive, critical thought. Desmarest's treatment of this theme – the position that a right-minded thinker seeks observable regularities in nature without requiring a thorough accounting of those regularities' underlying causes – associated good experimental philosophy with Newtonian ideas. He expressed frank admiration of Newton and Newtonian *attractionisme*, and matching disapproval of the Cartesians' *impulsionisme*.

Desmarest's Newtonian phenomenalism was to remain, more or less constantly throughout a long life in science, a characteristic feature of his empirical credo. One senses Desmarest's pride in belonging to an intellectual movement that established genuine (and therefore useful) knowledge of nature, not merely abstract knowledge, certainly not the illusory knowledge of wishful hypothetical systems that sought in backwards fashion to derive effects from causes. But this was a pride tempered by recognition that science's hard-won affirmations regarding natural regularities are, in a large perspective, of modest depth. Of the intellectual traits running consistently through Desmarest's writings of six decades, this surely is one of the most prominent: the conviction that natural science proceeds on the basis of discerning general relations (i.e., natural laws) among specifically known things, even if one is able only rarely to push much beyond this recognition to understand the relations' deeper causes. A price to be paid for true, secure knowledge is that it is often somewhat superficial. I believe it can be shown (though I will not attempt to do

---

claimed was other than what might generally be agreed upon, whether from common experience or reliable reports.

so here) that this attitude governed much of Desmarest's geological thinking throughout his life.[18]

The phenomenalistic perspective so clearly stated in Desmarest's additions to the French translation of Hauksbee's book on experimental physics is reiterated in the articles on "Fontaine" and (especially) "Géographie physique," published three years later in the *Encyclopédie*. These pieces focus not on experiment, but on observational science – indeed on the *principles* of how observational science is done, rather than particular observational results. In fact, a considerable part of "Géographie physique," situated in the zone of intersection between *physique* and *histoire naturelle*, is in effect an essay on the method of observing and of integrating observations. The conception of natural history manifested in both of these articles is distinctly Buffonian rather than descriptive. The emphasis is placed on finding general relations that may lead to explanatory principles: " . . . true philosopy consists in discovering the relations hidden from the shortsighted and from inattentive minds."[19] This is illustrated also in the succinct, concluding expression of Desmarest's idea of physical geography: "It is easy to see from this account that a system of physical geography is nothing other than a methodical scheme in which established and constant facts are presented, and in which these facts are brought near so as to draw general results from their combination."[20] Buffon's influence is

---

[18] Desmarest's Newtonian phenomenalism is especially clear in his "Discours historique et raisonné sur les principes & sur les expériences de M. Hauksbée" in *Expériences physico-méchaniques* (ref. n. 8 above), particularly in the opening section (1:xxi-xlix) and also in his remarks at 2:80–89, 134–136. An example of Desmarest's denigration of Cartesian impulsionism in favor of the attractionists' admirably candid disavowal of knowing what causes the force: "If I have had the good fortune of finding a *general measure*, to which I can refer exactly the effects that interest and make an impression on me, what does it matter whether I know the essence and cause of this measure." ("*Si j'ai eu le bonheur de trouver un module général, auquel je puisse rapporter exactement les effets qui me frappent & qui m'intéressent, qu'importe que je connoisse l'essence & la cause de ce module.*") (2:89). In his article "Fontaine" for the *Encyclopédie* (1757), Desmarest wrote of Descartes as one "who in physical matters imagined too much, calculated little, and applied himself even less to confining facts within definite limits, and availing himself of what was exposed to his eyes in order to succeed in solving obscure questions." ("*. . . qui dans les matieres de Physique imagina trop, calcula peu, & s'attacha encore moins à renfermer les faits dans de certaines limites, & à s'aider pour parvenir à la solution des questions obscures de ce qui étoit exposé à ses yeux*") (83b). To my knowledge, scholars who have studied the importation of Newtonian versions of experimental ideas into France have given no consideration to Desmarest's possible role in this process.

[19] "Géographie physique," 616b. ("*... la vraie Philosophie consiste à découvrir les rapports cachés aux vûes courtes & aux esprits inattentifs.*")

[20] "*Il est aisé de voir par cet exposé, qu'un système de Géographie physique n'est autre chose qu'un plan méthodique où l'on présente les faits avérés & constans, & où on les approche pour tirer de leur combinaison des résultats généraux: . . . .*" (626a).

clear both in overt references and in unattributed use of elements from, for instance, Buffon's *Théorie de la terre* (examples are: the presumed significance of geometrically-expressed symmetries in relations of continents and oceans on the globe's surface; orientations and configurations of features such as mountains and rivers). In "Fontaine" the great majority of references are to authors concerned with the theoretical interpretations of operations of springs (i.e., their mechanism), as distinct from natural-history reporting of particular springs. Like Buffon, Desmarest was inclined toward an idea of natural history that focussed on gaining a grasp of nature's general operations or processes.

Desmarest and the Antiquarian Scholars

I believe we must take note of one further aspect of the scholarly foundations of Desmarest's published work up through 1757. As I have already argued, his initial experience of the natural sciences was essentially literary, in that he at first learned about empirical science by reading about it. But in his erudition, he was a man of literary predilections in a deeper sense as well. His early writings tend to reflect the mind of a humanist scholar as well as a man of natural science. He often located the issues he discussed in a framework of tradition and authority, drawing out (not uncritically) information and perspectives from bygone authors. To some extent, his method was exegetical, centering on recitation and then examination of the opinions found in literature on a given subject. (That Desmarest never really abandoned this habit can be seen in the huge work of his old age, the volumes on *Géographie physique* done for the *Encyclopédie méthodique*.)[21]

This feature of Desmarest's approach is perhaps most obvious in his *Dissertation* (1753), which after all addressed a problem (England's former connection to the Continent) then considered to belong, first and foremost, in the domain of history and archaeology. Natural science's most evident contribution to this problem would be expected to come through discussion of how (i.e., by what natural processes) England might have been severed from France. But the more immediate question was whether indeed this topographic connection had once existed, and here the work of antiquarian scholarship was thought to be relevant, if not in fact paramount. It was Desmarest's concern to establish the separation of England "by different sorts of monuments," and to deal with physiographic change on the evidence of "authorities and medals."[22]

---

[21] Nicolas Desmarest, *Géographie physique*, 5 vols. (Paris: Chez H. Agasse, 1794–1828). In the series *Encyclopédie méthodique*; volume 5 was finished by other authors.

[22] *Dissertation* (ref. n. 1 above), 2, 98.

Works of ancient authors (such as Julius Caesar, Seneca, and Tacitus) and of antiquarian scholars of the previous two centuries (John Twyne, William Camden, William Somner, Pierre Borel, etc.) were consulted respectfully along with modern geographical authorities (Varenius, Halley, Buache). Desmarest's interest in this sort of problem, situating topographic issues within a temporal framework, did not end with the prize essay. For example, a question with both historical and physical dimensions that occupied Desmarest a few years later, in 1760, was whether French rivers may have been physically modified by natural means over the centuries, by filling in with silt. The navigability of the Seine in ancient and medieval times was the subject of a prize competition sponsored by the Rouen Academy. Desmarest believed that the Seine and its tributaries had been regularly filling in with transported material. Considering whether to prepare an entry for the competition (so far as I have been able to learn, he did not), for advice and assistance in addressing this question he turned to friends with expertise in antiquarian scholarship.[23]

Desmarest was acquainted with a number of antiquarian thinkers. Indeed, a striking fact about Desmarest's personal circumstances, during his first decade or so in Paris, is the extent to which his associations and friendships were with historical and archaeological scholars. It has already been mentioned that Desmarest had employment with Pierre-Nicolas Bonamy, editor of the *Journal de Verdun*. Bonamy was a member of the Académie des Inscriptions et Belles-Lettres. It may be that Bonamy helped to introduce Desmarest to others of that Academy, the most exalted institutional home in France for historical erudition. Probably a more important figure for Desmarest's insinuation into the Parisian network of scholars was Pierre-Jean Grosley (1718–1785). Several years Desmarest's senior and, like him, a product of the Collège de l'Oratoire at Troyes, Grosley was a literary and legal scholar resident in Troyes, who was to gain election in 1761 as *associé libre* of the Académie des Inscriptions. He needed eyes and ears in Paris, a function fulfilled by Desmarest. What survives from their correspondence, mostly from the mid-1750s through the 1760s, is an invaluable resource for a view into Desmarest's life (and Grosley's) during this time. Much in their letters concerns the details of Desmarest's involvement in Grosley's literary and intellectual transactions in Paris. The letters also make plain that Desmarest was, at least partly through Grosley, integrated into a society of erudite thinkers dedicated to antiquities, history, and the fine arts. These included Étienne Lauréault de Foncemagne (1694–1779), C.-J.-F. Hénault (1685–1770), the abbé Claude-Pierre Goujet (1697–1767), J.-B. Lacurne de

---

[23] Desmarest to Grosley, 3 March 1760, BN/803, fol. 98–99; Desmarest to Grosley, [April 1760], BN/803, fol. 127–128.

II

Beginnings of a Geological Naturalist

Sainte-Palaye (1697–1781), the comte A.-C.-P. de Tubières de Caylus (1692–1765), the doctor Camille Falconet (1671–1762), and (perhaps the one other than Grosley who was then closest to Desmarest) Claude-Henri Watelet (1718–1786). It is noteworthy that many in these circles – including Foncemange, Hénault, Goujet, and Lacurne de Sainte-Palaye, as well as Grosley and Desmarest – had backgrounds in the Oratorian schools, known for inculcating in students a taste for rationalism and critical historical inquiry.[24]

These scholars were not men of natural science as that term is now understood. Yet we may need to remind ourselves of the different configuration of the cognitive map during the eighteenth century; there was a far weaker distinction then than now, for instance, between natural science and historical inquiry. In *La Géographie des philosophes*, Numa Broc has pointed to the tendency among scholars of the Académie des Inscriptions to mingle issues of ancient and contemporary geography, in keeping with the Enlightenment's extension and deepening of the classical era's humanism.[25] Several historians of our time have helped us awaken to the methodological affinities in the activities of early modern thinkers trying to interpret relics of nature and human texts or artifacts within a single conceptual framework.[26] It was not by accident that some humanist-naturalists saw fit to collect both human antiquities and natural antiquities, and to display them in parallel (even overlapping) arrangements in their *cabinets*. In the assembly, examination, and interpretation of sacred and historical texts; of ancient medals, monuments, and inscriptions; or of fossil objects of various kinds, it was thought to be legitimate and necessary to bring to bear comparable tools for identification and authentication, to use similar critical methods to expose falsehood and myth, and to recognize how indirect and fragmentary is our access to the past through relics both human and natural.

Within the group of scholars of Desmarest's acquaintance, there was no strong sense of distance between philosophical scrutiny of nature on one

---

[24] See Lionel Gossman, *Medievalism and the Ideologies of the Enlightenment: The World and Work of La Curne de Sainte-Palaye* (Baltimore: The Johns Hopkins Press, 1968).

[25] Numa Broc, *La géographie des philosophes: Géographes et voyageurs français au XVIIIe siècle* (Paris: Éditions Ophrys, 1975), 20, 257.

[26] See, for example, Cecil J. Schneer, The Rise of Historical Geology in the Seventeenth Century, *Isis*, 1954, 45:256–268; Michael Hunter, *John Aubrey and the Realm of Learning* (New York: Science History Publications, 1975); Joseph Levine, *Dr. Woodward's Shield: History, Science and Satire in Augustan England* (Berkeley, Los Angeles, and London: University of California Press, 1977); Rhoda Rappaport, Borrowed Words: Problems of Vocabulary in Eighteenth-Century Geology, *The British Journal for the History of Science*, 1982, 15:27–44; Paolo Rossi, *The Dark Abyss of Time: The History of the Earth & the History of Nations from Hooke to Vico* (Chicago and London: University of Chicago Press, 1984).

hand and scholarly, literary, and aesthetic interests on the other. (Among the subjects raised in Desmarest's letters to Grosley, often through allusions to scholarly inquiries undertaken by their friends and acquaintances, were the classification and significance of prehistoric tells [*tombelles*] in Holland, the location in Champagne of Attila's defeat in battle, the technology of water distribution during the thirteenth century, and the interpretation of primitive fetishes [*fétiches*] or idols.) When, in due course, Desmarest took up collecting mineral objects, this was not likely to be seen as an activity incommensurate with the collection of monuments or inscriptions; indeed, as Rhoda Rappaport has reminded us, a common language applied to both.[27]

Most importantly for purposes of this essay's argument, even before he began to collect mineral objects or to make field observations, Desmarest had already acquired critical faculties common in antiquarian scholarship, skills that would have meaning for study in the natural sciences. He cultivated habits of critical examination of textual evidence. Desmarest had not just read; he had learned something of the importance of comparing facts, as reported in the texts, to assess their credibility. It is reasonable to suppose that, before he became involved in serious empirical investigations of his own, the young Desmarest gained from his association with the world of erudite scholarship some of his characteristic respect for the difficulty of clear and exact establishment of facts, some of his suspicion of theoretical systems, and some of his irreverence regarding conventional understanding of tradition and authority.[28]

Desmarest Becomes a Naturalist

I will try to sketch out only very briefly the beginnings of Desmarest's own activities as a field naturalist and natural history collector. Such evidence as is available about his new exercise of empirical practices is found in his personal

---

[27] Rappaport, "Borrowed Words" (ref. n. 26 above).

[28] On this point – the influence of scholarly erudition on Desmarest's convictions and attitudes – there is perhaps tangenital relevance in another of Desmarest's projects during the 1750s. He was editor of the posthumously-published compilation of the scholar Louis Dufour de Longuerue, *Longueruana, ou receuil de pensées, de discours et de conversations, de feu M. Louis du Four de Longuerue*, 2 vols. in 1, compiled initially by Jacques Guijon, then edited by Desmarest (Berlin [i.e., Paris]: [n. p.], 1754). While Desmarest apparently contributed nothing to the anecdotal texts themselves, other than perhaps to have helped select them from Longuerue's writings, one might suppose that he could not have been completely indifferent to some of the leading themes they convey: comparison of the former world with the present one, historical changes in the Earth's surface, antipathy toward abstract systems, enthusiasm for firm establishment of facts, and a critical attitude toward sacred and classical texts.

Beginnings of a Geological Naturalist 15

correspondence and surfaces for the first time in 1758. It happens that, while there are a few letters dating before 1758, it is only in that year that surviving letters from or to Desmarest become very abundant. So it must be admitted candidly that manuscript evidence from 1757 or earlier, which could either support or undermine the case that has been made for Desmarest's literary scientific habits and 'empiricism in principle,' is scant indeed; my argument on this point has to rest very largely on what we can see in print. For well over a decade after 1757, on the other hand, Desmarest published rather little, and our knowledge of his activities through the 1760s comes disproportionately from his correspondence (no other decade of his life is so well documented by extant letters).

The signs in his correspondence (especially with Grosley and with Gonthier, another *troyen* friend) that he began to collect mineral objects and to study them in the field start to appear in 1758. Some of the earliest references concern the examination of quarries near Troyes, to which Desmarest evidently made frequent journeys (he probably went also to Soulaines, where he retained property). By 1759 Gonthier and others in Troyes had been enlisted in gathering mineral samples (*terres et pierres*) for shipment to Desmarest in Paris, where he was assembling his personal *cabinet d'histoire naturelle*. (Desmarest's *cabinet* eventually grew sufficiently to be included in published lists of the finest collections to be seen in Paris.)[29] In 1759 he was carrying out field studies in the Marne valley, and in August he presented a report on some petrified wood he found at Montmirail to the Académie Royale des Sciences.[30] By early 1760 Desmarest apparently was participating in the *promenades savantes* organized by Bernard de Jussieu (1699–1777) in the countryside around Paris and perhaps in comparable *voyages lithologiques* arranged by Guillaume-François Rouelle (1703–

---

[29] See the lists of *cabinets d'histoire naturelle*, both institutional and personal, in Antoine-Joseph Dezallier d'Argenville, *La conchyliologie, ou histoire naturelle des coquilles de mer, d'eau douce, terrestres et fossiles*, 3rd edition, 2 vols. (Paris: Chez Guillaume de Bure fils aîné, 1780), 1:796; and Luc-Vincent Thiéry, *Almanach du voyageur à Paris* (Paris: Chez Hardouin, 1783), 124.

[30] It would be pointless here to try to cite every part of Desmarest's correspondence where one finds evidence of his mineralogical collecting and observing during 1758–1760. Some of the relevant letters are (a) Desmarest to Grosley in BN/803, fol. 98–99, 102, 103–104, 125–126, 127–128, 129–130; and Lettres inédites de Grosley (ref. n. 8 above), 320–322; (b) Desmarest to Gonthier in BN/803, fol. 121–122, 123–124; and Bibliothèque Municipale de Troyes, MS 2764.I(2) [correspondence of Desmarest, Malesherbes, and Gonthier; hereafter referred to as BMT/2764.I(2)]; (c) Gonthier to Desmarest in Bibliothèque Municipale de Troyes, MS 2890(7) [correspondence of Gonthier and Desmarest], nos. 7 to 10, 13, 16, 18, 20. For Desmarest's report to the Academy, Archives de l'Académie des Sciences, Procès-verbaux, vol. 78B, 11 août 1759, fol. 668r.

1770).³¹ It may have been on one such excursion that Desmarest discovered he shared interests with Anne-Robert-Jacques Turgot (1727–1781); Desmarest told Grosley that together with Turgot he had begun "the study of the environs of Paris in a project which will overthrow Guettard's entire system."³² Later that summer Desmarest, accompanied by Watelet, made an extended tour of the Vosges, with much emphasis on lithological and geographical observation and on the collection of mineral specimens crated and shipped to Paris.³³

Quite early in this expansion of his experiences, Desmarest compared and contrasted his investigatory interests with those of his humanist friend Grosley, then travelling in Italy:

> The excursions that I have made and that have kept me busy longer than I expected have shortened my stay at Troyes. You study men during your journeys, whereas I apply myself to stones. Each of these two studies has as its object things very difficult to comprehend: man's dissimulation hardly surpasses that of nature.³⁴

There is also much among the scattered references to Desmarest's early mineralogical collecting and observing to suggest that knowledge of useful mineral resources for agricultural or industrial purposes (*marnes, glaises, craies, pyrites*, and so forth) was much on his mind. Indeed, Desmarest remarked in 1760 to Grosley, in discussing the possible consequences of his studies for

---

³¹ The term *promenade savante* was used by the Marquis M.-J.-A.-N. Caritat de Condorcet in his "Éloge de M. de Jussieu," *Histoire de l'Académie Royale des Sciences*, Année 1777 (1780), 94–117, on 97. Desmarest spoke of Rouelle's *voyages lithologiques* in his *Géographie-physique* (ref. n. 21 above), (1794), 1:410.

³² Desmarest to Grosley, 28 April 1760, BN/803, fol. 102. ("*Nous avons commencé nous deux M. Turgot le Maître des Requetes l'examen des environs de Paris dans un plan qui culbutera tout le Systeme de Guettard.*")

³³ Desmarest later gave a lengthy account of his journey through the Vosges, in a letter to Malesherbes, 1 March 1764, BMT/2764.1(2), no. 33. Details also are seen in Desmarest's letters to Grosley, 21 June 1760, BN/803, fol. 105–106; and 18 July 1760, Lettres inédites de Grosley, 335–336.

I omit consideration here of some "Observations géologiques" made in 1760, on a journey through parts of Champagne, Burgundy, and the Lyonnais, notes which Desmarest claimed he had written, after they were published as part of the *Oeuvres de Mr. Turgot*, 9 vols., ed. P.-S. Dupont de Nemours (Paris: A. Belin & Delance, 1808–1811), 3:376–447. Although I once accepted Desmarest's claim at face value (Nicolas Desmarest and Geology in the Eighteenth Century, 1969, ref. n. 2 above, 344), I have come to think that evidence points toward Turgot as the real author (an article on this is in preparation).

³⁴ Desmarest to Grosley, Nov. 1758, BN/803, fol. 125–126. ("*Les courses que j'ai faites et qui m'ont occupé plus que je ne pensois ont abrege le sejour de Troyes. Vous etudiez les hommes en voiageant et moi je mattache aux pierres. Lun et lautre etude ont pour objet des choses tres difficiles a saisir: la dissimulation de l'homme ne surpasse gueres celle de la nature.*")

improved and less expensive sources of mineral fertilizers: "Thus natural history, which at first holds out only objects of curiosity, can offer useful ones."[35] This utilitarian motif is reaffirmed in the evidence from extensive natural history observations he made in Guienne in 1761 and 1762 at the behest of his patron of the moment, Charles-Robert Boutin, the Intendant of the *Généralité* of Bordeaux.[36]

The theme of utility brings us to an historical circumstance, as regards Desmarest's emerging active involvement in natural history, that cannot be ignored. For this important transformation in his scientific character coincided very closely with his commitment to an official role in the rational comprehension, control and stimulus of the useful arts. Starting in 1757 Desmarest took up employment with the royal government's regulatory supervision of industry, putting him on a path toward becoming an inspector of manufactures. His first responsibilities under the authority of the Intendant des Finances and Directeur du Commerce Daniel-Charles Trudaine (1703–1769) were to examine the textiles industries (*la draperie*), and within a few years he was analyzing techniques and trades, such as cheese manufacture, tanning, papermaking, and diverse agricultural arts. His journey through the Vosges in 1760 with Watelet was no doubt organized principally to visit certain industries. By about that time it was evident that he was coming to look on this sort of employment as his primary occupation. Not the least of its attractions was the opportunity to travel, which as he stated to Grosley "is the only way to learn precise details that are not found in books."[37]

---

[35] Desmarest to Grosley, 3 March 1760, BN/803, fol. 98–99. (*"Ainsi l'histoire naturelle qui ne presente dabord que des objets curieux peut en offrir d'utiles."*)

[36] Two sets of notes are known to survive from Desmarest's excursions in the southwest of France in the early 1760s: (a) Voyage dans une partie du Bordelois et du Périgord, Bibliothèque Municipale de Bordeaux, MS 721, and (b) Remarques de Mr. Desmarest (de l'Académie des Sciences) sur la géographie physique, les productions & les manufactures de la généralité de Bordeaux, lors de ses tournées depuis 1761 jusqu'en 1764, Archives départementales de la Dordogne (Périgueux), MS 26. Both sets are rationalized, reworked, and recopied notes, rather than moment-by-moment notations; this is particularly true of the latter document, which is the more interesting from the point of view of the geological ideas it expresses. Nonetheless these documents more nearly reflect the condition of notes on observations than of formal expositions abstracted from observations.

[37] Desmarest to Grosley, 9 May 1760, BN/803, fol. 103–104. ([In speaking to A.-R.-J. Turgot and, it would appear, his brother Étienne-François Turgot] *". . . je leur avois montré le desir de m'interesser en voiageant qui est le seul moien d'apprendre des precisions qui ne se trouvent point dans les livres."*) The context of this particular statement about Desmarest's interest in travel went beyond the study of industry: he had been given indications he would be selected as part of a group to travel, either to Siberia under Royal patronage, or to Germany at the expense of a group of "amateurs," evidently for observations of the transit of Venus; but no such plan worked out, as

For this essay's purposes, the important point is that in this emerging professional role as analyst of techniques and crafts, Desmarest rather abruptly had thrust upon him abundant opportunities to travel, as well as obligations to acquire and assess new information about the practical arts through his own faculties of observation. It is apparent that at almost precisely the same time that Desmarest began to exploit these opportunities and fulfill these obligations, he began also to engage in the activities of a naturalist. That is, within a short time after beginning his new career in the government control of industry, Desmarest began to collect mineral objects and to study them in the field. I do not think it excessively speculative to view Desmarest's new professional duties, which required him to exercise his empirical capabilities by observing artisans and craftsmen at work, as a critical catalyst in Desmarest's conversion into a naturalist-observer.

By the early 1760s Desmarest was making himself into a competent geological observer. There are opportunities for deepening our understanding of how this happened, but which cannot be explored in this short essay. For example, a detailed consideration of Desmarest's field notes of 1761 and 1762 could yield some interesting reflections on Desmarest's emerging observational character. Another problem of significant interest is whether Desmarest's early experiences in field observation, and particularly his interest in study of the Marne valley, was connected with (perhaps inspired by) his knowledge of Nicolas-Antoine Boulanger's *Anecdotes de la nature*. Yet another question, perhaps not resolvable without finding new manuscript evidence, concerns the reasons why Desmarest began around 1760 to speak disparagingly of the observations and ideas of Jean-Étienne Guettard (1715–1786), the naturalist-academician he had treated with apparent respect in the later 1750s. These and other possible lines of study are for other essays. For the moment, it suffices to recognize that in 1763, when Desmarest made his way into Auvergne to fulfil assignments to investigate certain industries, the skills in field observation that he would apply there, with unexpectedly important results for geology, were relatively newly-won, but established upon the foundations of prolonged and clearly directed scholarly study.

---

far as Desmarest was concerned. As for so much in tracing Desmarest's life, the biographical notes written by his son Anselme-Gaëtan (ref. n. 7 above) are a critical source of information on his father's early involvement in service to Trudaine and the Bureau du Commerce. See also Franc Bacquié, *Les inspecteurs des manufactures sous l'Ancien Régime*, vol. 11, 1927, of *Mémoires et documents pour servir à l'histoire du commerce et de l'industrie en France*, dir. J. Hayem, 12 vols. (Paris: Hachette & Marcel Rivière, 1911–1929).

Conclusion

Like many others who have taken an interest in trying to understand geology's early development, I think that an important factor in geology's establishment as a science toward the end of the eighteenth century was a growing sensibility on the part of geologists regarding cultivation of appropriate empirical methods of investigation, symbolized by (but not confined to) the act of geological fieldwork. The very notion of establishing a science of the earth upon the basis of field investigation was itself an innovation in progress during Desmarest's lifetime. This process was in turn part of a wider movement focussing on matters of fact, established critically by direct experience, and on fitting such critically-founded evidence into intelligible structures of explanation. The attitudes and skills needed to engage successfully in the process did not come automatically; they had to be developed, learned, refined, and internalized in the habits and language of practitioners. Desmarest's case affords us a chance to learn something about how an individual living through the emergence of geological science grappled with certain aspects of these new things. There is good reason to doubt that Desmarest's experience was peculiar to him; in the absence of a clearly established tradition of field-based geological science, it is highly likely that others too came to geological fieldwork from a background of text-based learning.

I have tried to show that Desmarest encountered the direct experiences of geological investigation in the course of (rather than from the start of) developing his interests in knowledge of the natural world. He evidently began his intellectual adventures in a world of books, in studies of texts. Only in his 30s did he begin extensive efforts to investigate nature directly. This was a major shift of orientation; yet through this change there were continuities in the ways he approached and understood the matters of fact and interpretation that concerned him. In particular, in embarking on a course of direct investigation of nature, Desmarest had already settled on a clear position about how observations must be treated: their significance depended upon their integration into a framework of generalization – observations must be used to find general relations, which in turn offer some prospect of yielding explanatory principles. His scholarly background, far from being irrelevant to his new empirical activities, provided important elements of their success. His work as an investigative naturalist, I believe, was strongly affected by patterns of thinking already well established before he seriously took up active observation.

Perhaps through his education, and probably through the most important of his early associations with other thinkers – many of whom were antiquarian scholars – he learned initially to approach scientific problems through procedures oriented around critical study of texts. We should not forget that in Desmarest's time the modernization of antiquarian scholarship had already established its own special emphasis on matters of fact and on critical examination of proper grounds for acceptance of factual truths; indeed, the focus of factual truth among antiquarians was on knowledge of a past inaccessible to direct scrutiny, ascertainable only through its relics, both verbal and material. To the extent that Desmarest's attitudes were formed through the lens of antiquarian scholarship, these would have furnished Desmarest – even in advance of any opportunities for direct investigation of the natural world – with understanding of a critical approach to the establishment of specific facts, and especially of facts bearing a temporal dimension, concerned with past events.

There is an apparent affinity between Desmarest's early scholarly experiences and the character of his scientific publications during his first decade in Paris. Toward the end of that decade, in 1757, Desmarest began what may be seen as his second lifelong career, in government service for the rationalization and regulation of industry. While his involvement in this government work certainly can properly be viewed as a distinct line of activity, parallelling his science and scholarship, in this essay I hope to have indicated one of the ways the two lines intersected. In Desmarest's particular case, initiation into government service, with its imperatives to travel and observe, probably was the occasion of his first intensive experience of deriving information less from the written word than from his own direct inquiry. Desmarest's emergence as a naturalist observer and collector began, so far as we can tell, very soon following the start of his technological occupation. I am inclined to think that this is more than chronological coincidence: perhaps confidence in the successful application of empirical procedures to industrial inspection helped to inspire commitment to a comparable approach to the earth sciences.[38]

***

This essay appeared originally in French, under the title "La Genèse d'un naturaliste: Desmarest, la lecture et la nature," in the volume presented to François Ellenberger for his 80th birthday: *De la Géologie à son histoire*, edited by Gabriel Gohau (Paris: Comité des Travaux Historiques et Scientifiques, 1997). I

---

[38] If this suggestion has merit, it would not be the only way that Desmarest's technological identity entered significantly into his scientific pursuits. In another essay that I hope soon to publish, I try to show that Desmarest's long campaign to gain election to the Académie Royale des Sciences finally met with success, in 1771, in some measure because by then he had established his credentials as an expert in the rationalization of the practical arts.

am grateful to M. Gohau and to the editors of CTHS for permission to publish this English version.

François Ellenberger died in January 2000. I was honored to know François, a man of rare and remarkable qualities, for over a quarter of a century. Just as this essay in its French form was offered in tribute to François, the present English version is dedicated to his memory.

My thanks to Jean-Marc Kehres; our discussions while he worked on formulating the French version forced me to think through several points anew. I also wish to express thanks to Rhoda Rappaport, Kennard B. Bork, and Martin J. S. Rudwick for helpful comments and suggestions, and to Jean Gaudant and Gabriel Gohau for their close reading of the essay. The research was supported in part by grants from the National Science Foundation (SES-8719713) and the University of Oklahoma Office of Research Administration; I gratefully acknowledge this support.

ARCHIVES

Bordeaux
Bibliothèque Municipale de Bordeaux
  MS 721: Desmarest, Nicolas, "Voyage dans une partie du Bordelois et du Périgord."
Paris
Archives de l'Académie des Sciences
  Procès-verbaux des séances
Bibliothèque de l'Institut de France
  MS 3199, Fonds Cuvier: Desmarest, Anselme-Gaëtan, "Notes et renseignements sur la vie et les ouvrages de mon père."
Bibliothèque Nationale [= BN]
  MS 803, Fonds français, nouvelles acquisitions françaises [= BN/803]
  [Correspondence: Desmarest, Grosley, Gonthier]
Périgueux
Archives Départementales de la Dordogne
  MS 26: Desmarest, Nicolas, "Remarques de Mr. Desmarest (de l'Académie des Sciences) sur la géographie physique, les productions & les manufactures de la généralité de Bordeaux, lors de ses tournées depuis 1761 jusqu'en 1764."
Troyes
Bibliothèque Municipale de Troyes [=BMT]
  MS 357: *Catalogus Scholasticorum, Collegii Treoensis*, 3 vols.
  MS 2764.I(2) [=BMT/2764.I(2)]
  [Correspondence: Desmarest, Malesherbes, Gonthier]
  MS 2890(7) [=BMT/2890(7)] [Correspondence: Gonthier, Desmarest]

REFERENCES

Bacquié, Franc, *Les inspecteurs des manufactures sous l'Ancien Régime* (Paris: Hachette & Marcel Rivière, 1927). [Vol. 11 of *Mémoires et documents pour servir à l'histoire du commerce et de l'industrie en France*, ed. J. Hayem, 12 vols. (Paris: Hachette & Marcel Rivière, 1911–1929).]

Broc, Numa, *La géographie des philosophes: Géographes et voyageurs français au XVIIIe siècle* (Paris: Éditions Ophrys, 1975).

Buffon, Georges-Louis Leclerc, Comte de, *Oeuvres philosophiques de Buffon*, ed. Jean Piveteau (Paris: Presses Universitaires de France, 1954).

Condorcet, Marie-Jean-Antoine-Nicolas Caritat, Marquis de, Éloge de M. de Jussieu, *Histoire de l'Académie Royale des Sciences*, 1777 (pub. 1780), 94–117.

Cuvier, Georges, Éloge historique de Nicolas Desmarets [sic], lu le 16 mars 1818, in *Recueil des éloges historiques lus dans les séances publiques de l'Institut Royal de France*, 3 vols. (Strasbourg & Paris: F. G. Levrault, 1819–1827), 2:339–374. Volume 2 appeared in 1819.

Desmarest, Nicolas, *Theses philosophicae. Has Theses, Deo duce & auspice Dei-parâ propugnabit Nicolaus Desmarets Trecensis, In Aula Collegii Treco-Pithoeani Sacerdotum Oratorii Domini Jesu, die Lunae Decimâ Sextâ Anno Domini millesimo septingentesimo quadragesimo-sexto, horâ post meridiem tertiâ* (Troyes: Bouillerot, 1746). [8 pages; copy in the Bibliothèque Municipale de Troyes].

Desmarest, Nicolas, Remarques sur les nouvelles expériences de l'électricité pour se préserver des effets du Tonnerre, *Suite de la clef, ou journal historique sur les matières du tems*, 1752, 72:50–61. [Desmarest's authorship of this article, and the three listed next below, is stated in the journal's *Table générale* for 1697–1756 (Paris: Chez Ganeau, 1759), 3:374, and 4:31.]

Desmarest, Nicolas, Nouvelles expériences sur l'Electricité des nuages orageux, *Suite de la clef, ou journal historique sur les matières du tems*, 1752, 72:134–137.

Desmarest, Nicolas, Lettre à l'auteur du journal historique, sur l'Histoire générale & particulière de l'électricité, &c., *Suite de la clef, our journal historique sur les matières du tems*, 1753, 73:276–291.

Desmarest, Nicolas, Observations sur la réponse à la lettre du 10 décembre 1752, touchant l'histoire générale & particulière de l'électricité: Insérée dans le journal de septembre, pag. 196., *Suite de la clef, ou journal historique sur les matières du tems*, 1753, 74:434–447.

Desmarest, Nicolas, *Dissertation sur l'ancienne jonction de l'Angleterre à la France, qui a remporté le prix, au jugement de l'Académie des Sciences, Belles-Lettres & Arts d'Amiens, en l'année 1751* (Amiens: Chez la Vve Godart; Et se vend à Paris, Chez Ganeau, Chaubert, Lambert, 1753).

[Desmarest, Nicolas], *Conjectures physico-méchaniques sur la propagation des secousses dans les tremblemens de terre, et sur la disposition des lieux qui en ont ressenti les effets* ([No place, no publisher], 1756). [Published anonymously.]

Desmarest, Nicolas, Fontaine, in *Encyclopédie, ou dictionnaire raisonné des sciences, des arts et des métiers, par une société de gens de lettres* (1757), 7:81b–101a.

Desmarest, Nicolas, Géographie physique, in *Encyclopédie, ou dictionnaire raisonné des sciences, des arts et des métiers, par une société de gens de lettres* (1757), 7:613b–626a.

Desmarest, Nicolas, *Géographie physique*, 5 vols. (Paris: Chez H. Agasse, 1794–1828). [Part of the *Encyclopédie méthodique* series; Vol. 5 completed by others following Desmarest's death.]

Dezallier d'Argenville, Antoine-Joseph, *La conchyliologie, ou histoire naturelle des coquilles de mer, d'eau douce, terrestres et fossiles*, 3rd ed., 2 vols. (Paris: Chez Guillaume de Bure fils aîné, 1780).

Ellenberger, François, *Histoire de la géologie*, 2 vols. (Paris: Technique et Documentation [Lavoisier], 1988–1994).

Encyclopédie, ou dictionnaire raisonné des sciences, des arts et des métiers, par une société de gens de lettres, 17 vols. (Paris: Chez Briasson, David l'aîné, Le Breton, Durand [Vols. 1–7]; Neufchastel: Chez Samuel Faulche & Compagnie [Vols. 8–17], 1751–1765).

Geikie, Archibald, *The Founders of Geology*, 2nd ed. (London and New York: Macmillan, 1905).

Gossman, Lionel, *Medievalism and the Ideologies of the Enlightenment: The World and Work of La Curne de Sainte-Palaye* (Baltimore: The Johns Hopkins Press, 1968).

Grosley, Pierre-Jean, Lettres inédites de Grosley et de quelques-uns de ses amis. Recueillies par M. Truelle Saint-Evron . . . et annotées par M. Albert Babeau, in *Collection de documents inédits relatifs à la ville de Troyes et à la Champagne méridionale, publiés par la Société Académique de l'Aube*, Vol. 1 (Troyes: Librairie Dufey-Robert, Léopold Lacroix successeur, 1878), 219–433.

Hauksbee, Francis, *Expériences physico-méchaniques sur différens sujets, et principalement sur la lumière et l'éléctricité, produites par le frottement des corps. Traduites de l'anglois de M. Hauksbée, par feu M. de Brémond, de l'Académie Royale des Sciences. Revûes et mises au jour, avec un discours préliminaire, des remarques & des notes, par M. Desmarest*, 2 vols. (Paris: Chez la Veuve Cavelier, & Fils, 1754).

Hunter, Michael, *John Aubrey and the Realm of Learning* (New York: Science History Publications, 1975).

Lallemand, Paul, *Histoire de l'éducation dans l'ancien Oratoire de France* (Paris: Ernest Thorin, 1888).

Levine, Joseph M., *Dr. Woodward's Shield: History, Science and Satire in Augustan England* (Berkeley, Los Angeles, and London: University of California Press, 1977).

Longuerue, Louis DuFour de, *Longueruana, ou receuil de pensées, de discours et de conversations*, 2 vols. in 1 (Berlin [i.e., Paris]: [n. p.], 1754). [Compiled by Jacques Guijon, edited by Nicolas Desmarest.]

Rappaport, Rhoda, Borrowed Words: Problems of Vocabulary in Eighteenth-Century Geology, *The British Journal for the History of Science*, 1982, 15:27–44.

Rossi, Paolo, *The Dark Abyss of Time: The History of the Earth & the History of Nations from Hooke to Vico*, transl. Lydia G. Cochrane (Chicago and London: University of Chicago Press, 1984).

Schneer, Cecil J., The Rise of Historical Geology in the Seventeenth Century, *Isis*, 1954, 45:256–268.

Suite de la clef, ou journal historique sur les matières du tems. Contenant quelques nouvelles de littérature & autres remarques curieuses, 120 Vols. (Paris: Chez Ganeau, 1717–1776). [Also known as the *Journal de Verdun*.]

Table générale alphabétique et raisonné du journal historique de Verdun, sur les matières du tems, depuis 1697. Jusques et compris 1756, 9 Vols. (Paris: Ganeau, 1759–1760).

Taylor, Kenneth L., Nicolas Desmarest and Geology in the Eighteenth Century, in *Toward a History of Geology*, Cecil J. Schneer, ed. (Cambridge, Massachusetts: MIT Press, 1969), 339–356.

Taylor, Kenneth L., Desmarest, in *Dictionary of Scientific Biography*, 16 vols., ed. Charles C. Gillispie (New York: Charles Scribner's Sons, 1970–1980), 4:70–73. [Vol. 4 was published in 1971.]

Thévenot, Arsène, Notice historique sur l'ancien collége et le lycée de Troyes, *Annuaire administratif et statistique du département de l'Aube*, 1877, 79–132.

Thiéry, Luc-Vincent, *Almanach du voyageur à Paris, année 1783* (Paris: Chez Hardouin, 1783).

III

Nicolas Desmarest and Italian Geology

An Italian Journey

In 1765, in his fortieth year, the French naturalist Nicolas Desmarest set out for Italy on a journey that lasted nearly a year. This voyage through Italy came at a decisive point in Desmarest's growth as a scientist, when his scientific activities were coming to focus successfully on problems of a geological nature. The experience of a year of travel in several parts of Italy played a significant role in Desmarest's subsequent work and thought. In this paper I will try to explore what Italy meant to Desmarest's geological outlook, both from the standpoint of his own Italian observations and, to a lesser extent, with respect to his awareness of the work of Italian scientists and of other travelers in Italy.

The occasion of Desmarest's trip to Italy was what we might consider the advanced education of the young Duc Louis-Alexandre de La Rochefoucauld (1743–1792). Son of the Duchesse d'Enville, who was a friend and patron of scientific scholars, the Duke needed reliable and learned companionship for his tour. The Duke had scientific interests, which Desmarest could help to nourish. Indeed, the Duc de La Rochefoucauld became a noted scientific amateur, and through his eventual appointment as one of the Paris Academy's *honoraires* he participated in French science at a high level. He continued, until his brutal death in the Revolution, to show Desmarest particular interest and encouragement.

The party left Paris in summer 1765, and after spending much of August in Geneva and the nearby Savoy Alps, proceeded by the Mont Cenis Pass to Turin and Parma in September. They spent a month or more in Tuscany, before going on to Rome near the end of October. Soon (after an audience with the Pope) they were off to Naples and Campania, then returned to Rome for the winter season. Departing Rome finally in mid-March, the travelers crossed the central Apennines into the Marches, and went on to Ravenna, Ferrara, Padua, Venice, Verona, and Milan. Following a side trip from Milan to Lago Maggiore, they made their way to Genoa, and back to France through Provence in early May.

III

2   *Desmarest and Italian Geology*

Evidence to Reconstruct Desmarest's Italian Experience

I think of the main historical documentation regarding Desmarest's Italian journey as classified in three basic groups, descending in reliability (so far as the journey itself is concerned) as the documents depart in time from the events of 1765–1766. First are the documents connected immediately with the trip, including what survives from Desmarest's correspondence, La Rochefoucauld's travel diary, and the letters and sketches of the accompanying artist, Boissieu.[1] Second are Desmarest's published references to his Italian observations during the succeeding dozen years or so.[2] Third, and sometimes problematic, are the many things Desmarest had to say about Italy much later, mostly in his compendious *Géographie physique* of the *Encyclopédie méthodique* series.[3]

The volumes of *Géographie physique* for which Desmarest was responsible appeared between 28 and 45 years after the journey ended, so one may question how accurately the details of what Desmarest had done and thought in the 1760s were remembered here. One needs to distinguish between two types of article in *Géographie physique*. On one hand is material which explicitly reports on Desmarest's personal activities in Italy (and there is a fair amount of this first-

---

[1] Letter from Desmarest, Geneva, to Mlle Collot, Soulaines, 20 August 1765, in possession of Desmarest's descendant, M. Chartrin of Blois; letter from Desmarest, Rome, to Grosley, 12 November 1765, Bibliothèque Nationale, F.F., N.A., MS 803, fol. 117–118; La Rochefoucauld notebook, "Voyage d'Italie," 1766, Archives de l'Académie des Sciences. La Rochefoucauld's progress in his travels was reported from time to time in *Gazette de France*: For 1765, no. 75 (20 September), p. 299; no. 80 (7 October), p. 319; no. 84 (21 October), p. 335; no. 86 (28 October), p. 343; no. 93 (22 November), p. 371; no. 96 (2 December), p. 383; no. 100 (16 December), p. 397; no. 102 (23 December), p. 406; and for 1766, no. 16 (24 February), p. 62; no. 28 (7 April), p. 114; no. 42 (26 May), p. 171. Some information on Desmarest's Italian journey, apparently compiled from original documents, is found in the "Notes et Renseignements sur la vie et les ouvrages de mon père," prepared by Desmarest's son Anselme-Gaëtan, Bibliothèque de l'Institut de France, Fonds Cuvier, MS 3199. Desmarest took a letter of introduction from d'Alembert to Lagrange in Turin; for the letter and Lagrange's response see J.-A. Serret, 1867–1892, vol. 13, pp. 41, 48–49. See also the draft of a letter from Turgot, Limoges, to Desmarest, 7 January 1766, partially published in G. Schelle, 1913–1923, vol. 2, pp. 603–604. A member of the Duke's party was the artist Jean-Jacques de Boissieu (1736–1810), from Lyon; among Boissieu's papers are letters with details about the journey. See Pérez, 1979 and plates; also Pérez, 1984. De Boissieu's sketches and paintings are, of course, testimony to what the Duke and his entourage saw. Another of the Duke's companions for the Italian trip, together with Desmarest, was a younger brother of André Morellet, acting as 'secretary.'

[2] Desmarest, 1771; 1773; 1779a; 1779b.

[3] Desmarest, 1794–1828. The fifth volume (1828) and atlas (1827) were published posthumously, and in the case of the last text volume appear not to incorporate Desmarest's own work to any significant extent; the last volume Desmarest put out was the fourth (1811). Also Desmarest, 1806.

person reportage, oddly enough for an encyclopedia). On the other hand are parts of the work relating to Italy which bear no such personal stamp, or which are clearly derivative or discuss places we know Desmarest did not see. Not surprisingly, there exists in *Géographie physique* an extensive middle zone between these two categories: articles where he discusses topics pertaining to Italy with which we know he had (or could have had) familiarity first-hand, but where his terms do not explicitly claim this direct experience. In this paper I have drawn on the sometimes tenuous *Géographie physique*, along with sources that are chronologically more proximate, to develop my account of Desmarest's Italian activities. But I have done so with, I hope, a cautious respect for the limitations of the collection's use for this purpose. On the other hand, I have also wanted to avoid neglect of Desmarest's encyclopedia as a valuable reflection on the rather different issue of Desmarest's views about Italian geology late in his life, around the turn of the century; for this purpose *Géographie physique* is not quite so problematic a resource.[4]

From the diverse kinds of sources I have mentioned, it is possible to assemble at least a partial account of Desmarest's scientific activities during his Italian travels. The account that emerges is quite consistent with what is known on other grounds about Desmarest's career line at this time in his life. That is, the details of Desmarest's journey reflect his ongoing assimilation of the still novel idea that study of the earth's features and the processes shaping them could be a suitable focus for his scientific energies. One sees in various ways his excitement and enthusiasm, his special interest in the volcanic phenomena upon which his recently-started Auvergne investigations were centered, his belief that a bright new field of endeavor lay before him, and his desire to sort critically through the work of other naturalists and draw selectively from it in the task of formulating the firmly-grounded, general earth science he thought hardly yet existed.

---

[4] Even in this respect – the use of Desmarest's idiosyncratic, highly personal encyclopedia as a window onto the opinions he held late in his career – the existence of some difficulties must be acknowledged. In several places Desmarest, 1794–1828 essentially reproduces parts of others' works without attribution; certainly this puts one on guard. So does Desmarest's habit of reproducing here and there sections of his own writings, sometimes decades-old. One does not always know whether to read an article in Desmarest, 1794–1828 as an up-to-date expression of Desmarest's thinking, or as the slapdash result of haste to fill a volume out. One instance of a disappointing article is "Italie," which is drawn to an alarming extent from an essay of Guettard's published in 1768 (Desmarest, 1794–1828, vol. 4, pp. 435–440).

4

The Alps and Apennines

Even before arriving in Italy, early in the journey Desmarest experienced a close-up view of Alpine terrain that was eye-opening, if not almost revelatory. Although he had seen mountainous country in the Vosges and in Auvergne, and had even visited a limited part of the western Pyrenees, Desmarest was not prepared for what he encountered at Chamonix in August 1765. To his good friend Mademoiselle Collot, he expressed astonishment at the enormous scale of the mountains, and the apparent instability and constant change in the glaciers, ice crashing and tumbling violently everywhere "with a horrible fracas that resounds in all the mountains round about."

> What struck me most was the horror of the bearing of mountains that change at every instant, of precipices, of torrents and cascades that vanish into a fine rain after having begun as considerable flows of water. These are frightful beauties ... where the operations of nature are seen on a large scale ... [5]

He went on to contrast nature's engagement "in continual destruction," in the Savoy Alps, with the "smiling and tranquil appearance" of nature in his native Champagne village of Soulaines.

We should not be surprised that Desmarest, in so many ways a man of the Enlightenment, was not possessed of a Romantic sensibility. Possibly the expression "these are frightful beauties" might tempt us to discover in him some forerunning thread of Romantic feeling. But more central to Desmarest's comments is uneasiness bordering on terror in the presence of the destruction peculiar to high mountains, destructive power which he opposes to the character of places fit for human habitation. He saw the processes acting in the high mountains more nearly as specimens of the exceptional than as sublimely intensified examples of the normal.

Some two months after first arriving in Italy, Desmarest wrote an almost equally revealing letter to another close friend. To Pierre-Jean Grosley he confided that the mineralogical and geographical natural history of both the Alps and the Apennines was essentially "new and virgin: neither the forms nor the materials of these ranges are known ..." So, he said, "none of the frameworks of explanation heretofore imagined fits the phenomena." He

---

[5] Letter from Desmarest to Collot, 20 August 1765 (see note 1). A letter written by the artist Boissieu describes the group's activities among the "Monts Maudits"; published in M.-F. Pérez, 1984, pp. 81–83. Of a number of mountain landscape drawings done by Boissieu in the Chamonix region, one showing the Aiguille du Dru is reproduced by Pérez, 1984, p. 78. Although this was Desmarest's first view of high Alpine scenery, the Duke had already been to Chamonix three years earlier (La Rochefoucauld, 1893).

thought that the similarities he recognized between the characters of the Alps and Apennines and those of the western Pyrenees warranted confidence in a scheme for a general ordering of important facts about mountain ranges and about the earth as a whole. We may lament the fact that Desmarest's letter omits any specification of this scheme of ordering facts.[6]

Desmarest also told Grosley of his sustained contacts with the Florentine physician and naturalist Giovanni Targioni Tozzetti (1712–1783). Targioni allowed Desmarest to examine his collections, and the two had lengthy discussions. Desmarest read in Targioni's writings, which the French visitor considered to be closely reliant on the views of Steno. Desmarest implied that his excursions in Tuscany were organized to a substantial degree with a view toward applying Targioni's descriptions and analyses, in which Desmarest found both "exactitude and defects." The paths of these trips included the trajectories between Florence and Livorno, from Pisa to Lucca and Carrara, from Lucca back to Florence, and from there to Siena and Volterra.[7]

---

[6] Letter from Desmarest to Grosley, 12 November 1765 (see note 1). Grosley, incidentally, had been to Italy twice, and on his second journey there in the late 1750s, Desmarest had acted as his Paris agent, serving to coordinate the dispatch of appropriate letters of introduction.

[7] *Ibid.* See also Desmarest, 1773, p. 644; and Desmarest, 1794–1828, vol. 1, pp. 525–573 ("Targioni"); vol. 2, pp. 687–697 ("Apennin"); vol. 2, pp. 801–808 ("Arno"); vol. 3, pp. 301–302 ("Carrare"); vol. 4, pp. 298–300 ("Golfoline"); vol. 4, pp. 650–653 ("Montamiata"); vol. 4, p. 656 ("Monte-Rotondo"). In a eulogy of Targioni, Félix Vicq-d'Azyr attributed to the Italian scientist a minor role (and to the Duc de La Rochefoucauld a rather more favorable one) in normalizing French and Italian mineralogical communication in the later part of the 18th century. La Rochefoucauld purchased, and had Targioni classify and label, an entire collection of minerals and fossils of Tuscany, which he then had brought back to Paris. According to Vicq-d'Azyr this permitted French specialists to untangle the unmanageable Italian nomenclature, in which Targioni had according to general Italian usage applied local Tuscan names, and to substitute a set of terms convenient to French and Italians alike (Moreau, 1805, vol. 3, pp. 322–323).

Evidently Desmarest read Targioni in Italian – he told Grosley he was "translating" some volumes by the Tuscan naturalist (one of whose travel accounts was published in French only a good many years later).

Desmarest's own previous publications display his familiarity with the work of some Italian authors, some written in Italian but much in Latin and sometimes French. For example amidst the voluminous material Desmarest added to the French translation of Francis Hauksbee's *Physico-Mechanical Experiments* (1754) he repeatedly cited the work of the Florentine Accademia del Cimento, as well as the acoustical research of the Bolognese Francesco Maria Zanotti (1692–1777) published in 1731 (Zanotti, 1731). And in his article "Fontaine," for example (Desmarest, 1757), Desmarest discussed critically the defense presented by Niccolò Gualtieri (1688–1744) of the theory of springs' derivation of water through subterranean connections with the sea, opposing Vallisnieri's meteoric theory. (Desmarest referred to the lengthy review of Gualtieri's book in *Journal des Sçavans*, June 1725.) Desmarest also cited information, in "Fontaine," from works

III

6 *Desmarest and Italian Geology*

Italian Volcanic Phenomena

On the way from Tuscany to Rome, Desmarest and La Rochefoucauld had their first opportunity to observe Italian volcanic phenomena. Because of his preoccupation during the previous two years with the extinct Auvergne volcanoes, and especially with his identification of prismatic basalts as volcanic products, examination of Italian volcanic districts no doubt lay highest among Desmarest's priorities. He clearly counted heavily on this chance to compare Italian with Auvergnat conditions, and to move toward stronger generalization of conclusions he had already begun to draw from the French region familiar to him.[8]

Desmarest's subsequent discussions of Italian volcanic effects tended to mark out his experiences in four geographical regions: (1) from western Umbria – around Radicofani and Acquapendente – southward to the vicinity of Rome; (2) the Roman neighborhood, especially to the east and southeast, particularly the Albano Hills; (3) the Neapolitan region, including of course Vesuvius and the Phlaegrean Fields; and (4) the Vicentine and Paduan districts, including particularly the Euganean Hills, the Monti Berici, and the Alpine foothills near Vicenza.

Naturally Desmarest was intensely interested in finding that in Italy, just as in Auvergne, there were abundant proofs that prismatic basalts – indeed all basalts of any degree of organization – are distinctly associated with volcanic lavas and must therefore be considered volcanic in origin. Among the places where this was most clearly seen, he said, were the areas around and to the north of the Lake of Bolsena, and in the Vicentine-Paduan districts. He particularly singled out, in the latter area, the basalts to be seen in the Roncà Valley, and in the Alpone Valley between San Giovanni Ilarione and Bolca.[9]

---

by Giambattista Riccioli (1598–1671), Domenico Guglielmini (1655–1710), Luigi Ferdinando Marsigli (1658–1730), and Giovanni Poleni (1683–1761).

[8] Desmarest's anticipation of seeing Italian volcanic terrain is clear in a letter from Turgot to Desmarest, 17 May 1765, Bib. Nat., F.F., N.A., MS 10359, fol. 3; also in letters from Pasumot to Desmarest, 7 May and 31 May 1765, Bibliothèque Municipale de Beaune, MS 310; and in the above-mentioned letter from Desmarest to Grosley, 12 November 1765. See also Desmarest, 1771, pp. 746–748, 758–760; and Desmarest, 1773, pp. 626, 639, 644, 647, 658, 661–663, 665, 667. In his "Lettre à M. l'abbé Bossut" (Desmarest, 1779b), written in a spirit of practical justification for science, Desmarest stressed the concern he had in Italy to inform himself fully on natural cements found in volcanic districts that could be useful as building materials.

[9] Desmarest, 1771, pp. 746–748, 760; Desmarest, 1773, pp. 617, 658, 662, 666. See also Desmarest, 1794–1828, vol. 2, p. 308 ("Alpes du Vicentin, du Véronois & du Brescian"); vol. 2, pp. 689, 694 ("Apennin"); vol. 3, pp. 58–65 ("Basalte"); and vol. 3, pp. 169–170 ("Bolca").

Desmarest and Italian Geology

Of the only visit Desmarest was ever to make to the site of an active volcano, toward the end of 1765, we know surprisingly few details. That the Duke and his companions climbed Vesuvius and viewed the summit crater is reported in two of Boissieu's letters.[10] Desmarest shipped home several crates of rock specimens from the volcano. They also toured the hills, solfataras, and fumaroles near Pozzuoli, and went to the island of Ischia. But for some reason Desmarest left (or we have discovered) relatively little commentary on these excursions. It is notable that, in common with many other 18th-century naturalists, Desmarest apparently thought of Vesuvius as a mountain that pre-existed the volcanic action which he presumed altered it and its constituent materials. This view was in keeping with the concept of volcanic action as a denaturing process, seen more nearly as cooking and transforming mineral substances than as generating them. It is probably relevant that in other contexts Desmarest sometimes exhibited to some degree a conflation of the notions of morphology and structure: a hill (*colline*) might be at once both a topographic feature and a structural entity. Another point claimed by Desmarest, incidentally, on what authority it is not clear, was that Vesuvian lavas decay rapidly, reducing to arable soil in only twenty years.[11]

Of his later passage through the districts of Padua, Vicenza, and Verona we know somewhat more, since appeals to observations in this region, including the Euganean Hills and Monti Berici, became rather frequent in his later writings. One judges from the abundant mention of these northeastern Italian parts that his imagination warmed more easily to them than to the sights around Vesuvius. The most readily apparent reason for this is that Desmarest was impressed by local evidence for alternating depositions of volcanic and sedimentary materials. For at least part of their Paduan-Vicentine tour the travelers were

---

[10] Pérez, 1984, pp. 87–90.

[11] Desmarest, 1771, pp. 739, 748ff.; Desmarest, 1773, pp. 645–646, 649, 665–669; Desmarest, 1779b, pp. 193–196. Also Desmarest, 1794–1828, vol. 2, pp. 204–205 ("Agnano"); vol. 2, pp. 889–890 ("Averno"); vol. 3, pp. 268–269 (" Campanie heureuse"); vol. 3, p. 328 ("Cendres"); vol. 4, pp. 654–656 ("Monte-Nuovo"); vol. 4, pp. 684–688 ("Naples"). For an example of Desmarest's incomplete distinction between morphology and structure, Desmarest, 1794–1828, vol. 3, pp. 437–439 ("Collines"). Some articles in Desmarest, 1794–1828 contain tangential information on Desmarest's ideas about Vesuvius and the Vesuvius district, sometimes through comparison of samples found there with volcanic materials seen elsewhere—e.g., vol. 2, pp. 255–256 ("Albano"); vol. 2, p. 344 ("Alun"); vol. 3, pp. 445–446 ("Communications souterraines"); vol. 4, pp. 394–395 ("Incrustations"); vol. 4, pp. 502–503 ("Lave").

Because Desmarest never got near the end of the alphabet in Desmarest, 1794–1828, we lack his account of "Vésuve," "volcan," and the like. The fourth volume terminates amidst the N's. Following an entire volume (the second) of 896 pages given to the letter A, coverage thinned out rapidly; the next (third) volume treats the letters B, C, and D in 704 pages.

in the company of the young naturalist Giovanni Battista (Alberto) Fortis (1740/41–1803), whose interests in the same geological features may already have been connected with the opportunity they afforded for an approach toward a periodization of geological stages of development.[12]

One other aspect of Desmarest's program of volcanically oriented observations that I wish to mention, before summing up on this dimension of his journey, depended not on field studies but rather on antiquities preserved in Rome. He wanted to sort out the types of stone known by the name basalt. This inquiry included checking the material nature of Roman monuments and other ancient works of art, made from what was considered in antiquity to be basalt. In this effort Desmarest had the help of, among other scholars, the antiquarian Johann Joachim Winckelmann (1717–1768), with whom Desmarest formed a cordial friendship. Desmarest concluded eventually that the term basalt was applied rather indiscriminately in antiquity, so that little was to be gained from reliance on ancient commentators.[13]

Desmarest's volcanological observations in Italy, as he wrote about them later, served by and large to address a fairly clearly delineated set of ideas, which may be summarized as follows:

(1) Basalt's volcanic origin is confirmed by a wide variety of field observations, and this igneous rock assumes diverse forms owing not only to its composition but also to the conditions of its cooling.[14]

(2) Volcanic products in different places display different degrees of destruction through natural processes. So in certain localities it is relatively easy to identify specific lava flows and their centers of eruption, while in others the situation is more complex and confusing, owing to the more advanced state of the rocks' degradation.[15]

(3) Signs are frequently found, in volcanized areas, of volcanic and aqueous geologic processes succeeding one another. Volcanic layers are sometimes seen overlying calcareous or other water-formed strata, and sometimes are covered in turn by similarly aqueous sediments. Desmarest took pains also to describe

---

[12] Desmarest, 1771, pp. 746–747; Desmarest, 1773, pp. 617, 647, 658–661; Desmarest, 1806, pp. 256–260. In Desmarest, 1794–1828, see especially vol. 1, pp. 3–7 ("Arduino"); vol. 2, pp. 155–159 ("Adige"); vol. 2, pp. 306–308 ("Alpes du Vicentin, du Véronois & du Brescian"); vol. 3, pp. 169–170 ("Bolca").

[13] Desmarest, 1773, pp. 607ff. See Winckelmann, 1952–1957, vol. 3, pp. 134–136, 186–187, 216–217, 236–237, 309–311.

[14] See note 9.

[15] Desmarest, 1771, pp. 746–747; Desmarest, 1773, pp. 663, 665; Desmarest, 1806, pp. 256–260; Desmarest, 1794–1828, vol. 1, p. 7 ("Arduino"); vol. 2, pp. 203–204 ("Agnani"); vol. 2, p. 308 ("Alpes du Vicentin, du Véronois & du Brescian").

volcanic rocks that showed the results of aqueous infiltration and mineralization subsequent to their cooling.[16]

(4) Finally, all volcanic products show a range of alteration of various rock materials presumed to be the pre-volcanic stuff upon which volcanic fires acted. These primordial or source materials included especially rocks of granitic type, but not to the exclusion of others, including calcareous materials. Like a great many other naturalists of the age, Desmarest thought it improbable that crystals could grow from an igneous melt, so their presence in lavas was thought to show the incomplete volcanic reduction of the crystalline source rock. Examination of the minute parts of volcanic matter yielded clues, Desmarest believed, both to the types of rock present in the source volcanic site and to the extent of what we might call the petrological transformation effected by heat acting with different intensities or for different lengths of time. As he put it in an article on varieties of pozzolanes, the differences within a class of volcanic product result from "different primary materials, altered or melted by fire, as well as from principles joined to these materials upon contact with the blaze (*flamme*) within the vast crucible of volcanoes."[17]

These four ideas tended to dominate what Desmarest had to say of Italian geology during the decade or so following the end of his journey in 1766. He may well have been convinced of each of these views before ever setting foot in Italy, but if so his Italian experience enlarged their support. And these represented firm opinions from which he did not stray, and which remained central to his volcanological thought. However, volcanoes and their effects were far from Desmarest's sole geological interests in Italy. One can be misled by the fact that Desmarest's main publications of the 1770s, when he was establishing his scientific reputation, dealt with basalts and epochs of volcanism. To the extent that these publications made use of his Italian observations – and this was substantial – it might be natural to conclude that Desmarest's geological vision during his Italian travels was pretty much confined to things volcanic. But this was not the case.

---

[16] Desmarest, 1773, pp. 647, 658–662, 666, 668; Desmarest, 1806, pp. 256–260; Desmarest, 1794–1828, vol. 1, pp. 6–7 ("Arduino"); vol. 1, pp. 270–271 ("Lazzaro Moro"); vol. 2, p. 308 ("Alpes du Vicentin, du Véronois & du Brescian"); vol. 3, pp. 169–170 ("Bolca"); vol. 3, pp. 638–640 ("Dessiccation"); vol. 4, pp. 394–395 ("Incrustations"); vol. 4, pp. 650–653 ("Montamiata").

[17] Desmarest, 1771, pp. 724, 749, 757–761, 764; Desmarest, 1773, pp. 624–649, 658–669; Desmarest, 1779b, p. 194.

III

Desmarest's General Geological Interest in Italy

In fact, Desmarest put to good use his chances to study Italian geological phenomena of all kinds, not just those he would consider volcanic. In brief defense of this claim, I will mention three considerations arising out of his writings postdating the 1770s, a decade in which, as I have said, practically all of his geological publications had a volcanological focus.

First, he did eventually find occasion to report on his visits to Italian localities whose interesting features had little or nothing to do with volcanism. Thus, for example, he rendered accounts in *Géographie physique* of his observations in the drainage of the lower Adige, of his travels near Rome and across the central Apennines from Foligno into the March of Ancona, and of other crossings of the Apennines between Genoa and Turin, between Pavia and Genoa, and by two routes between Bologna and Florence.[18]

Secondly, one finds in Desmarest's later writings a wealth of attention to varied types of geological features or processes, his understanding of which he illustrates at least in part through his Italian experiences. When he came to Italy, for instance, he brought with him a conviction that the study of river configurations – their geometric orientation and the disposition of confluences, sinuosities, and opposing banks – could provide important clues to the generation of physiographic regularities. These ideas he applied, and perhaps developed further, as he looked on the Italian scene.[19] He found evidence to support the anti-Buffonian opinion that running waters, not marine currents, have been the primary agents of physiography.[20] In the Adige he saw waterborne rocks at great heights above the river's modern course, which he took as proof of the water's power to excavate to depths of hundreds or even thousands of feet.[21] Taking note of a dry valley meeting the main valley of the Topino, filled with torrential deposits, he regarded this as evidence for the

---

[18] Desmarest, 1794–1828, vol. 1, pp. 535–536 ("Targioni"); vol. 2, pp. 155–159 ("Adige"); vol. 2, pp. 514–517 ("Ancone"); vol. 2, pp. 635–638 (Anio"); vol. 2, pp. 687–697 ("Apennin"); vol. 3, p. 438 ("Collines"). See also vol. 2, p. 25 ("Abri").

[19] Desmarest, 1794–1828, vol. 1, pp. 125–132 ("Ferber"); vol. 1, pp. 528–530, 541–542 ("Targioni"); vol. 2, pp. 515–517 ("Ancone"); vol. 2, pp. 687–697 ("Apennin"); vol. 2, pp. 702–703 ("Aquino"); vol. 2, pp. 801–808 ("Arno"); vol. 4, pp. 509–511 ("Ligurie"); vol. 4, pp. 638–639 ("Montagne"); vol. 4, pp. 650–653 ("Montamiata").

[20] Desmarest, 1794–1828, vol. 1, p. 5 ("Arduino"); vol. 1, pp. 525–573 ("Targioni"); vol. 2, pp. 155–159 ("Adige"); vol. 2, p. 177 ("Affaissement"); vol. 2, pp. 203–204 ("Agnani"); vol. 2, pp. 306–308 ("Alpes du Vicentin, du Véronois & du Brescian"); vol. 2, pp. 515–517 ("Ancone"); vol. 2, pp. 801–808 ("Arno").

[21] Desmarest, 1794–1828, vol. 1, pp. 5–6 ("Arduino"); vol. 2, pp. 155–159 ("Adige"); vol. 2, pp. 307–308 ("Alpes du Vicentin, du Véronois & du Brescian").

great variability in the rate of operation of geological agents.[22] He was on the lookout for erratic rocks, signs of agents capable of transporting materials considerable distances from their source localities.[23] He was sensitive to the presence of extraneous fossils as witnesses to conditions in which organisms formerly lived – whether these were the fine fish impressions at Monte Bolca or remains of elephants dug up in relatively recent sediments.[24] He wanted to see caves, springs, brine pools, and mofettes or gas vents.[25] Perhaps most earnestly of all he sought evidence to confirm his suspicions that repeated incursions and departures of the sea had resulted in alternating production of marine and non-marine geological effects. For Desmarest, pertinent signs of sea incursions and retreats included not only marine sediments and attendant organic remains, but also rounded stones which he presumed were formed by sustained wave action. Non-marine geological action would be attested to by both freshwater and volcanic deposits, and also by erosive effects in deposits of any kind.[26]

A striking aspect of this last point, concerning Desmarest's interest in periodic advances and retreats of the sea, is its implicitly historical character. Geological remnants were here being looked at as keys to events which were to be put in sequential order. The same must be said, of course, of the distinctions Desmarest made during the 1770's in relative ages of lava flows based on their degree of degradation as well as their relative position. There are grounds for

---

[22] Desmarest, 1794–1828, vol. 2, pp. 514–515 ("Ancone").

[23] Desmarest, 1794–1825, vol. 1, p. 5 ("Arduino"); vol. 2, pp. 155–159 ("Adige").

[24] Desmarest, 1794–1828, vol. 2, pp. 306–308 ("Alpes du Vicentin, du Véronois & du Brescian"); vol. 2, p. 703 ("Aquino"); vol. 2, pp. 804, 806 ("Arno"); vol. 3, pp. 169–170 ("Bolca"); vol. 3, pp. 173–174 ("Bolonois"); vol. 4, p. 387 ("Ichtyopètres"). In his "Notes justificatives des faits rapportés dans les Epoques de la nature," Buffon made mention of elephant remains brought back to Paris by La Rochefoucauld and Desmarest from a site near Rome (Piveteau, 1954, p. 201).

[25] Desmarest, 1794–1828, vol. 1, pp. 444–446 ("Sénèque"); vol. 2, pp. 342–343 ("Aluminières"); vol. 2, p. 697 ("Apennin"); vol. 2, p. 776 ("Arezzo"); vol. 3, p. 193 ("Bouches d'Eole"); vol. 3, pp. 240–245 ("Bullicames"); vol. 3, pp. 315–318 ("Castelnuovo"); vol. 3, p. 331 ("Cesi"); vol. 3, pp. 329–330 ("Cerboli"); vol. 3, pp. 564–565 ("Crissolo"); vol. 4, p. 183 ("Foligno"); vol. 4, p. 477 ("Lagonis"); vol. 4, pp. 497–498 ("Latera"); vol. 4, pp. 508–509 ("Libbiano"); vol. 4, p. 624 ("Modène"); vol. 4, pp. 628–629 ("Mofettes"); vol. 4, pp. 629–631 ("Molfetta"); vol. 4, p. 652 ("Montamiata"); vol. 4, pp. 656–657 ("Monte-Rotondo"); vol. 4, pp. 687–688 ("Naples").

[26] Desmarest, 1794–1828, vol. 1, pp. [493], 506–508 ("Stenon"); vol. 1, p. 534 ("Targioni"); vol. 2, pp. 155–159 ("Adige"); vol. 2, pp. 169–174 ("Adriatique"); vol. 2, pp. 203–204 ("Agnani"); vol. 2, pp. 306–308 ("Alpes du Vicentin, du Véronois & du Brescian"); vol. 2, pp. 514–517 ("Ancone"); vol. 2, pp. 687–697 ("Apennin"); vol. 3, pp. 169–170 ("Bolca"); vol. 4, pp. 28–31 ("Embouchures des fleuves"); vol. 4, p. 176 ("Fleuve"); vol. 4, pp. 478–481 ("Lagunes"); vol. 4, p. 585 ("Méditerranées ou mers intérieures"); vol. 4, pp. 650–653 ("Montamiata"); vol. 4, pp. 684–688 ("Naples").

thinking that by the time of his Italian journey, or at least by its end, Desmarest was already beginning to grow beyond the rather two-dimensional orientation toward physical geography to which he had been mainly drawn by Guillaume-François Rouelle.

Italian Geological Figures

The third and final consideration I wish to raise, regarding the breadth of geological interest Desmarest exhibited while he traveled in Italy, has to do with the other scientists he knew, or of whose work he took cognizance. Here again we know less than I would like, but the late *Géographie physique* gives some help. As I have mentioned, while in Italy Desmarest evidently had significant contacts with Targioni Tozzetti and Fortis. He respected both, but evidently Targioni the more – not surprisingly since Targioni was an accomplished observer in the prime of life, while Fortis was at that time a quite young naturalist.[27] There is nothing I know of to suggest that Desmarest met Giovanni Arduino (1714–1795), but in due course at least Desmarest became an admirer of his work.[28] Anton Lazzaro Moro (1687–1764) had recently died, and in any case Desmarest was to depict Moro, in *Géographie physique*, as a somewhat misguided doctrinaire whose system relied excessively on global volcanism.[29] At what point Desmarest came to know of Lazzaro Spallanzani (1729–1799) is not clear – perhaps only some time after his Italian travels. In later years Desmarest's work contains only scanty (if implicitly complimentary) references in passing to Spallanzani.[30] The fact that this Italian scientist was still living was sufficient, on Desmarest's terms, to disqualify him (as it did Fortis as well) from being accorded a full article in *Géographie physique*'s first volume, a selective gallery of heroes and rogues in earth science.

Besides Arduino, Moro, and Targioni Tozzetti, the only other Italian appearing among the forty figures Desmarest selected for this volume was Luigi Ferdinando Marsigli (1658–1730). Although he came a generation even before Moro, Marsigli received Desmarest's warm praise on the strength of his combination of broad geographical vision and careful topographic observation,

---

[27] Besides the fact that Desmarest, 1794–1828 includes a long article on Targioni (vol. 1, pp. 525–573), his name arises far more often elsewhere in these volumes than does that of Fortis. For Targioni, e.g., vol. 1, pp. 156, 489–491, (493); vol. 2, pp. 689, 788, 804; vol. 3, pp. 241, 438; vol. 4, pp. 395, 650–651; for Fortis, vol. 1, pp. 3, 7; vol. 4, pp. 629–630.

[28] Desmarest, 1794–1828, vol. 1, pp. 3–7 ("Arduino").

[29] Desmarest, 1794–1828, vol. 1, pp. 267–271 ("Lazzaro-Moro").

[30] E.g., Desmarest, 1794–1828, vol. 3, p. 301 ("Carrare").

as shown in his physical studies of the sea and of the Danube. Desmarest looked with a forgiving spirit on Marsigli's excessive enthusiasm for the hypothesis of a worldwide structure of interconnecting mountain ranges – the so-called *charpente du globe* or global scaffolding – perhaps because Desmarest had once found the theory highly attractive himself.[31]

Two non-Italian scientific personages loom large in Desmarest's *Géographie physique* through their Italian-based observations. One is the Swedish naturalist Johann Jacob Ferber (1743–1790), who first traveled in Italy a half-dozen years after La Rochefoucauld and Desmarest. Desmarest saw Ferber as developing a more detailed knowledge of different Italian rock massifs on the basis of distinctions established by Arduino, applied especially in northern Italy. Particularly approved by Desmarest were Ferber's inclinations toward interpretation of topography with reference to distribution of river systems, and toward extraction of conclusions about successions of rocks from their dispositional order.[32]

The other foreign observer in Italy chosen for special recognition by Desmarest was Steno. In *Géographie physique* a translation of a large part of Steno's *Prodromus* is presented. Desmarest claims to have excised some of the language he found overly metaphysical, by which I suppose he meant too comprehensively formulated, seemingly too much aimed at the general nature of things and not enough at the immediate problem of interpreting rock forms. But Desmarest gave Steno honorable credit for establishing much of the framework of thinking that permitted people like Arduino and Targioni Tozzetti to proceed further.[33]

---

[31] Desmarest, 1794–1828, vol. 1, pp. 328–[336] ("Marsigly"). One sees throughout Desmarest, 1794–1828 Desmarest's recognition and use of the work of a number of other Italian authors. Among these were the Bolognese mathematician and hydraulics specialist Eustachio Manfredi (1674–1739); the Veronese naturalist Giovanni Giacomo Spada (1680–1744?); Vitaliano Donati (1713–1763), the authority on the Adriatic Sea; the Piedmontese botanist and fossil-describer Carlo Allioni (1728–1804); and Paolo Mascagni (1755–1815), the anatomist and expert on mineral springs.

[32] Desmarest, 1794–1828, vol. 1, pp. 125–132 ("Ferber"). See also Desmarest, 1773, pp. 642–644; and Desmarest, 1806, pp. 276–277. Desmarest participated, with Macquer and Lavoisier, in a favorable report for the Academy of Sciences on the 1776 French translation of Ferber's book on Italian natural history.

[33] Desmarest, 1794–1828, vol. 1, pp. 489–510 ("Stenon"). An "observer in Italy" whom I ignore here is Seneca (Desmarest, 1794–1828, vol. 1, pp. 436–489). Seneca's presence among the forty selected major figures in the history of earth science is a bit of a puzzle. The only ancient personage of the lot, Seneca's inclusion may possibly be partly understood by noting Desmarest's favorable attitude toward that sage's views on the moral neutrality of nature, and on the explanation of the world by recourse to natural causes, especially causes seated in cyclically-repeating processes. And Desmarest was not the only 18th-century geologist who showed interest

III

14 Desmarest and Italian Geology

In both Arduino and Targioni, Desmarest saw allies in conceptions of geological dynamics and temporal ordering. They were enlisted against Buffon's marine currents, in favor of running water's primacy as the landscape's excavator. Arduino's authority was brought forward on the alternating succession of volcanic and sedimentary rocks, especially in northern Italy. A special talent of Targioni's was the ability mentally to reconstruct original geological structures from which the visible ones had been lately carved by water, an exercise of the imagination that Desmarest had already shown sympathy with early in the 1760s.[34] And both Arduino and Targioni appeared in *Géographie physique* with high marks on one of Desmarest's main scales of evaluation, that of functioning as an accurate observer with due suspicion of hypothetical systems.

While Desmarest's remarks often permit us to gauge his opinions about his predecessors and contemporaries, there are cases where we are left in uncertainty. Throughout Desmarest's work, I find a notable shortage of references to local Italian observers of Vesuvius, such as Giovanni Maria della Torre (1713–1783). Neglected practically as much is the noted Vesuvius amateur William Hamilton (1730–1803). During their stay in Naples in December 1765, the Duc de La Rochefoucauld and his companions were hosted by Hamilton, who arranged musical performances.[35] But nothing suggests that Hamilton and Desmarest – each in his way seen in hindsight as one of the more celebrated volcanological figures of the 18th century – established a relationship based on their common

---

in the Senecan dictum that nature must be investigated in terms of its normal, regular features, without regard to its extraordinary, irregular manifestations. It must be remembered, too, that scientists like Desmarest saw themselves as advocates of a new empirical method for investigating the earth's past, a method making possible for the first time independent determination of the accuracy of statements made by ancient writers (figures whose traditional authority in scientific questions still seemed in need of restraint). It was in this light that Desmarest explained his attention to Seneca in the Desmarest, 1794–1828 article "Anciens (Auteurs)," vol. 2, pp. 513–514. Also pertinent, perhaps, in reducing to size the puzzle of Seneca's presence in Desmarest, 1794–1828 are the similarities between Desmarest's lengthy piece on Seneca and the notes he had contributed as part of the commentary for a French translation of the *Natural Questions*. It may be that Desmarest simply had all too readily available some remotely relevant padding for the volume.

[34] Desmarest, 1794–1828, vol. 1, pp. 525–573 ("Targioni"). Desmarest's notions about imaginative restorations of landforms that had been denuded can be seen in his notes, "Remarques de Mr. Desmarest (de l'Académie des Sciences) sur la géographie physique, les productions & les manufactures de la généralité de Bordeaux, lors de ses tournées depuis 1761 jusqu'en 1764," Archives Départementales de la Dordogne, MS 26.

[35] Boissieu letters in M.-F. Pérez, 1984, pp. 85, 90. Desmarest acknowledged Hamilton's observations of Etna and in the Naples region at least once in Desmarest, 1794–1828, vol. 3, p. 446 ("Communications souterraines"). Similarly, I have found one citation in Desmarest, 1794–1828 of della Torre's studies of Vesuvius: vol. 3, p. 445 ("Communications souterraines").

interest.[36] Nor do Desmarest's publications cite very frequently his French predecessors in Italy, such as Charles-Marie de La Condamine (1701–1774), who a decade earlier than the Duke's journey had been quite attentive to the Italian volcanic terrain.[37]

Conclusion

Desmarest's Italian journey of 1765–66, and his experience of Italy and Italian science, were important elements in his scientific career. The opportunity to see a different and geologically varied land, and to widen his scientific acquaintance, came at an auspicious time for Desmarest. It enabled him to rest his general geological ideas on a wider basis. He could hardly have avoided enlarging his intellectual horizons, during this fortunate journey. What he secured from Italy and Italians for himself, and by extension in some measure for the French scientific community, possibly he was able to repay in small part to Italians like Fortis or even Targioni Tozzetti, or their successors – but this is a question I have not looked into. It does bear noting, however, that in the early 19th century Desmarest's accounting of Italian geology and geologists, in *Géographie physique,* became a public source of at least modest significance.[38]

Let me end with a point that is more nearly a question than a conclusion. Of the sum of Desmarest's published writings about Italy, the ones closest in time to his journey dwell on problems relating to volcanic geology, where his early successes in geological research had begun prior to his Italian trip. In the

---

[36] After this paper was written I discovered an anonymous manuscript, which I believe is in Desmarest's hand, entitled "Copie de plusieurs notes et réflexions isolées sur l'ouvrage de M. Hamilton" (Bibliothèque Municipale de Rouen, Collection Montbret, no. 22). These notes appear to discuss, rather critically, Hamilton's *Campi phlegraei.*

[37] References to La Condamine that I have found in Desmarest, 1794–1828 are at vol. 2, p. 637 ("Anio"); vol. 2, p. 694 ("Apennin"); vol. 2, p. 776 ("Arezzo"); and vol. 3, p. 540 ("Cratère"). A French observer upon whom Desmarest relied to some extent was the illustrator Jean Houel (1735–1813), especially for his travels in Sicily (e.g., Desmarest, 1771, p. 749; Desmarest, 1794–1828, vol. 4, pp. 103–106, 268, 434). It happens that during the same year as La Rochefoucauld and Desmarest, the French scientist Lalande was also traveling in Italy. Lalande soon published a travel account that acquired some notoriety as a guide for French travelers. I have found no evidence of Lalande's path crossing that of La Rochefoucauld's party. But La Rochefoucauld did note that on the road one March day near Spoleto they met the abbé de Condillac ("Voyage d'Italie," see note 1.) From La Rochefoucauld's journal it would appear, incidentally, that they used the venerable old guide by Maximilien Misson, with the supplement by Joseph Addison. They had maps by d'Anville.

[38] Lyell is known to have drawn from *Géographie physique* as he prepared the historical introductory sections of his Principles.

mid-1770s he made public his approach toward a historical interpretation of geological formations in Auvergne, analyzing volcanic deposits into sequential epochs. In the later part of his life, when he published more expansively than ever before, we find that his focus, so far as Italian-related topics go, seems to have shifted so that issues having to do with distinctions among successive formations and with the agents that progressively destroy those formations, are at least as prominent as volcanological questions. One cannot altogether set aside the possibility that this simply reflects a real shift in the center of Desmarest's interests and ideas during his middle years and beyond. But in this paper I have been suggesting otherwise. We have quite firm evidence of Desmarest's convictions about the power of subaerial denudation from a quite early date. But also there are in many parts of Desmarest's late publications, chiefly the *Géographie physique*, signs that some of their contents may be authentic indicators of both the activities and the thoughts he experienced many years earlier. If this sort of reading of Desmarest's *Géographie physique* has any merit – that is, looking at these writings as an accumulative repository of scientific expressions deriving over a lengthy period, with some parts bearing marks that carry back even to the 1750s – then I think we can ask this question: may it not be that Desmarest's *Géographie physique* points toward his Italian experiences, observations, and associations as significant factors in his *early* adaptation, already during the 1760s, to an increasingly historical stance in geology?

***

Research support for this paper was provided by the University of Oklahoma Office of Research Administration.

REFERENCES

DESMAREST, N. (1757). *Fontaine*. In: *Encyclopédie, ou Dictionnaire Raisonné des Sciences, des Arts et des Métiers*. Vol. 7, 81–101, Briasson, David, Le Breton, Durand, Paris.

DESMAREST, N. (1771). *Mémoire sur l'origine & la nature du Basalte à grandes colonnes polygones, déterminées par l'histoire naturelle de cette Pierre, observée en Auvergne*. Histoire et Mémoires de l'Académie Royale des Sciences, (publ. 1774), 705–775.

DESMAREST, N. (1773). *Mémoire sur le basalte. Troisième partie, où l'on traite du basalte des anciens; & où l'on expose l'histoire naturelle des différentes espèces de pierres auxquelles on a donné, en différens temps, le nom de basalte*. Histoire et Mémoires de l'Académie Royale des Sciences, (publ. 1777), 599–670.

DESMAREST, N. (1779a). *Extrait d'un Mémoire sur la détermination de quelques époques de la nature par les produits des volcans, & sur l'usage de ces époques dans l'étude des volcans*. Observations sur la physique, sur l'histoire naturelle et sur les arts, vol. 13, 115–126.

DESMAREST, N. (1779b). *Lettre à M. l'abbé Bossut, de l'Académie des Sciences, sur les différentes sortes de pozzolanes, & particulièrement sur celles qu'on peut tirer de l'Auvergne*. Observations sur la physique, sur l'histoire naturelle et sur les arts, vol. 13, 192–199.

DESMAREST, N. (1794–1828). *Géographie-physique*. In: *Encyclopédie méthodique*. 5 vols., atlas. H. Agasse, Paris.

DESMAREST, N. (1806). *Mémoire sur la détermination de trois époques de la nature par les produits des volcans, & sur l'usage qu'on peut faire de ces ces époques dans l'étude des volcans*. Mémoires de l'Institut des Sciences, Lettres et Arts. Sciences mathématiques et physiques, vol. 6, 219–289.

HAUKSBEE, F. (1754.) *Expériences physico-méchaniques sur différens sujets, et principalement sur la lumière et l'électricité, produites par le frottement des corps. Traduites de l'anglois de M. Hauksbée, par feu M. de Brémond, de l'Académie Royale des Sciences. Revûes & mises au jour, avec un discours préliminaire, des remarques & des notes, par M. Desmarest*. 2 vols. Veuve Cavelier & Fils, Paris.

LA ROCHEFOUCAULD, L.-A. DE (1893). *Relation inédite d'un voyage aux glacières de Savoie en 1762 par le Duc de La Rochefoucauld d'Enville, avec introduction et notes par Lucien Raulet*. Annuaire du Club Alpin Français, vol. 20, 458–495.

MOREAU, J.-L. (ed.) (1805). *Oeuvres de Vicq-d'Azyr*. 6 vols. L. Duprat-Duverger, Paris.

PÉREZ, M.-F. (1979). *The Italian Views of Jean Jacques de Boissieu*. Master Drawings, vol. 17, 34–43.

PÉREZ, M.-F. (1984). *Lettres du voyage d'Italie de Jean Jacques de Boissieu (1765–1766)*. In: Chomer, G., Pérez, M.-F., & Ternois, D. (eds.), *Lyon et l'Italie: Six Études d'Histoire de l'Art*, 75–100. C.N.R.S., Paris.

PIVETEAU, J. (ed.) (1954). *Oeuvres philosophiques de Buffon*. Presses Universitaires de France, Paris.

SCHELLE, G. (ed.) (1913–1923). *Oeuvres de Turgot et documents le concernant*. 5 vols. Paris, Alcan.

SERRET, J.-A. (ed.) (1867–1892). *Oeuvres de Lagrange*. 14 vols. Gauthier-Villars, Paris.

WINCKELMANN, J.J. (1952–1957). *Briefe* (edited by W. Rehm & H. Diepolder). 4 vols. Walter de Gruyter, Berlin.

ZANOTTI, F.M. (1731). *De sono.* De Bononiensi Scientiarum et Artium Instituto atque Academia Commentarii, vol. 1, 173–181.

IV

New light on geological mapping in Auvergne during the eighteenth century: the Pasumot-Desmarest collaboration

A landmark in the early history of geological mapping was the volcano-geomorphological map of part of Auvergne published in 1774 with an article on basalts by Nicolas Desmarest (1). The map was identified as the work of two *ingénieurs-géographes du Roi,* Pasumot and Dailley. Prepared under Royal government sponsorship, this map set a new standard for precise graphic representation of the positions of distinct kinds of rock, and played a role in emerging conceptions about geomorphological change and about the distribution of volcanic rocks (2).

While historians of science have been aware that cartographic innovations were important in the early development of geological science, relatively little detailed information has been available about the process of mapping for geological purposes so early as the 1760s. Contemporary documentation does exist, however, that sheds light on the preparation of this map, and particularly on the working relations between Nicolas Desmarest (1725-1815) and François Pasumot (1733-1804). A large part

---

(*) An earlier version of this paper was presented at the 12th International Conference on the History of Cartography, Paris, in September, 1987.

(1) Mémoire sur l'origine et la nature du basalte à grandes colonnes polygones, déterminées par l'histoire naturelle de cette pierre, observée en Auvergne, *Mémoires de l'Académie Royale des Sciences,* 1771 (1774), 705-775. The map's full title is "Carte d'une partie de l'Auvergne où sont figurés les courants de laves, dans lesquels se trouve le basalte en prismes, en boules, etc. pour servir à l'intelligence du mémoire de M. Desmarest sur ce basalte."

(2) For a discussion of the 1774 map, and of the much larger and more complete Auvergne map at which Desmarest worked for many more years (finally published by his son in 1823), see François Ellenberger, Recherches et réflexions sur la naissance de la cartographie géologique, en Europe et plus particulièrement en France, *Histoire et Nature,* 22-23 (1983), 3-54, especially 26-29. While historically significant as thematic maps of a geological character, the maps with which Desmarest was associated were different in conception from the structurally-oriented "geological map" that emerged in the nineteenth century. It is Ellenberger who refers to Desmarest's maps, appropriately, as "*cartes volcano-géomorphologiques.*"

of the pertinent documentation is a collection of letters from Pasumot to Desmarest, most of them written during the years the mapping was in progress (3). Here one finds much to help form a picture of the enterprise that yielded this novel scientific map.

The idea of mapping the volcanic terrain of part of Auvergne was put into effect following Desmarest's first reconnaissance of the Puy-de-Dôme district in 1763. A dozen years earlier the naturalist Jean-Etienne Guettard, of the Royal Academy of Sciences, had discovered the volcanic character of the region's mountains (4). Now Desmarest made the additional find that the abundant prismatic basalts, whose configurations had long been considered indicative of an aqueous formation, were associated with the congealed lavas and must therefore be volcanic as well. An aspirant to the Academy, Desmarest perceived an opportunity in the largely unstudied Auvergne volcanic landscape. He hoped to prepare a general work on the extinct volcanoes of Auvergne, and a map which would locate the limits of the lavas. His exploitation of that opportunity led to the production of the remarkable map under consideration, and to an original interpretation of the Auvergne basalts and of the region's geological history. It also undoubtedly contributed substantially to Desmarest's election to the Academy in 1771 (5).

Desmarest and Pasumot had a good deal in common. Both were educated in provincial Oratorian schools (Desmarest at Troyes, Pasumot at Beaune), both went to Paris to seek their professions (in 1747 and 1750, respectively), and both had interests in antiquities as well as in natural science (6). Desmarest and Pasumot became acquainted, in fact, several

---

(3) Bibliothèque municipale de Beaune, MS 305. Hereafter, this collection of letters is referred to as BMB/305. I wish to thank the *bibliothécaire,* Mme Bernadette Blandin, who kindly provided me with photocopies of the letters. There are twenty letters, nineteen of them to Desmarest. The other one was written to Desmarest's close friend Mlle Collot, while Desmarest was away in Italy; it also has a bearing on Pasumot's and Desmarest's joint endeavor. The first letter is dated 31 August 1763, the last 29 December 1770. (A year lacking in the date of one letter can be estimated, from internal evidence, at 1764.) Sixteen of the letters date from the period, some two and one-half years in length, from immediately before fieldwork began to just after its completion.

(4) François Ellenberger, Précisions nouvelles sur la découverte des volcans de France : Guettard, ses prédécesseurs, ses émules clermontois, *Histoire et Nature,* 12-13 (1978), 3-42.

(5) On Desmarest, see Kenneth L. Taylor, Nicolas Desmarest and Geology in the Eighteenth Century, in *Toward a History of Geology,* ed. by Cecil J. Schneer (Cambridge, Mass., and London : MIT Press, 1969), p. 339-356; and *Id.,* Desmarest, in *Dictionary of Scientific Biography,* ed. by Charles C. Gillispie, vol. 4 (New York : Charles Scribner's Sons, 1971), p. 70-73.

(6) Desmarest's attention to antiquarian matters remained amateur, but Pasumot in due course earned a reputation as an antiquarian author. Biographical sources on Pasumot, other than in standard biographical dictionaries, include C.-M. Grivaud, Notice biogra-

New light on geological mapping in Auvergne

years before they were brought together for work on the Auvergne map: Pasumot was linked to a network of Desmarest's friends in Troyes, perhaps as a result of his preparation of a map of the Troyes district in 1759, and one of those friends, the Oratorian geographer David Anselme de Bardonnanche, had been Pasumot's teacher in Beaune before moving on to Troyes (7).

During the winter of 1763-1764 it was arranged that Pasumot would be responsible for the mapping in Desmarest's Auvergne researches (8). Pasumot worked in the field for three summers, 1764 through 1766, aided during the second and third summers by another *ingénieur-géographe*, Dailley. Desmarest joined in the work in 1764 and 1766, although rather late in the season, well after the engineers had begun. In 1765 Desmarest was away on a year-long journey in Italy, as companion to the young duc Louis-Alexandre de La Rochefoucauld d'Enville. Fortunately for historians, one consequence of Desmarest's absence from the scene for at least part of each year's fieldwork is that Pasumot's letters reported to Desmarest on events in the field as they progressed (9).

The scheduling of the cartographic fieldwork and the pace of its accomplishment are among the points illuminated by Pasumot's letters to Desmarest. In 1764 Pasumot began work in June, and was joined by Desmarest around the end of August. They finished late in October (10). In 1765 Pasumot got a late start, in August. With Dailley and another engineer (Pezet) he worked again until about the end of October (11). In 1766 Pasumot and Dailley got started once more in August, and kept at the task until near the end of October, joined in September and October by Desmarest (12). The seasons in the field, then, lasted from under three to as much as five months, and extended well into autumn.

phique sur M. Pasumot, in Pasumot's *Dissertations et mémoires sur différents sujets d'antiquité et d'histoire*, ed. by C.-M. Grivaud, 1 vol. in 7 parts (Paris: no publisher, 1810-1813), part 1, p. 1-14; and Charles Aubertin, *Notice sur François Pasumot* (Beaune: Impr. Blondeau-Dejussieu, 1852).

(7) Letter from Desmarest to Gonthier, 10 May [1759], Bibliothèque nationale, F. F., N. A., MS 803, fol. 121-122 (hereafter referred to as BN/803); letters from Gonthier to Desmarest, 15 September 1761, 30 January 1762, and 29 January 1764, Bibliothèque municipale de Troyes, MS 2890, no. 7 (hereafter referred to as BMT/2890). Pasumot's "Carte des environs de Troyes" is in *Éphémérides troyennes*, 1760. Père Bardonnanche (alternatively Bardonnenche or Bardonenche) is referred to frequently in the correspondence among Desmarest, Gonthier, and Grosley mentioned above (BN/803 and BMT/2890), often as someone sharing Desmarest's interests in collecting rocks, minerals, and fossils.

(8) Letter from Gonthier to Desmarest, 29 January 1764 (see note 7 above). The first of Pasumot's letters to Desmarest in the Beaune collection (BMB/305, 31 August/1 September 1763) makes no reference to the Auvergne project.

(9) Six of Pasumot's letters to Desmarest are dated at Clermont, five before Desmarest's arrival in Auvergne for the season's fieldwork, one after his departure.

(10) BMB/305, 28 July 1764, 23 October 1764.

(11) BMB/305, 26 July 1765, 22 November 1765, 7 June 1766.

(12) BMB/305, 17 August 1766, 23 November 1766.

The tasks for the remaining part of the year included calculations from data, drafting of the results, and comparison and discussion of notes and their outcome. Some gestures were also made toward cataloguing and description of rock specimens. In due course, after the fieldwork was done, there were map proofs to be scrutinized and corrected. Throughout, securing commitments of assistance and support from patrons and supervisors, and getting these superiors to take desired action, were matters of sustained concern. But neither Desmarest nor Pasumot could devote more than a fraction of his time between late fall and summer to the Auvergne project. Each had other occupations on his mind. From 1763 to 1771 Desmarest was inspector of manufactures for the intendant Turgot in Limoges. In 1764 Pasumot assumed the physics professorship in the *collège* at Auxerre. Indeed, Pasumot required help from ministerial authorities in Paris to get free from responsibilities at the *collège* for his summer fieldwork (13). Nonetheless, since Desmarest's obligations in Limoges allowed for a winter period in Paris, and in Auxerre Pasumot was close to the capital, the two were able to confer there on occasion.

Pasumot's initial hope was that the fieldwork could be done in a single year. After some preliminary scouting in 1764, Pasumot informed Desmarest that the area to be mapped would be a good deal larger than Desmarest had estimated. Part of the problem lay in confusion over the value of the Auvergne league, which was one-third greater than the Paris league. As executed for publication the map displays an approximately square area, each side about seven Auvergne leagues of 3,000 toises each. This is roughly equivalent to forty by forty kilometers. But a considerable fraction of this space was left uncharted. Pasumot estimated that an able engineer could, on flat terrain, work up in one month an area of at most 6,000 toises squared, or four square Auvergne leagues (14). Thus, even though the map as prepared within the next few years was thorough only for the Mont-Dore district, and for selected areas to the northeast, additional help was necessary. With the collaboration during the succeeding two years of another qualified *ingénieur-géographe* (Dailley), the fieldwork was essentially done in three seasons. Pasumot and Dailley worked separately for the most part, meeting periodically to connect their respective sections.

Conditions for work were not always favorable. Fieldwork was frequently impeded by bad weather. Physical health, as well, was a topic of more than casual concern in Pasumot's letters. Illness not uncommonly afflicted those involved in the mapping, keeping even the healthy

(13) BMB/305, 15 December 1764, 7 May 1765.
(14) BMB/305, 28 July 1764.

New light on geological mapping in Auvergne

away from work to care for the sick. Pezet, who accompanied Dailley from Guyenne for the fieldwork in 1765, fell ill with a "violent malignant fever" in September and never recovered, dying finally in January without having digested or plotted out his observations. So the area of Pezet's responsibility had to be redone (15). Even when everyone was well enough to work, the competence of assistants could not be depended upon. It was also necessary to take precautions against the suspicion and hostility of the local citizenry, who were apt to think that men laden with instruments and apparatus for signals could be up to no good. To put local officials in a hospitable temper and certify that the fieldworkers were not practitioners of sorcery, the intendant Balainvilliers was asked periodically to send out circular letters to his *subdélégués* attesting to the legitimate activity of these men (16).

Equipped with a horse, the engineer might hope to return to the comfort of a proper inn in a sizable town like Clermont or Rochefort as frequently as once each week (17). The innkeeper through whom Pasumot received his mail might be able to direct Desmarest where to look in the field for his colleague (18). Rooms were expensive at Mont-Dore, where the baths were popular, so sometimes the *subdélégué* was prevailed upon to provide a bed (19). Apart from indirect information of this sort, however, Pasumot's letters are not eloquent regarding either the privations or the pleasures of daily work in the open air. If feeling ran strong about the hardships of the work or the beauty of the setting, little of it is expressed in Pasumot's letters.

We do learn such details as that the mapping proceeded at double the scale of the *carte de l'Observatoire* (20); that the engineers were able to use a newly-manufactured graphometer with telescopic sights, made by Canivet (21); and that Desmarest had a portable barometer that could be used in the levelling process (22). Illustrations were made at first by an inexperienced young assistant from Limoges, Dupain, but were later

---

(15) BMB/305, 7 June 1766.
(16) BMB/305, 31 May 1765. A letter, written in the name of the bishop, to be carried by Pasumot to secure the aid of local *curés* (and no doubt to forestall difficulties among parishioners) is reproduced in [L. Welter], Carte des volcans d'Auvergne préparée par Pasumot, *Bulletin historique et scientifique de l'Auvergne*, LXIV (1944), 215-216.
(17) BMB/305, 28 July 1764, 14 August 1764, 17 August 1766.
(18) BMB/305, 28 July 1764, 24 August [1764].
(19) BMB/305, 14 August 1764.
(20) BMB/305, 23 October 1764, 13 November 1764; also letter from Desmarest to Dailley, 26 March 1765, Arch. de l'Acad. des sciences, "Minéralogistes et géologues français," vol. 3. It seems that Pasumot had a change of mind about scale, having thought at first that the *carte de l'Observatoire* scale would serve (BMB/305, 28 July 1764).
(21) BMB/305, 7 May 1765.
(22) BMB/305, 28 July 1764.

done by a more practiced artist, de Boissieu (23). Some *auvergnat* enthusiasts for natural history took an interest in what the visitors were doing, and were helpful in boxing up and shipping crates of specimens that had been collected (24). These and many other minor points that help fill out an image of the enterprise are here. Largely missing from the correspondence, on the other hand, is technical information concerning the calculations, or discussion of graphic techniques that could be used on the map.

One would like to know about the relationships between the scientific mappers and their patrons in government. In Pasumot's letters, and elsewhere, there are suggestive pieces of information that contribute to our knowledge. The Auvergne intendant Balainvilliers had some authority over the project, and needed to be dealt with diplomatically for the practical assistance he could provide. But Pasumot implied disdain for Balainvilliers' pretensions of control (25). Greater responsibility and authority evidently lay elsewhere. Daniel-Charles Trudaine, the former Auvergne intendant under whose protection Desmarest had begun to work for the Bureau du Commerce in the 1750s, and who had special interest in cartographic engineering, was one Paris authority. But the supervisory patrons who are mentioned most often by Pasumot, and who appear personally to have counted the most, were members of the Boutin family. Charles-Robert Boutin, the reforming intendant in Bordeaux, had employed Desmarest for cadastral observations in Guyenne in 1761 and 1762. It was through him that Dailley and Pezet were made available to assist in the Auvergne mapping in 1765 (26). He and his older brother, the *receveur-général* Simon-Charles Boutin, evidently held in some manner the bureaucratic purse strings for the Auvergne mapping project. To judge from Pasumot's letters, it was in regard to one or the other of the brothers Boutin (mainly Simon-Charles, one senses, although the references are usually unclear) that the most serious presentation of accounts and requests had to be made. Records of expenses, reports of work completed and promises of future progress, requests for reimbursement and honorarium payments *(gratifications),* and presentations of basalt prisms — all went to the Boutins. Pasumot frequently saw

---

(23) BMB/305, 17 August 1766, 23 November 1766; also letter from Desmarest to duc de La Rochefoucauld, 29 August 1766, Bibliothèque municipale de Mantes, collection Clerc de Landresse, MS 1170. Two views of Auvergne basalts by de Boissieu appeared among the engravings in the *Recueil de planches* (vol. 6, 1768) of the *Encyclopédie* together with Desmarest's first, brief exposition of columnar basalt's volcanic origin. The existence of additional drawings done in Auvergne by de Boissieu is reported by Marie-Félicie Perez and Madeleine Pinault, Three New Drawings by Jean-Jacques de Boissieu, *Master Drawings*, XXIII-XXIV (1985-1986), 389-395.

(24) E. g., BMB/305, 17 August 1766, 23 November 1766.

(25) BMB/305, 6 August 1764.

(26) BMB/305, 7 May 1765, 31 May 1765, 26 July 1765; also letter from Desmarest to Dailley, 26 March 1765 (see note 20 above).

New light on geological mapping in Auvergne

the Boutins as responsible for frustrating bureaucratic delays, but their genuine commitment to the project is also apparent.

Government support for the mapping was of course motivated mainly by belief in the practical worth of the project. Desmarest was careful at critical junctures to pay lip service to utilitarian considerations, and must have been convinced that the knowledge they were generating would have useful value (27). But it is noteworthy that in Pasumot's letters the practical dimension of the project has quite a low profile. The letters are, to be sure, communications between men sharing an implicit commitment to the merit of what they were undertaking; perhaps for this reason their motives and objectives required little mutual discussion. Whatever the cause, in these private communications the rhetoric of technocratic dedication to improvement practically vanishes, in favor of the technical task at hand, along with the myriad of questions, comments, opinions, bits of news, and tangential personal remarks (on such things as employment opportunities, or points of antiquarian interest) passing between fellow professionals.

Indeed, not the least interesting thing about Pasumot's letters is what they tell us of the professional relationship that he had with Desmarest during the 1760s. Pasumot acted not at all like a subservient technician under orders of a client. One judges from his letters that Pasumot, without quite putting himself forward as an altogether equal partner to Desmarest, adopted the manner of a scientific collaborator. Pretty clearly, Desmarest accepted him cordially enough on those terms.

The evidence for this is spread through the correspondence, and must be alluded to here quite generally and incompletely. Perhaps what counts most is the clear impression one gets of Pasumot's deep involvement in the geological aspects of the work. Pasumot, the cartographic engineer, did not just help define the locations of volcanic peaks and flows, or aid in collecting specimens. He also participated extensively in the interpretation of the terrain. It is obvious that at least some of this he did at Desmarest's request, and that the two were accustomed to open discussion of the physiographic and lithological issues to which the mapping was supposed to lend itself (28). Quite possibly Pasumot on occasion assumed a wider role as a lithological naturalist than Desmarest might have intended for him, and Pasumot's letters do reveal his geomor-

---

(27) Letter from Desmarest to Balainvilliers, 9 August 1766, Arch. départementales du Puy-de-Dôme, C 7044. See also Desmarest, Lettre à M. l'Abbé Bossut, de l'Académie des Sciences, sur les différentes sortes de pozzolanes, et particulièrement sur celles qu'on peut tirer de l'Auvergne, *Observations sur la physique,* XIII (1779), 192-204.

(28) E. g., BMB/305, 28 July 1764, 13 November 1764, 23 November 1766. The most immediate surviving account written by Desmarest about the 1764 fieldwork implies Pasumot's thorough participation in this *expédition lithologique* (letter to Grosley, 11 December 1764, BN/803, fol. 115-116; see note 7 above).

phological limitations more than once (29). On the other hand, Pasumot may have had a point in his self-assured appraisal of his own qualifications for this job: Pasumot said in effect that his possession of a geographer's way of seeing, in addition to a naturalist's sense, gave him more nearly than Desmarest the optimal qualities for this mapping project (30). To put it this way may make Pasumot sound, somewhat misleadingly, presumptuous or aggressively competitive. It is more accurate to say that Pasumot's posture toward Desmarest was that of a candid and confident junior partner.

Pasumot consistently acknowledged Desmarest's authority in geographical-mineralogical knowledge of the extinct volcanoes, and seems never to have challenged Desmarest's status as the scientific investigator around whose work the mapping project was oriented. But Pasumot was never obsequious, nor did he underestimate the value of the services his own learning and judgement could render. He gave Desmarest advice on terminology (suggesting, for example, that the map title omit the word *minéralogique* in favor of *litho-géographique*) (31). Perhaps more tellingly, Pasumot's main rationale in seeking assistants for the mapping project was to turn the work of running triangles over to subordinates so that he could attend more fully to the natural history problems involved (32).

It has not escaped others' attention that the cartographic engineer Pasumot was also an accomplished naturalist (33). Yet, if left to assess his role in early geological research in Auvergne only from printed sources, without the letters commented upon here, we might not realize how fully intertwined the cartographic and geological facets were in the investigations of the 1760s. And we would not be in a position to recognize fully the highly collaborative relationship that existed between Pasumot and Desmarest (34).

(29) BMB/305, 23 October 1764 (Pasumot imagines past lava sources as equal in altitude to existing mountains); 13 November 1764 (Pasumot reports abundant signs of volcanism in the Morvan region).
(30) BMB/305, 28 July 1764.
(31) BMB/305, 29 December 1770.
(32) BMB/305, 28 July 1764.
(33) See Ellenberger, art. cité in n. 2, 26; also François Poplin, Pasumot, Buffon et la dent de mammouth d'Auxerre, *Bulletin de la Société des Sciences Historiques et Naturelles de l'Yonne*, CXX (1988), 81-95, and *Id.*, Buffon, Pasumot et le sommeil paradoxal du chat, *Mémoires de l'Académie des Sciences, Arts et Belles-Lettres de Dijon*, CXXX (1991), 297-308.
(34) It has long been known that Desmarest and Pasumot came to be on poor terms in the later 1770s, in a priority dispute over the identification of zeolite among volcanic products. The final letters of the Beaune correspondence (10 November, 29 December 1770) reveal that already in that year a deterioration in their relations was developing. Desmarest had evidently accused Pasumot of an act of plagiarism.

V

Buffon, Desmarest and the Ordering of Geological Events in *Époques*

Between 300 and 200 years ago, questions about the extent of time through which the Earth's natural operations have functioned were often on the minds of articulate naturalists. However, many of these thinkers seem *not* to have set explicit formulations about such questions at the center of their work. So, to oversimplify the research problem a little, one faces a choice between two possible approaches to historical study of scientific ideas about terrestrial time. One way is to search through what authors said directly about the matter. The other is to focus on what the authors appear to have thought was most important in their efforts to comprehend the Earth (or perhaps what they believed they could speak sensibly about as responsible savants), and then see what inferences might be drawn out regarding how they dealt with time.

Both approaches are important. It is not surprising, if only because of the need to keep the problem within manageable proportions, that many of the most prominent historical studies concerned with the Earth's age are organized around the first of these conceptions (Haber 1959; Albritton 1980; Dean 1981; Hallam 1989, ch. 5; Dalrymple 1991). This paper, however, takes the second, less direct path. At least two related reasons justify this approach. First, we know that there were cultural and institutional constraints on what many reputable naturalists of the eighteenth century felt they could say about the extent of natural time. This is one consideration, although not the only one, in understanding why some of the scientific writers of the period who disavowed a religiously orthodox timescale of several thousands of years were cautious in expressing their views, often resorting to the use of ambiguity. (Quite effective for this purpose, in the French language, was the double meaning of the word *siècle*, which in addition to 'century' might alternatively signify 'age' or 'period.') So time is one of those topics where we have reason to suppose certain eighteenth-century writers did not say all that was on their mind. Second, the constraints on candid expression about time need not always have been external; they might result from personal intellectual integrity and self-discipline. Scientific thinkers who did not want to give utterance to what they saw as mere speculation might have good reason to be circumspect on an issue where they were keenly aware it was so difficult to find solid footing. Time conceptions, or explorations of new possibilities for moving beyond conjecture with regard to the Earth's duration, often might be only implicit or latent in the ways astute writers addressed what they thought to be the central questions, or issues where there seemed reasonable prospect of making headway.

An empirical spirit was highly valued, and was still treated as something rather new, in the natural sciences of the eighteenth century. This was an age when great enthusiasm was expressed about the revitalization of philosophy owing to revived consultation of the book of nature, in place of reading texts inherited from traditional authorities. It was a time when Francis Bacon's reputation was at its peak, as an originator and publicist of inductive and experimental methods, on the Continent as well as in England. In part as a result of applying this Baconian spirit to features observable over the Earth's surface or enclosed within its accessible parts, many eighteenth-century naturalists had begun to conclude that the Earth in its present condition must be far more greatly changed from its earlier state than had been generally thought by their predecessors. If the Earth had truly undergone vast changes, the time required for their accomplishment might very well be much more than was allowed by conventional chronologies. Leading natural philosophers, guided by precepts emphasizing experience as the source of knowledge, were aware that serious difficulties lay in the way of efforts to know the past, particularly a remote past which no human being could have witnessed. It was apparent to those committed scrupulously to empirical methods that the best chances for acquiring secure knowledge of nature's past lay in cultivating the means of interpreting natural 'monuments,' natural relics of times and events inaccessible to direct observation. Among those best informed about such methods were antiquarian and historical scholars, many of whom were just as keen about pursuing historical understanding of 'natural antiquities' as they were about human history (Rappaport 1982, 1997). By the close of the eighteenth century it was widely felt that a far stronger grip had been gained on the means to extend knowledge of the Earth's past, if not actually to determine its age, than had been the case at the century's beginning. I will try here to trace a part of the developments within geological science that lent justification to that opinion.

Buffon, Desmarest and Natural *Époques*

During the last third of the eighteenth century several scientific thinkers and investigators advanced ideas for interpretation of terrestrial features in terms of distinct stages or sequences of the Earth's past. A concept that came to play a very significant role in efforts to construct periods in the Earth's past was expressed through the French word *époque*, or epoch. It is a story that can be told largely through the work of two notable French naturalists, Georges-Louis Leclerc, the Comte de Buffon (1707–1788), and Nicolas Desmarest (1725–1815) (Figures 1 and 2).

Figure 1. Portrait of Buffon, 1761, by Drouais, engraved by Baron. Frontispiece, *Histoire Naturelle*, vol. 1 (1749), from the set held in the History of Science Collections, University Libraries, University of Oklahoma. (Evidently this frontispiece, identified as a 1761 Drouais portrait, was added before binding of the 1749 volume.)

The first linked his interest in *époques* directly with notions about the Earth's age, while the second evidently preferred to avoid making pronouncements on a problem he saw as unavoidably conjectural. While their approaches to establishing a vocabulary of *époques* reflect great differences in their scientific sensibilities, I suggest here that behind such differences there were important intellectual affinities that can be understood through their mutual attachment to ideas found within the prevailing geological idiom, the Theories of the Earth. I will further suggest that Desmarest's important contribution toward the establishment of scientific procedures for placement of geological phenomena in relative order is best understood as arising out of guiding principles found within the tradition of Theories of the Earth, rather than as opposed to them.

Figure 2. Portrait of Desmarest, from an engraving by Tardieu. This presumably shows Desmarest at an age of at least 70 years, since he is shown wearing the apparel of the Institute (founded 1795) designed by David. From a photograph given to the author in 1966 by Desmarest's descendant, Marthe Chartrin.

Buffon and Desmarest each put the term *époque* forward publicly during the 1770s. For both, the term had reference to successive periods and geological events in the Earth's history. Similarly, each of them attached to the notion of *époques* an expansive conception of the time through which the Earth's processes have operated. But an initial review of their respective uses of an *époques* terminology reveals a number of striking differences that seem to outweigh the similarities between them.

The Burgundian Buffon, by far the more famous of the two, centred his 1778 masterpiece *Époques de la Nature* around a series of stages in the Earth's development, from its supposed generation around 75 000 years ago (Buffon 1778, p. 67) (Figure 3).

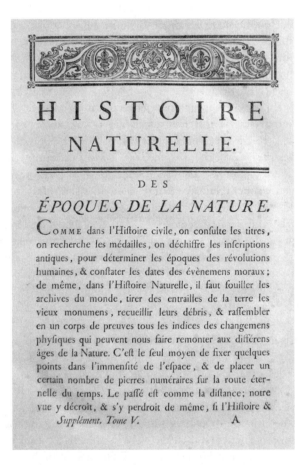

Figure 3. Title and opening text of Buffon's *Époques de la Nature*, from *Histoire Naturelle*, *Supplément* vol. 5 (1778), p. 1.

Desmarest, considerably less renowned than Buffon both in his own day and later (but certainly not an obscure figure to those familiar with geology's history), utilized a conception of distinct geological epochs in an analysis of volcanic features he had been studying since 1763 in the region of Auvergne, in south-central France. Although he had made brief indication of such an analysis into epochs in his earlier work on the Auvergne basalts (Desmarest 1774, 1777),[1] it was in a 1775 memoir, read for a special meeting of the Paris Academy of Sciences, that Desmarest formally presented his views on the epochs of nature in these volcanic remnants. A much less grandiose exposition than that of Buffon, Desmarest's paper was only published in truncated form a few years later, in 1779, the same year that Buffon's great treatise became available for purchase (Desmarest 1779). A longer version was presented anew for the French Institute a quarter of a century later, in 1804, and published in 1806 (Desmarest 1806).

The close coincidence in timing between the theory of Buffon and Desmarest's summary has fostered suspicions among some commentators (e.g. Roger 1962, pp. xl–xli; Ellenberger 1994, p. 240) of contention between Buffon and Desmarest over rights to the term *époque* in scientific treatment of the Earth. If there was in fact a contest over rightful possession of the word, it was a competition between two men of quite unequal status in the French scientific world. Buffon, nearly two decades older than Desmarest, was no less greatly his social and cultural senior, as a patrician who had long been the superintendant of the King's garden and keeper of the royal natural history collections (Roger 1989). Desmarest, by contrast, was of humble origins, and in the 1770s was a newcomer to the Paris Academy, to which he had been elected in 1771 with patronage assistance from reform-minded ministers and officials who appreciated his work in support of rational improvements in crafts and industry (Taylor 1969, 1971).

Jacques Roger, the foremost interpreter of Buffon's life and work, found that the term *époque* was used with reference to the Earth's history by other naturalists besides either Buffon or Desmarest at least as early as 1750 (Roger 1962, pp. xl–xli). Yet at the time that Buffon and Desmarest began to cultivate their respective conceptions of *époques*, such occasional antecedent uses of the term had not been incorporated within any systematic schema for analysing the Earth. Buffon himself first used the word in print, as a means to designate a distinct period of the Earth's history, in a 1766 passage of his monumental

---

[1] Desmarest's initial perceptions about the volcanic terrain in Auvergne – specifically his determination that columnar basalt is a volcanic product – were presented briefly in 1765. His subsequent and more thorough investigations of the region, published in two parts during the 1770s, were presented in 1771.

Histoire Naturelle where he envisioned setting the problem of animal life's history within the broad framework of the terrestrial globe's creation and development (Buffon 1766, p. 374; Roger 1962, p. xxvi). Buffon's first presentation of his *Époques de la Nature* came in August of 1773, when he read a chapter of the text at a meeting of the Academy of Dijon (Roger 1962, p. xxxi). While the *Supplément* volume of *Histoire Naturelle* containing the *Époques* has an imprint of 1778, it was not placed on sale until April of 1779 (Roger 1962, p. xxxvii). If Desmarest became annoyed, as may well have been the case, at Buffon's apparent appropriation of the terminology of *époques* as the learned world was abuzz about the Royal naturalist's splendid new treatise, his pique may be understandable in that he had been using that term in his private notes and correspondence since the early 1760s.[2]

Regardless of their respective entitlements to priority regarding use of the term *époque*, in the final two decades of the eighteenth century both Buffon and Desmarest were acknowledged as significant exponents of an intensified sense of sequential developments in the Earth, and of the operations of natural processes over periods greatly surpassing the time of human record and memory. Both friends and enemies of Buffon agreed that his *Époques de la Nature* did much to strengthen conviction in a natural timescale dwarfing the extent of human history, and to recast scientific thinking in terms of development or a periodized series of phases of change (Gohau 1987, ch. 7; Ellenberger 1994, pp. 215–217; Oldroyd 1996, pp. 89–92). Although with a much more restricted and less varied audience, Desmarest too gained respect, within a comparatively exclusive community of scientific scholars, as a pioneering advocate for interpreting geological phenomena in a way that allowed determination of successive operations and effects.[3]

Attentive observers saw differences as well as similarities in the deployment of an 'epochs' concept, on the part of Buffon and Desmarest. For Buffon, epochs represented vast stages in an unfolding story boldly narrated, situating

---

[2] Desmarest wrote about viewing 'les traces de chacune des époques des trois révolutions principales que le globe a essuyées' – evidence of three main epochs of global revolution – in his notes from observations in Périgord in 1761: 'Remarques de Mr. Desmarest (de l'Académie des Sciences) sur la géographie physique, les productions & les manufactures de la généralité de Bordeaux, lors de ses tournées depuis 1761 jusqu'en 1764' [the period concerned actually ended in 1762], Archives Départementales de la Dordogne, Périgueux, MS 26, part I, p. 27. On Desmarest's further private use of the term *époque* in the later 1760s see Taylor 1992a, p. 374, n. 10.

[3] Ezio Vaccari has kindly pointed out to me that during the same period under discussion here, several Italian naturalists were also adopting a terminology of *epoca* to define periods of global alteration. In particular, Dr Vaccari informs me that a form of geological periodization or change was discussed in terms of *epoca* by Giambattista Passeri in 1753, by Alberto Fortis in 1766 and by Giovanni Arduino in the early 1770s.

a panorama of natural events and effects in a temporal order, the overall plausibility of which was not meant to depend critically on any narrow body of data. Indeed, Buffon's story rationalized the facts, rather than being constructed out of them (Taylor 1992a). For Desmarest, however, the account of the Auvergne volcanic epochs involved no dramatically comprehensive narrative. In fact, Desmarest's regionally restricted story was expressly incomplete. It made no claims to give an account of changes on a global scale, and it was confined to discussion of effects deriving from discrete periods, separated from one another by time gaps about which he indicated nothing could be said, at least for the moment (Figure 4).

Each of these Enlightenment naturalists plainly required a natural timescale that disregarded traditional chronologies. But they differed markedly in how overtly this departure from convention was expressed. Buffon's chronological framework was explicitly stated in terms of a developmental sequence where the present moment stands some 75 000 years from the Earth's origins as a piece of incandescent matter torn from the sun by a passing comet (Figure 5). This computation was ostensibly informed by Buffon's experiments, already reported in earlier volumes of the *Histoire Naturelle*, on the cooling of molten iron in the forges at his Burgundy estate (Buffon 1774, 1775). Buffon's knowledge of how radically variable the results of such an extrapolation might be is suggested in the fact that his private calculations of the Earth's duration were measured in millions of years instead of a few scores of thousands as in his published account (Roger 1962, pp. lx–lxvii; Roger 1989, pp. 537–543).

From a later perspective it is easy to underestimate the novelty of a deliberate narration of the Earth's history through a period of less than a hundred thousand years. But we should consider that a ten-fold enlargement of orthodox figures for the Earth's age could be a genuine challenge to the imagination. Even if many astute readers discounted the pretended precision of Buffon's chronology, they saw the unalloyed naturalism in his disregard for orthodox time-reckoning. There is reason to believe that a number of Buffon's progressive contemporaries, even among naturalists who disdained Buffon as an armchair naturalist and an undisciplined system-builder, registered grudging respect because this prominent office-holder had pushed the envelope confining open expression of what serious thinkers might say about the natural time frame. They of course knew that Buffon was read – not only because he held a scientific position of unexcelled prestige and authority, but above all because he was readable.

Desmarest was more discreet than Buffon, less rashly explicit about time. In his account of the Auvergne epochs, and for that matter in all his writings public or private, to my knowledge Desmarest never ventured to attach numbers to

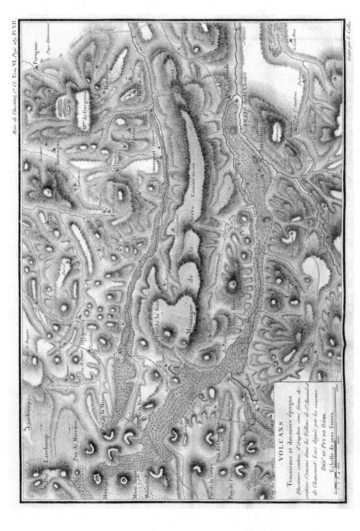

Figure 4. Desmarest's map (Plate VII) showing volcanic remnants from distinct epochs, from his 1806 publication, 'Mémoire sur la détermination de trois époques de la nature par les produits des volcans.' The Auvergne locality centers on the valleys of Chanonat (N) and Saint-Amand (S), separated by the long, sloping plateau of the Montagne de la Serre. This map features the third and last epoch, the products of which are the volcanic cones on the left and the stippled lava flows in the two valleys. The Montagne de la Serre was identified as the remnant of a lava flow from the much earlier second epoch, left high and dry by subsequent erosion. (Map dimensions: 24 x 34.5 cm. Scale: 2000 *toises* [fathoms] = 75 mm. Engraved by E. Collin.)

Figure 5. Plate from *Histoire Naturelle*, vol. 1 (*Théorie de la Terre*), 1749 (opposite p. 127, at the start of 'Preuves de la Théorie de la Terre'). By N. Blakey, engraved by St. Fessard. An allegorical representation of the origins of the planets from solar stuff, a result of a collision of a comet with the Sun. While this image accompanied Buffon's *Théorie de la Terre*, it was in the later *Époques de la Nature* (1778) that Buffon elaborated more extensively on the idea of the planets' generation from solar matter.

the periods of time his geological interpretations required. It seems probable that he did not believe such reckoning to be possible. Nonetheless there was no escaping the fact that Desmarest's distinctions among the volcanic epochs of Auvergne depended on recognition of the progressive destruction of the volcanic productions by erosive processes working at more or less constant rates. For Desmarest, the positions of volcanic remnants from the earliest epoch, capping crests overlooking sizeable valleys, spoke of the operation of geomorphic processes – excavation by moving water – in slow and regular fashion. So while Desmarest in his public declarations avoided sensationally direct or precise assertions about the duration of geological processes, neither did he shrink from explaining his views on how natural operations produced the observed effects in ways clearly necessitating enormously long periods of time.

To summarize, then, Buffon popularized an 'epochs' concept in a form something like that of a likely story about the Earth's formation and progressive alteration. And Desmarest helped to introduce the possibility – or at least strengthened the possibility – of working by empirical analysis of geological phenomena to a genetic or developmental account of particular geological operations and effects.

Most of what I have said so far about Buffon and Desmarest and their place in late-eighteenth-century developments regarding geology and time is fairly well agreed upon – an abbreviated form of a standard narrative, one might say. Notice that while this story entails the approximately simultaneous presentation by Buffon and Desmarest of accounts of sequential geological *époques* conceived as periods of substantial duration – and thus a form of geological explanation organized around placement of events in a chronological series – it also emphasizes differences in the approaches taken by the two naturalists in establishing the events they respectively addressed, and in the contrasting degrees of continuity and completeness of the resulting sequences. It also makes no reference to possible interdependence between Buffon and Desmarest as regards adoption and use of a vocabulary based on *époques*.

Contemporary Theories of the Earth, and an Undifferentaited Past

I propose now to enlarge on this picture, to sketch out a relevant part of the culture of geology in the period when Buffon and Desmarest were preparing themselves to make public presentations about Epochs of Nature. The result should not require any major retractions from the story that has just been rehearsed, as far as it goes. It will, however, draw attention to some common ground held by Buffon and Desmarest, as one looks more closely at their apparently contrasting applications of the *époques* concept. I will maintain that Desmarest's mode of thinking about geological problems drew significantly from some Buffonian doctrines set forth long before either Buffon or Desmarest talked in public about natural epochs. And I will suggest also that this means we must be cautious about drawing any stark contrast between Buffon as exclusively a system-building theorist and Desmarest as an altogether field-based empiricist thoroughly disengaged from Theories of the Earth.

Buffon's *Époques de la Nature* is viewed as one of the culminating entries in the genre of scientific writing called Theories of the Earth. Even as it was being discussed and debated, in the years around 1780, there were voices saying that this style of treatment was of limited use, that it was in the process of being replaced by less comprehensive, more empirically-based studies of restricted areas without pretensions of global application (Taylor 1992a). But for us to heed *only* those voices risks allowing the term 'Theory of the Earth' to be hijacked as a pejorative by a few late Enlightenment advocates of scientific reform. Through most, if not the whole, of the eighteenth century, the expression 'Theory of the Earth' was generally used rather flexibly, and by and large *not* pejoratively, to refer to scientific understanding of the Earth – including ongoing but so far largely inconclusive efforts at such understanding (Roger 1973; Taylor 1992b; Magruder 2000).

This usage is exemplified in a standard reference work of the third quarter of the century: the *Rational Universal Dictionary of Natural History* produced by the French mineralogical naturalist Jacques-Christophe Valmont de Bomare (1731–1807). Valmont de Bomare, from Rouen, first put out this multi-volume compendium in the 1760s, following years of travelling at Royal expense throughout Europe – including Iceland – as an official *naturaliste-voyageur du gouvernement*. The *Dictionnaire* went through five successively enlarged editions by the end of the century. As a teacher of public courses, as well as through his encyclopedic exposition, Valmont de Bomare was a successful digester and popularizer of scientific knowledge (Burke 1976).

In a moderately lengthy article in the *Dictionary* entitled 'Theory of the Earth,' Valmont de Bomare addressed geological knowledge broadly (Valmont de Bomare 1768–1769, vol. 11, pp. 222–245). The article exemplifies a formula valuable for historians of geology proposed several years ago by our late colleague François Ellenberger, to the effect that we can profitably distinguish *specific* or *individual* Theories of the Earth from the *ideal* or *generic* Theory of the Earth enterprise (Ellenberger 1994, pp. 12–16). Specific Theories, according to this distinction, were systems of explanation offered with more or less assurance as an author's True Answer to problems relating to the Earth's configuration, origins, development, purposes, or future prospects. Generic theories by contrast represented the large and frequently critical and skeptical enterprise in which attainment of a true Theory, or at least elements of it, was viewed as highly desirable but a work still in progress. In other words, a particular use of the phrase 'Theory of the Earth' might refer to aspirations for a complete system, or alternatively it might refer to something close to what later came to be thought of as geological investigation. One needs to look at each case to know where it fits along the spectrum between these two points.

Valmont de Bomare's account of the theory of the Earth belongs in this latter, generic category. It exhibits doubts about any possibility for serious resolution of questions about the Earth's beginnings, declaring that rather than try to determine origins what we should do instead is to study the Earth in its present state and arrangement. It pursues such themes as: the contrast between sudden, violent causes, and slow and general ones; the inadequacy of any single cause to account fully for the evident transformations the Earth has undergone; and disparity between superficially apparent confusion and disorder in the Earth, as distinct from order and regularity discernible on close inspection.

Quite significantly, relative to the concerns of the present paper, Valmont de Bomare's article on the theory of the Earth treats as more or less conventional a distinction between the *present* or *new* state of the world and an *old* or *former* condition from which the present is demarcated by one or multiple *revolutions*. Valmont is somewhat unclear about the accessibility of satisfactory knowledge regarding the revolution or revolutions separating present from past, as he is also about the extent of time that may be at stake. For our purposes, however, the notable feature of Valmont's treatment is the largely opaque divide between the old and new worlds, as regards any prospect for articulation of the past into discrete parts. Past conditions in the Earth are tantalizingly suggested by various sorts of terrestrial evidence – and indeed the Theory of the Earth enterprise here has, as perhaps the major part of its agenda, to explore that evidence and illuminate that past. But Valmont de Bomare puts forward no protocol for setting past conditions in any kind of temporal order or sequence, let alone for developing a chronology. Valmont was apparently a scrupulous empiricist, uninterested in merely speculative discussion of time and the past. As far as he could see, there existed no reliable means to establish chronological sequences in the Earth's past, which is to say for doing history. Thus what might be called the 'chronlogical duality' in Valmont's demarcation, distinguishing simply between the *present* world and a *former* one, is perhaps mostly the result of a lack of perceived options. The former world, from this perspective, is in a sense monolithic, undifferentiated, impervious to scientific analysis into a refined set of parts placed in relative sequence. This is a view of nature's remote past that would persist in nineteenth-century popular conceptions, as Martin Rudwick has shown, long after geological specialists had done a great deal to establish a highly differentiated series of periods in the Earth's history (Rudwick 1992).

Desmarest and Buffon's Two Theories of the Earth

A decade after publication of this article in Valmont's *Dictionary*, anyone giving the summary of Desmarest's 1775 memoir close attention, or anyone reading Buffon's *Époques de la Nature*, would find in one or the other of these pieces denials of an opaque divide between present and past. Buffon's scheme of continuous epochs offered an appealingly complete account, but little in the way of tools for confident determination that the story was told correctly. Desmarest, on the other hand, laid his emphasis on precisely those analytical tools for access on a limited basis to an ordered past. An observer at the end of the 1770s whose attitudes were formed by – or conformable to – a dependable reference like Valmont de Bomare's *Dictionary* would be justified in regarding both Desmarest and Buffon as breaking through a barrier between present and past, although in rather different ways. And that observer would probably consider each of their two achievements as part of the Theory of the Earth endeavor.

In general, modern historians of science have not wanted to categorize Buffon's and Desmarest's 'epochs-undertakings' alike, in contrast to a contemporary outlook for which I have just made a claim. Such reluctance is readily understandable, indeed in some measure I share it. Unlike Buffon, Desmarest in the 1770s outlined interpretive procedures for information generated by investigation that showed a way to translate observations of a landscape into a historical narrative. It represented a fulfilment, or rather a promising beginning of fulfilment, of an aspiration expressed by forebears such as Leibniz and Fontenelle, to establish a reliable history of nature derived out of its analogues to archival documents in civil history – natural monuments and inscriptions (Roger 1968; Rossi 1984; Rappaport 1991, 1997; Leibniz 1993; Cohen 1998).

All the same, without retreating on maintaining the difference between Buffon's narrative *époques* and Desmarest's analytical ones, I support a respectful observance of their contemporaries' habit of linking both with Theories of the Earth. Beyond the fact that, as I believe, users of Valmont de Bomare's *Dictionary* would speak that way, an additional consideration is that important elements in the path that brought Desmarest to his historic formulation of *époques* had roots in Theories of the Earth, in fact especially in the original

Théorie de la Terre with which Buffon opened his monumental *Histoire Naturelle* in 1749.

That Buffon produced not just one but two theories of the Earth – the first so titled and the second the *Époques de la Nature*, three decades later – has elicited much comment. The two theories had in common mainly the same hypothetical origin of planet Earth as generated from a chunk of solar matter drawn out by a passing comet. Perhaps because of Buffon's reputation as a broad-canvas naturalist and literary synthesizer, Buffon's Earth-theorizing in both instances has sometimes been treated as spun out of the imagination, and thus as an encouragement to a conjectural kind of thinking. (If many contemporaries had reservations about Buffon's standing as an active scientific observer, few of his readers are likely to have been misled on this point. They know he did not personally perform all the things he said should be done in the name of good science.) But in fact the two theories had quite different characters. Of the two theories, the *Époques de la Nature* is a good deal better known and understood, certainly in part because it has long been regarded as a masterpiece of French literature, which the *Théorie de la Terre* has not. Jacques Roger gave us a finely commented edition of *Époques*, and nothing of the sort has ever happened with Buffon's 1749 *Theory*.

Buffon's 1749 *Theory of the Earth* – from which by the way it is apparent Valmont de Bomare drew for his article of the same name – was in considerable measure an appeal for a program of judicious observational study as the correct path toward understanding the Earth (Figure 6). The greater part of the 1749 *Theory* reflects a spirit of 'empiricism in principle,' for which Buffon was a master spokesman. It deserves notice that the *Théorie* was identified as the 'second discourse' of the *Histoire Naturelle* project (Buffon 1749b), immediately after the initial discourse entitled 'On the manner of studying and expounding natural history' (Buffon 1749a). A consistency in methodology, as well as didactic tone, is sustained through both discourses. Buffon presented himself to be sure as an interpreter of the Earth, but no less as a teacher of how to interpret the Earth. The central interpretive principles are empirical in character and tend to emphasize assimilation of observations to laws, not to history (Gohau 1990, 1992).

In the *Théorie* Buffon advocated careful and patient observation especially of a wide range of configurational regularities in terrestrial features at various levels, ranging from continental in scale to a matter of inches. The underlying idea, of which Buffon was not so much originator as eloquent expositor, was that generalizations arising out of combinations and comparisons of these dispositional regularities are the most promising avenues to as-yet unknown pieces of a proper Theory of the Earth (Taylor 1988). There is a kind of epistemological modesty in this Buffonian program that may surprise those familiar only with Buffon's later theory, the *Époques de la Nature*. Although certainly not without its own evidences of confidence – indeed overconfidence – in the extent of human knowledge about certain terrestrial features and their meanings, Buffon's outlook in the *Théorie* was strongly investigatory. The *Théorie de la Terre* was not a result in hand, it was an empiricist's adventure in progress. Perhaps not incidentally, in his *Théorie* Buffon evaded direct discussion of the question of how long terrestrial operations have been working.

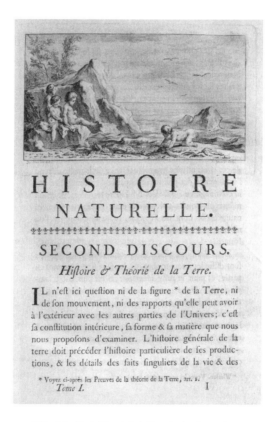

Figure 6. Opening page (65) of Buffon's *Théorie de la Terre*, in the first volume (1749) of *Histoire Naturelle*. Vignette by DeSeve, engraving by St. Fessard.

Buffon, Desmarest and the Ordering of Geological Events

Buffon's emphasis on examination of dispositional regularities was part of a style much respected among francophone naturalists throughout the eighteenth century – and I believe others outside the French-speaking world as well. The promise it seemed to hold out for its adherents was in discovery of useful laws, generalizations in the spirit of natural philosophy, rather than in the form of historical formulations. In retrospective view, we may be tempted to think that investigators of the late eighteenth century would have been well advised to seek historical formulations, in preference to philosophical generalizations; we know that during the early nineteenth century geological science was to achieve an unprecedented coherence in part because of a focus on stratigraphy, oriented precisely around a concern for sequence rather than law. But we must remember that it was not until the nineteenth century that scientists began to be comfortable with the idea that placement of events in historical sequence constituted a wholly legitimate form of scientific explanation. For our eighteenth-century protagonists, it came naturally to expect that a proper resolution of the puzzles of Earth science – attainment of a satisfactory Theory of the Earth – must take the shape of laws or fixed generalizations.

It is perhaps ironic that the principal hero of this discussion, Nicolas Desmarest, whose achievement it was to articulate a method for translating observations of apparently static structures of a volcanized landscape into empirically-determined chronological phases or epochs, was brought up geologically on a diet of timeless configurational generalizations. For so he was formed as a beginning investigator, as one finds on close examination of his early manuscripts and publications (Taylor 1997). In both indirect and direct ways, Desmarest acknowledged that his principles of geological investigation were largely the same as those found in Buffon's 1749 *Théorie de la Terre*.

But maybe this is not quite so ironic after all. For much more was involved in Desmarest's Auvergne research than the insight that the volcanic structures he studied there could be distinguished in accord with discrete epochs. Desmarest had first to gauge the configurational limits of these structures, by fieldwork. In what is no doubt an oversimplified formula, Desmarest's geochronological achievement with his *époques* required that he use two distinguishable sorts of cognitive faculties in an integrated manner: one of them spatial, and the other temporal or historical. For the first of these, the school to which Desmarest belonged, the approach he adopted in seeking to chart out dispositional regularities, was – as he acknowledged – nowhere better expressed than in Buffon's 1749 *Théorie de la Terre*.

***

In preparing this paper I have benefited from the kindness of Martin J. S. Rudwick in allowing me to read in draft parts of his forthcoming book, *Bursting the Limits of Time*. I am grateful to K. V. Magruder for his invaluable assistance with the illustrations. I thank S. Knell, M. Rudwick and E. Vaccari for a number of constructive suggestions for revision of the manuscript. Figures 1, 3, 4, 5 and 6 courtesy of the History of Science Collections, University of Oklahoma Libraries.

References

ALBRITTON, C.R., JR. 1980. *The Abyss of Time: Changing Conceptions of the Earth's Antiquity after the Sixteenth Century*. Freeman, Cooper, San Francisco.

BUFFON, G.L. LECLERC, COMTE DE. 1749a. Premier Discours. De la manière d'étudier & de traiter l'histoire naturelle. *In: Histoire Naturelle, générale et particulière*. Vol. 1. De l'Imprimerie Royale, Paris, 3–62.

BUFFON, G.L. LECLERC, COMTE DE. 1749b. Second Discours. Histoire & Théorie de la Terre. *In: Histoire Naturelle, générale et particulière*. Vol. 1. De l'Imprimerie Royale, Paris, 65–124. [Preuves de la théorie de la terre: 127–612.]

BUFFON, G.L. LECLERC, COMTE DE. 1766. De la dégénération des animaux. *In: Histoire Naturelle, générale et particulière*. Vol. 14. De l'Imprimerie Royale, Paris, 311–374.

BUFFON, G.L. LECLERC, COMTE DE. 1774. Expériences sur le progrès de la chaleur. *In: Histoire Naturelle, générale et particulière*. Supplément, Vol. 1. De l'Imprimerie Royale, Paris, 145–300.

BUFFON, G.L. LECLERC, COMTE DE. 1775. Recherches sur le refroidissement de la Terre & des planètes. *In: Histoire Naturelle, générale et particulière*. Supplément, Vol. 2. De l'Imprimerie Royale, Paris, 361–515.

BUFFON, G.L. LECLERC, COMTE DE. 1778. Des Époques de la Nature. *In: Histoire Naturelle, générale et particulière. Supplément*, Vol. 5. De l'Imprimerie Royale, Paris, 1–254. [Additions et corrections aux articles qui contiennent les preuves de la Théorie de la Terre, 255–494; Notes justificatives des faits rapportés dans les Époques de la Nature, 495–599.]

BURKE, J.G. 1976. Jacques-Christophe Valmont de Bomare. *In*: GILLISPIE, C.C. (ed) *Dictionary of Scientific Biography*, Vol. 13. Charles Scribner's Sons, New York, 565–566.

COHEN, C. 1998. Un manuscrit inédit de Leibniz (1646–1716) sur la nature des 'objets fossiles.' *Bulletin de la Société Géologique de France*, 169, 137–142.

DALRYMPLE, G.B. 1991. *The Age of the Earth*. Stanford University Press, Stanford.

DEAN, D.R. 1981. The Age of the Earth Controversy: Beginnings to Hutton. *Annals of Science*, 38, 435–456.

DESMAREST, N. 1774. Mémoire sur l'origine & la nature du basalte à grandes colonnes polygones, déterminées par l'histoire naturelle de cette pierre, observée en Auvergne.

Mémoires de l'Académie Royale des Sciences, Paris, Année 1771, 705–775 [presented to the Academy 3 July 1765 (in part) and 11 May 1771].

DESMAREST, N. 1777. Mémoire sur le basalte. Troisième partie, où l'on traite du basalte des anciens; & où l'on expose l'histoire naturelle des différentes espèces de pierres auxquelles on a donné, en différens tems, le nom de basalte. *Mémoires de l'Académie Royale des Sciences*, Paris, Année 1773, 599–670 [presented to the Academy 11 May 1771].

DESMAREST, N. 1779. Extrait d'un mémoire sur la détermination de quelques époques de la nature par les produits des volcans, & sur l'usage de ces époques dans l'étude des volcans. *Observations sur la physique, sur l'histoire naturelle et sur les arts*, 13, 115–126 [summary of paper presented to the Academy at the *Séance publique* on 15 November 1775, at its official autumn *rentrée* or reconvening after St. Martin's day].

DESMAREST, N. 1806. Mémoire sur la détermination de trois époques de la nature par les produits des volcans, et sur l'usage qu'on peut faire de ces époques dans l'étude des volcans. *Mémoires de l'Institut des Sciences, Lettres et Arts. Sciences mathématiques et physiques*, 6, 219–289 [presented to the Institut *1 Prairial an XII* (21 May 1804)].

ELLENBERGER, F. 1994. *Histoire de la géologie. Tome 2: La grande éclosion et ses prémices, 1660–1810*. Technique et Documentation (Lavoisier), Paris, London, New York [1999 English translation by M. Carozzi, *History of Geology: The Great Awakening and its First Fruits, 1660–1810*, Balkema, Rotterdam].

GOHAU, G. 1987. *Histoire de la géologie*. Editions La Découverte, Paris [1990 English translation and revision by A.V. Carozzi, & M. Carozzi, *A History of Geology*, Rutgers University Press, New Brunswick and London.]

GOHAU, G. 1990. *Les sciences de la terre aux XVIIe et XVIIIe siècles: Naissance de la géologie*. Albin Michel, Paris.

GOHAU, G. 1992. La 'Théorie de la Terre,' de 1749. *In*: GAYON, J. (ed.) *Buffon 88: Actes du Colloque international pour le bicentenaire de la mort de Buffon*. Vrin, Paris, 343–352.

HABER, F.C. 1959. *The Age of the World: Moses to Darwin*. The Johns Hopkins Press, Baltimore.

HALLAM, A. 1989. *Great Geological Controversies*. (Second edition). Oxford University Press, Oxford & New York.

LEIBNIZ, G.W. VON. 1993. *Protogaea: De l'aspect primitif de la Terre et des traces d'une histoire très ancienne que renferment les monuments mêmes de la nature*. Translated by B. de Saint-Germain, edited with introduction and notes by J.-M. Barrande, Presses Universitaires du Mirail, Toulouse.

MAGRUDER, K. V. 2000. *Theories of the Earth from Descartes to Cuvier: Natural Order and Historical Contingency in a Contested Textual Tradition*. PhD dissertation, University of Oklahoma.

OLDROYD, D.R. 1996. *Thinking About the Earth: A History of Ideas in Geology*. Athlone, London.

RAPPAPORT, R. 1982. Borrowed Words: Problems of Vocabulary in Eighteenth-Century Geology. *British Journal for the History of Science*, 15, 27–44.

RAPPAPORT, R. 1991. Fontenelle Interprets the Earth's History. *Revue d'histoire des sciences*, 44, 281–300.

RAPPAPORT, R. 1997. *When Geologists Were Historians, 1665–1750.* Cornell University Press, Ithaca and London.

ROGER, J. (ed.) 1962. *Buffon, Les Époques de la Nature, Édition critique.* Mémoires du Muséum National d'Histoire Naturelle, Série C, Sciences de la Terre, Tome 10. Éditions du Muséum, Paris.

ROGER, J. 1968. Leibniz et la théorie de la Terre. *In*: *Leibniz 1646–1716: Aspects de l'homme et de l'oeuvre.* Aubier-Montaigne, Paris, 137–144.

ROGER, J. 1973. La théorie de la Terre au XVIIe siècle. *Revue d'histoire des sciences*, 26, 23–48.

ROGER, J. 1989. *Buffon: Un philosophe au Jardin du Roi.* Fayard, Paris [1997 English translation by S.L. Bonnefoi, edited by L.P. Williams, *Buffon, A Life in Natural History.* Cornell University Press, Ithaca and London].

ROSSI, P. 1984. *The Dark Abyss of Time: The History of the Earth & and the History of Nations from Hooke to Vico.* Translation from the Italian by L.G. Cochrane, University of Chicago Press, Chicago and London.

RUDWICK, M.J.S. 1992. *Scenes from Deep Time: Early Pictorial Representations of the Prehistoric World.* University of Chicago Press, Chicago and London.

TAYLOR, K.L. 1969. Nicolas Desmarest and Geology in the Eighteenth Century. *In*: SCHNEER, C.J. (ed.) *Toward a History of Geology.* MIT Press, Cambridge (Mass.) and London, 339–356.

TAYLOR, K.L. 1971. Nicolas Desmarest. *In*: GILLISPIE, C.C. (ed.) *Dictionary of Scientific Biography*, Vol. 4. Charles Scribner's Sons, New York, 70–73.

TAYLOR, K.L. 1988. Les lois naturelles dans la géologie du XVIIIème siècle: Recherches préliminaires. *Travaux du Comité Français d'Histoire de la Géologie*, 3ème série, 2, 1–28.

TAYLOR, K.L. 1992a. The *Époques de la Nature* and Geology during Buffon's Later Years. *In*: GAYON, J. (ed.) *Buffon 88: Actes du Colloque international pour le bicentenaire de la mort de Buffon.* Vrin, Paris, 371–385.

TAYLOR, K.L. 1992b. The Historical Rehabilitation of Theories of the Earth. *The Compass: The Earth-Science Journal of Sigma Gamma Epsilon*, 69, 334–345.

TAYLOR, K.L. 1997. La genèse d'un naturaliste: Desmarest, la lecture et la nature. *In*: GOHAU, G. (ed.) *De la géologie à son histoire: ouvrage édité en hommage à François Ellenberger.* Comité des Travaux Historiques et Scientifiques, Paris, 61–74.

VALMONT DE BOMARE, J.C. 1768–1769. *Dictionnaire raisonné universel d'histoire naturelle. Edition augmentée par l'auteur* (second edition), 12 vols., Yverdon.

VI

Natural Law in Eighteenth-Century Geology: The Case of Louis Bourguet

From an eighteenth-century perspective, one of the ancestors of the science of geology was the speculative cosmogony of which Burnet, Woodward, and Whiston had been such famous exponents. Louis Bourguet belonged in many ways to this same cosmogonical tradition. His contribution was yet another outline of a "theory of the earth." Yet in some respects Bourguet reflected a mood that later came to characterize actualistic features of geological science. Something about the development of earth science during the eighteenth century can perhaps be learned from an examination of Bourguet's reputation in the fifty or seventy-five years following his work.

Bourguet, born in Nîmes in 1678, was the son of a wealthy Protestant merchant. His family moved to Switzerland following the revocation of the Edict of Nantes. With ample opportunity for education and travel, Bourguet became widely acquainted with philosophy, corresponded with Leibniz, and gained the friendship of Réaumur and entrance to the Academy of Berlin. Among his chief interests were languages (notably Etruscan), numismatics, and fossil-collecting. Settling in Neuchâtel, he taught philosophy and mathematics for a time, and died there in 1742.[1]

One of Bourguet's claims to originality was his principle of salient and re-entering angles. A statement of this principle appeared in his "Memoir on the Theory of the Earth," published in 1729 at the end of his *Philosophical Letters on the Formation of Salts and Crystals*.[2] Here he recorded that his numerous excursions through the Alps had enabled him to discover "the surprising regularity of

---

[1] On Bourguet's career consult *La France protestante*, 2nd ed., III (1881), cols. 2–7; St. Le Tourneur's article in *Dictionnaire de biographie française*, VI (1954), cols. 1507–1508; the biographical sketch by Vincent-Saint-Laurent and Du Petit-Thouars in Michaud's *Biographie universelle*, V (1812), 384–385; and the biographical article by Michel Nicolas in Hoefer's *Nouvelle biographie générale*, VII (1855), cols. 91–93.

[2] *Lettres philosophiques sur la formation des sels et des crystaux. Et sur la génération & le méchanisme organique des plantes et des animaux; A l'occasion de la pierre belemnite et de la pierre lenticulaire. Avec un mémoire sur la théorie de la terre* (Amsterdam: François l'Honoré, 1729). A second edition was published in Amsterdam by Marc-Michel Rey in 1762. References are to the first edition.

the structure of these great masses."[3] This regularity lay in the arrangement of the projections or buttresses emanating from mountains more or less at right angles to the trend of the ridges. The projections of mountains lying on opposite sides of a valley are always located, he believed, so that "the salient angles of each side correspond reciprocally to the re-entering angles which are alternatively opposed to them."[4] The effect of this topographic condition is to maintain a more or less constant width to the valley. In broad valleys these corresponding angles are less acute than in narrow and deep ones; on plains they are barely discernible. Bourguet believed this general physiographical relationship to be an empirical rule, applicable alike to mountain valleys, lake bottoms, and the ocean floor.

Bourguet did not conceal his pride in what he regarded as a discovery of great importance, "the principal key to the *Theory of the Earth*."[5] Nobody before him, he was certain, had made this observation.[6] Monsieur Du Fay of the Academy of Sciences, he reported proudly, viewed this "extremely fine and judicious" observation as one of those facts about the world that almost required a special ability at divination in order to be seen as a significant rather than fortuitous condition in the earth.[7]

Bourguet's apparently extravagant feeling about his analysis of the earth's structure deserves our attention because it reveals a special attitude toward the proper method of philosophizing about the earth's present condition and past history. The essay in which this appears was an outline for a projected (but never fulfilled) large-scale study of the theory of the earth. What he offered was a brief approach to the science of the earth. Bourguet believed that previous attempts at such a theory had failed for reasons that he understood and could remedy. A new era of earth-theorizing was arriving, and he would help to lead the way.[8]

One distinctive aspect of Bourguet's approach to earth history was his reference to empirical correlations in the earth's components. For example, aside from the question of angular harmonies in topography, Bourguet offered as "another capital phenomenon" the identical mineral content of fossil shells

---

[3] *Ibid.*, p. 181.
[4] *Ibid.*, p. 182. See also pp. 195–196.
[5] *Ibid.*, p. 182.
[6] *Ibid.*, p. 181.
[7] *Ibid.*, p. 183–184.
[8] An indication of Bourguet's frame of mind a decade and a half earlier than this 1729 definition of his geogonical ideas can be found in Bernhard Sticker, "Leibniz et Bourguet. Quelques lettres inconnues sur la théorie de la terre," *XIIe Congrès International d'Histoire des Sciences, Actes*, IIIB (Paris: Albert Blanchard, 1971), 143–147.

and of the strata in which they are found.[9] Empirically-founded generalizations such as these might or might not inspire immediate explanations, but Bourguet indicated that their recognition would assist in the task of rejecting erroneous theories of the earth.

Another important feature of Bourguet's line of argument was an extension to new phenomena of successful elements from sciences already firmly established, especially geometrical mechanics. The truths recently discovered in mechanics were, in his opinion, fixed and eternal verities, not subject to the sorts of change witnessed in the condition of the earth.[10] Why could not these permanent laws of nature bring forth knowledge as yet unknown about the globe's past vicissitudes? The solid parts of the earth, for example, should be expected to be most compact and massive in those regions where the earth's rotation is greatest. An understanding of centrifugal forces is thus of utility both in the search for precious materials[11] and in quest of knowledge about the creation of mountains, the largest of which Bourguet and his contemporaries placed between the tropics.[12]

Bourguet's sketch of the theory of the earth depended, then, on proper inference from mechanical facts to historical ones. The first order of business was a "description of the main phenomena"; a logical process then would lead to knowledge of their causes. He proposed a method of "ascent from the consideration of the present state of the Globe to the change that has certainly occurred in it. From there one will come to the manner in which this change took place. Next one passes to consideration of the primitive state of the Globe, knowledge of which depends on that of states derived from the first."[13] The path to a sound understanding of the series of stages by which the earth has become what it is begins with precise knowledge of the disposition and behavior of matter. Bourguet foresaw that this branch of philosophy would flourish with the as yet untried combination of knowledge of both mechanics and the details of the earth's structure.

The specific historical conclusions reached by Bourguet from his own procedures were rather conventional. The earth's present state was the result of

---

[9] *Lettres philosophiques*, p. 183.

[10] Bourguet's view that recent scientific endeavors had succeeded in producing sure knowledge of nature was later recorded in these words: "And experimental philosophy has been cultivated with so much care, since Galileo, that certitude has been achieved in the explanation of the largest part of Physics or of the phenomena of nature." ("Seconde lettre à Mr. Meuron . . . sur la philosophie de Mr. Le Baron de Leibnitz," *Journal helvétique*, [XV], July 1738, p. 18.)

[11] *Lettres philosophiques*, pp. 185–186.

[12] *Ibid.*, pp. 189–190, 195.

[13] *Ibid.*, pp. 218–219.

a revolution wrought upon an earlier primary condition, and demonstrates the power and wisdom of the Almighty. To the extent that he explained the great "key" of corresponding angles, he did so in terms of an *ad hoc* catastrophic event, the almost instantaneous condensation of the solid crust from a fluid state under the influence of the earth's changing planetary motion characteristics.[14] On the whole, later earth scientists who found Bourguet memorable may have done so not for his scheme of the earth's history but for some elements of his declared method, and specifically for the principle of corresponding angles as a specimen of observed order in the earth that called for physical explanation.

There can be little doubt that without Buffon, Bourguet's name would not have attained even the uncertain prominence in eighteenth-century geology that it did. In his *Théorie de la terre* (1749) Buffon adopted Bourguet's principle of salient and re-entering angles, and he reaffirmed his commitment in the *Époques de la nature* (1778).[15] Unlike Bourguet, Buffon attributed this regularity to the action of ocean currents that once covered the entire surface of the earth. He was critical of Bourguet's undisciplined speculation, but agreed with Bourguet that the principle of corresponding angles is "the key to the theory of the earth."[16]

It may safely be assumed that some of those who later commented on Bourguet's rule of salient and re-entering angles knew of it only through Buffon, and indeed failed to distinguish between the two expressions of it. For example, Jean-Claude de Lamétherie's mistaken criticism of Bourguet for maintaining that all valleys are water-carved is a misapprehension that seems likely to have derived from a confusion of Bourguet's and Buffon's ideas.[17]

During the second half of the eighteenth century, numerous references to Bourguet's principle of corresponding angles appeared in the French scientific literature. A sampling of these brings forth a variety of degrees of approval and disapproval. Daubenton, in the first volume of the *Encyclopédie*, discussed this "very important" observation in terms supporting Buffon's interpretation of its significance.[18] Nicolas-Antoine Boullanger, in his article "Déluge,"

---

[14] *Ibid.*, pp. 211–213.

[15] *Oeuvres complètes de Buffon*, ed. J. V. F. Lamouroux, I (Paris: Verdière et Ladrange, 1824), 76–77; II (1824), 80–83. Also *Les Époques de la nature*, ed. J. Roger. Mémoires du Muséum National d'Histoire Naturelle, Sér. C, Tome X (Paris: Éditions du Muséum, 1962), 80.

[16] *Oeuvres de Buffon*, II, 82.

[17] *Théorie de la terre*, III (Paris: Maradan, 1795), 344–345, 348. Lamétherie's confusion cannot have been based, however, on unavailability of Bourguet's work, which he partially reprinted. See note 24 below.

[18] "Angles correspondans des montagnes," *Encyclopédie, ou dictionnaire raisonné des sciences, des arts et des métiers*, I (Paris: Briasson, David l'aîné, Le Breton, Durand, 1751), 464.

approved Bourguet's generalization while doubting that submarine currents could be entirely responsible, and thus may have influenced Buffon's later emphasis upon action of waters retreating from continents.[19] In the 1768–1769 edition of Valmont de Bomare's *Dictionary*, the article "Théorie de la terre" supported the principle in terms of the action of both meteoric and marine water movement, without mentioning Bourguet;[20] the article "Montagne" also expressed a favorable opinion, although a note added by Haller reported that nature adheres to this principle quite irregularly.[21] Another critic of the accuracy of Bourguet's generalization was Peter Simon Pallas, who reported in 1777 that his observations in the Altai Mountains did not confirm the rule.[22] Horace-Bénédict de Saussure was even more emphatically negative, and said that in his experience reciprocally salient and re-entering angles were found only under special circumstances, where narrow valleys of recent origin run transverse to the trend of mountain ranges.[23]

Lamétherie devoted some eleven pages of his *Théorie de la terre* to an exposition, discussion, and criticism of Bourguet's principle.[24] Receiving it rather coolly, he nonetheless made it clear – and in this regard he had much in common with Bourguet's other critics – that this reaction was founded on Bourguet's failure to be accurate, not on opposition to the search for general correlations in the condition of the earth's crust. The comparatively slight attention Lamétherie gave to Bourguet's "system" stems from his agreement that accurate descriptive generalizations about valley structure, or any other geological feature, are welcome.[25]

One of the most attentive students of the problem of corresponding angles was Nicolas Desmarest. Early in his career, in 1757, he alluded admiringly in an *Encyclopédie* article to Bourguet's work as the sort of observational generalization that ought more often to typify the endeavor of physical geographers.[26] Decades later, in his *Géographie-physique*, he produced an annotated account of Bourguet's expression of his principle, as well as a broader analysis of the place of corresponding angles in geological science. Desmarest regarded

---

[19] *Encyclopédie*, IV (1754), 802.

[20] *Dictionnaire raisonné universel d'histoire naturelle*, XI (Yverdon: no pub., 1769), 232–233.

[21] Ibid., VII (1769), 148.

[22] *Observations sur la formation des montagnes et les changements arrivés au globe, particulièrement à l'égard de l'empire russe* (St. Petersburg: Académie Impériale des Sciences, [1777]), pp. 21–22.

[23] *Voyages dans les Alpes*, II (Neuchâtel: Louis Fauche-Borel, 1803), 339–340 (§577). [Original edition: *Voyages dans les Alpes*, I (Neuchâtel: Samuel Fauche, 1779), 511 (§577).]

[24] III, 344–355.

[25] "Système de Bourguet," III, 436–440.

[26] "Géographie physique," *Encyclopédie*, VII (1757), 616.

the generalization as only partially accurate, and took issue with Buffon for regarding marine and fluvial currents as similar in character and in their physical effects. Convinced that valleys are continually degraded by rivers, Desmarest argued that several factors are involved in determining a river's course within the boundaries of a valley floor, and so concluded that neatly corresponding salients and re-entrances cannot be expected everywhere.[27]

But since we are concerned here with Bourguet's reputation we need only note Desmarest's high respect for Bourguet's discovery (as it seemed to him) of a significant facet of physiography. Desmarest expressed disappointment not at Bourguet's failure to provide a reasonable explanation for the regularity of angles (something Buffon had partially done), but at his general inability to sense the proper link between observation and historical explanation in physical geography. Having transcribed (none too accurately) a substantial portion of Bourguet's account of the correspondence of angles, Desmarest pointedly omitted Bourguet's sketch for a theory of the earth, commenting that Bourguet's promising beginning at the phenomenological level was not matched by satisfactory results in the production of an account of the earth's physical history. This fine observer ended, Desmarest said, as a formulator of "absurd hypotheses" because he did not understand sufficiently well the methodological problems in finding the causes of observed effects.[28]

This somewhat impressionistic survey of the fate of Bourguet's principle of salient and re-entering angles up to the end of the eighteenth century is sufficient to suggest that, although there was no unanimity among French-language scientists regarding the value of Bourguet's geological studies, attention to the cosmogonical aspect of his work was overshadowed by interest in his more narrowly empirical observations, typified by the principle of corresponding angles. It is plausible that this orientation of later eighteenth-century attitudes toward Bourguet is not without significance. It seems to me that the lingering interest in Bourguet's topographic rule, even in cases where the aim was to refute it, was couched in recognition that Bourguet and later scientists shared at least one methodological belief.

Bourguet had perceived, perhaps a bit dimly, that a major difference between what he was attempting and the accomplishments of the successful mechanical sciences lay in the historical nature of his subject. Somehow one had to make the transition from constant, eternal truths (physics) to ephemeral events in time (theory of the earth). His hope was that the key to that transition

---

[27] *Encyclopédie méthodique, Géographie-physique*, II (Paris: H. Agasse, 1803), 590–593, article "Angles correspondans des vallées"

[28] *Ibid.*, I (*an* III [1794–1795]), 28–34, article "Bourguet."

lay in a clear apprehension of certain present physical relationships. It was a hope that was echoed repeatedly throughout the remainder of the eighteenth century by numerous Continental and British scientists. The goal was to find natural laws that would permit a scientific unravelling of the history of nature. Opinion gathered slowly and irregularly against the use of *ad hoc* hypotheses to accomplish this goal. Bourguet aroused the curiosity of some later observers because, in the middle of a cosmogony in the *ad hoc* tradition, he seemed to be trying to inject the seed of a natural law.

VII

Reflections on Natural Laws in Eighteenth-Century Geology

Introduction

The aim of this paper is to point out the importance, in eighteenth-century geological thinking, of ideas about natural laws specific to this science. I intend to discuss the ambition then held by some men of science to establish a number of general truths within the particular domain of geology. A varied group of geological investigators of the eighteenth century set out to organize knowledge of the Earth in the form of general principles or laws. While to some this may appear an obvious or banal claim, I think it is not. When we consider the consequences of this observation about geological science during the eighteenth century, it seems to me we are enabled to see a bit more clearly the conceptual framework in which many of the historical characters concerned with these problems worked and reasoned. And although this is only a preliminary study, I will maintain that even if the majority of the period's geological efforts at discovering general relations failed, it does not follow that these endeavors lack historical significance.

One task I am *not* undertaking, then, is an effort to find links between geology's prehistory and *the* "natural law" of ethical or legal theory. Such an idea would not even have occurred to me had it not been that our President, François Ellenberger, raised the question with me recently.

Nor is it my intention, furthermore, to examine eighteenth-century perceptions about the comprehensive laws of nature developed in the physical sciences, or the application of those laws to geological questions. Regarding this latter exclusion, it is worth saying that the prestige of universal laws in the corpus of natural philosophy was indeed enormously consequential for the formation of contemporaneous ideas on laws peculiar to geology. First, these natural and universal laws, whose formulation represented the greatest successes of the New Philosophy, served as methodological examples to all who might want to participate in the continued renewal of the sciences, a matter of intellectual pride during the century of Enlightenment. Secondly, the theorists of the Earth, the geologists and mineralogists of the eighteenth century, generally saw themselves (in common with geological scientists of practically any period, for that matter) as under an obligation to reconcile their assertions with accepted natural laws in the physical sciences and to avoid claims incompatible with established physical principles.

One last preliminary remark: I do not intend to insist too strongly on the word *law*, or at least I do not want to assign to it too strict or technical a meaning. The eighteenth-century savants with whom I am concerned usually used terms such as *regularities*, and *general or constant conditions* in referring to the stability of observable objects and operations that they studied and tried to explain. I might well have entitled this presentation "Regularities of Disposition in Eighteenth-Century Geology," or perhaps even better "The Search" for such regularities. For the moment, however, I retain the word *law*, for several specific reasons. I believe in fact that the recognition of regular dispositions, imagined or real, played the role in a sense of specific laws in geology. One encounters actual uses that were made of the word *law*, or *loi*, taken in this sense, in certain texts of the period; this is not an invention on my part.[1] No less important is the methodological weight carried in the expression *natural law*, tangibly greater than found in the terms *regularity* or *constancy*. Questions about adoption of the most appropriate methods, as seen by leading geological characters of the eighteenth century, are in fact part of my present concern. We will find, I am sure, that there were links between these issues of method and substantial geological claims or propositions that were put forward. But in broad historical perspective, the preoccupations of eighteenth-century geologists, developed within a framework of confidence in rules or laws special to geology, may have historical significance that is mainly methodological. That is to say, the form of investigation and expression can sometimes count for more than the concrete result of inquiry. It is to this matter of formal orientation in the eighteenth century's geological enterprise, founded on the hope of discovering valid laws, that I wish to draw attention, and in doing so I hope to suggest how much it is worthy of our historical consideration.

Buffon's Regularities

We begin by examining a remarkable feature of the *Théorie de la Terre* published in 1749 by Georges-Louis Leclerc, comte de Buffon. It is well known, I believe, that quite early in this discourse Buffon chose to contrast the Earth's apparent and superficial disorder with its deeper and meaningful organization.[2] If this aspect of his rhetoric is widely recognized, I have the impression that the particular details of the terrestrial organization upheld by Buffon are far less known. Since it seems to me that Buffon's *Théorie de la Terre* shows a marked fascination with natural regularities, I take this work as a point of departure. I have attempted to draw up a synoptic table of the main instances of order presented by Buffon in this work, that is in his *Théorie* and the accompanying *Preuves*. This summary, however incomplete, affords us an idea of Buffon's design of finding in constant relations, and especially in regularities of location or disposition, the keys to an understanding of the Earth. (See the Appendix.)

Let me emphasize that the taxonomy shown in this provisional inventory is my own, not Buffon's. (There is nonetheless some resemblance between the classification I have chosen and that encountered in the work of Louis Bourguet, prior to Buffon).[3] Partly for reasons of convenience in discussing them, I have divided the regularities identified by Buffon into four categories. First, there are the gross features in the division of continents and seas. Buffon sees broad significance in the positions of the continental masses and oceans, their dimensions and orientations on the globe's surface, and the common points in their forms. He believes one sees here abundant examples of natural order too pronounced to be the result of chance. For example, the axes of each of the great continental masses – the axes of maximum length – do not have precisely a North-South direction, but rather show an obliquity of about 30 degrees. These axes, according to Buffon, divide precisely in halves the surface areas on the East and West sides. And each main continental mass is set off with respect to the equator, by an amount from 16 to 18 degrees, the Old Continent (Europe, Asia, and Africa) toward the North, and the Americas toward the South.

In my tabulation, the second type of regularity applies to the parts of continents and oceans that constitute the leading features of the terrain and the sea bottom: the mountains and valleys, the river courses, and the ocean depths. Here as elsewhere, it should be noted that certain sorts of relations are to be seen only as tendencies – for example, the fact (as Buffon sees it) that mountain ranges are generally situated in mid-continent – whereas others more nearly approach a state of invariable rule.

The third category is comprised of regularities among the Earth's constitutive entities, on a scale we would recognize as stratigraphic, petrological, or mineralogical. These include, for example, stratigraphic beds, the materials composing such beds, their fissures, and the physical and chemical properties of these materials.

Finally, the fourth class consists of regularities – not of objects or materials – but rather of processes, natural operations discernible on the Earth. These include the motions of water, especially of marine waters but without exclusion of rain and river waters, and the action of winds.

Regularities of Disposition, and of Operation

Rather than rehearse the entire contents of this list (Appendix), which can be examined at leisure, I will limit myself to a few observations about it. It is of course impossible to discuss at length all that is found in so large a list of proposed regularities. I will pass over without discussion the ways Buffon treats the greater part of these relations: his certainty, for example, of the former ocean's existence over the present continents, the fashioning of topographic inequalities by under-sea currents, and the migration of the continents and seas around the globe in a Westerly direction as caused by the action of tides and winds. These convictions are, to be sure, not without importance. In general Buffon, in his characteristic self-confidence, was not lacking for explanations for a given phenomenon or an interpretation of its presence. Perhaps the most important element of Buffon's train of thinking about regularities was to exemplify the contention that regular effects are proofs of equally regular causes.[4] But I think it is notable that on some occasions Buffon indicates candidly that he cannot determine the specific significance of certain regularities. He makes known, for example, as regards the dispositions of the continents, that "These relations may be connected with something general that may perhaps one day be discovered, of which we are presently ignorant."[5] This goes to show that, in Buffon's view of things, knowledge of terrestrial regularities has a value practically unto itself: perceivable regularities should be sought and collected, in expectation of our being able one day to determine their hidden relations, bringing to light their hitherto unknown meaning.

Even if we accept that, in accord with Buffon's personal opinion, there exist many fewer obscurities about the established facts than would be accepted by an open-minded modern observer (one would hardly expect Buffon to overlook points he might use to support his own system[6]), it is true all the same that both the reality and the interpretation of a good number of these regularities were subject to contestation. The main point I wish to emphasize is that, in the middle part of the eighteenth century, the language used to discuss the Theory of the Earth (or essentially all of geological science) included assertions about a number of supposed regularities, which constituted a set of basic ideas open for discussion within the geological world.

Seen from a certain angle, perhaps all of this may appear quite conventional. Is it not the case that any modern reader of Buffon's *Theory of the Earth* must be struck – as I recall having been impressed myself when I first read it – by the fact that Buffon clearly shows himself here as an *actualist*? by the fact that he stresses observable operations, the natural processes and agents that generally act slowly and are responsible both for the Earth's organization and for its inequalities? This is indeed so, and I do not mean to minimize the point. But allow me to draw your attention to the fact that, among the four categories in my synoptic table of Buffon's regularities, only one is genuinely dedicated to natural *operations*. Leaving this (fourth) class aside, the three others do not address natural processes, rather they concern regularities that may be understood as *consequences* of certain operations, perhaps in certain cases even of operations not yet discovered or understood. I think it noteworthy that most of these consequences or regularities are those of disposition. Every example found in the two first categories are of this kind, as are nearly all those of the third. These are supposed observations with respect to situational relations, conditions of fixed spatial position among one or another class of geographical, geological, or mineralogical entity. It matters little, at least for the moment, if these putative regularities are real or imagined, significant or inconsequential. In my view it would be a mistake to overlook the fact that the majority of regularities pointed out by Buffon are regularities of static disposition, available in principle to direct, empirical perception. These dispositional constancies, as statements about perceived configurations, are of course to be distinguished from regular operations or processes.

In the remaining part of this presentation I will concentrate on these dispositional relations. We will see that in emphasizing such relations an Earth Theorist is permitted to begin with fixed, observable facts – *effects* seen in the present state of affairs – and move to their causes and to past conditions (the latter being, of course, distinguished from observed effects in their inaccessibility to direct observation). Thus the theoretician concerned with change and development in the Earth can nonetheless act in accord with a credo precious to Enlightenment savants, regarding sound procedures for gaining natural knowledge: true science arises only out of sense experience and is best expressed in terms of general or universal rules or laws.

If this predilection for general dispositions were a specific quality found only in Buffon's writings, it would no doubt be little more than a historical curiosity. But while Buffon certainly represents a case of an exceptionally stubborn seeker of regularities, I think this preoccupation was quite broad and durable, at least in francophone circles. If this is true, a number of questions spring to mind. Where did this intellectual habit come from, and how was it perpetuated? What is its historical significance? What relations obtained between this fascination with regularities and the other elements of geological reasoning and activity? What variations existed in this way of thinking, and what changes and developments did it undergo? If one can speak of a tradition of constant dispositions and of regularities, what was its historical legacy?

Clearly it is not possible to answer all these questions here. I intend for the present to leave aside questions about the origins of this preoccupation, in order to focus on the second half of the eighteenth century. During the period from 1750 to 1800, the Theory of the Earth tradition drew back gradually in the face of new ways of doing geology, and it seems possible to me that the predisposition to speak in terms of regularities had a role in this change. My main historical task here is to show that this mental habit was in fact widespread, and to suggest how it may have played a role in the transformation of geological thinking and practice.

Bourguet

Having renounced study of the earlier part of the century, nonetheless I cannot refrain from taking one retrospective look, toward Louis Bourguet. A case could be made for Bourguet as the paradigmatic example of the thinker placing natural regularities at the center of geological reasoning. The principle of salients and re-entrants that he announced in 1729 is certainly among the most-discussed candidates for standing as a dispositional geological law in the eighteenth century. Buffon did not conceal his debt to this Huguenot author, even if the use Buffon made of the rule of correspondence was not in keeping with Bourguet's intentions.[7] But Buffon was far from being the only one to exploit or comment upon the idea of constant salients and re-entrants. My own first encounter with this rule came through reading Desmarest's remarks about it, which led me to a number of other comparable discussions during the second half of the eighteenth century. This brought to my notice for the first time the curious strength of dispositional regularities in the geological literature of the period. It seemed strange, and perhaps indicative of an obscurely important element in contemporary thought patterns, to find traces of a persistent debate over a principle that was repeatedly rejected by most of its commentators.[8] If they did not believe in it, why did they keep talking about it?

Given that the putative rule on the correspondence of salients and re-entrants evoked an interest disproportionate to the number of its adherents, I would like to add two comments before leaving Bourguet. Accepting the dominant place of the principle of correspondences among Bourguet's regularities, in his sketch of a Theory and in his subsequent reputation, nonetheless we should take note of the fact that in Bourguet's work there is mention of many different sorts of regularities, and that most of them reappear in Buffon's work (those taken up by Buffon from Bourguet – whether unchanged or slightly modified – are marked with an asterisk in the list presented in the Appendix). Furthermore, Bourguet expresses himself forthrightly on the extension of "rules of mechanics and of geometry" to the globe's structure, and on the right way to improve the Theory of the Earth through comparison of repeated observations, for example "on the heights of mountains, on the disposition of their different strata, on the tilt of these strata or beds relative to the horizontal, on the specific gravity of the materials composing them," and so forth.[9]

Let us note in passing that this allusion to specific gravity in strata refers implicitly to John Woodward's book on the natural history of the Earth, in which a great deal is made of the sequence of strata correlating with specific gravity.[10] The problem of understanding the extent to which such rules are present in the Theories of the Earth from the late seventeenth century onward requires study, but I venture to suggest that a tendency toward formulation of dispositional regularities is more a natural than a fortuitous feature of such theories. This tendency, I imagine, is entirely compatible with (and perhaps even called forth by) the fundamental nature of Theories of the Earth, as defined and interpreted by our colleague Jacques Roger.[11] And at this point let me state that I am in full agreement with a point emphasized by Rachel Laudan in her recent book on the foundations of geology, namely that from the late seventeenth century on, the basic geological problem was not one of choosing between theories and facts (since sensible authors realized that both are important), but rather of determining the right relations between theories and facts.[12] In any case, I think that in studying the eventual outcome of the preoccupation for regularities, we can follow the unfolding of a characteristic feature of Theories of the Earth, and in a fashion investigate the actual relations between treatments of the observed facts and their theoretical comprehension.

Boulanger

Let us now return to the second half of the eighteenth century. I promised to show the more or less continual and general presence in the period's geological discourse of an interest in regular dispositions. My fulfillment of this promise will be a bit feeble, I fear, not just for lack of time in this exposition but also because of my provisional and incomplete review of the period's texts from the viewpoint of this issue. Allow me, however, to attempt a summary of my research's results so far.

A leading example is found in the work of Nicolas-Antoine Boulanger. Let us consider briefly a little-known work by this engineer, his *Nouvelle mappemonde* published in 1753 with its accompanying *Mémoire* (Figure 1).[13] Here one finds an echo of one of Buffon's conceits about the positions of the continents and oceans. Boulanger believes he sees our world divided naturally into two parts, one essentially continental and the other oceanic. The natural center of

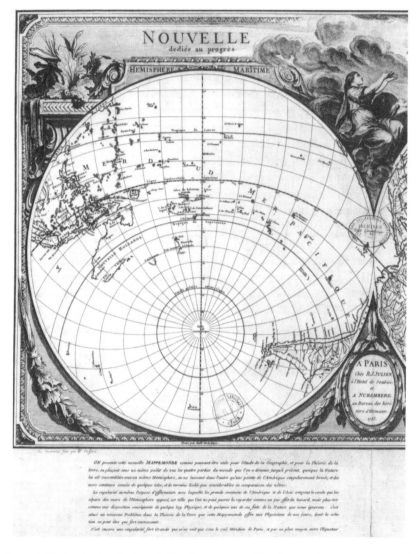

Figure 1. Nicolas-Antoine Boulanger's *Nouvelle mappemonde dédiée au progrès de nos connoissances*, 1753. By permission of Bibliothèque nationale de France. Boulanger contended that the distribution of the continental land masses and of the oceans over the globe's surface constitutes a regular disposition arising out of some as-yet unknown physical law. He thought it was not accidental, but indicative of some deep organizational principle, that these two hemispheres display a predominance of ocean in one case and of land in the other.

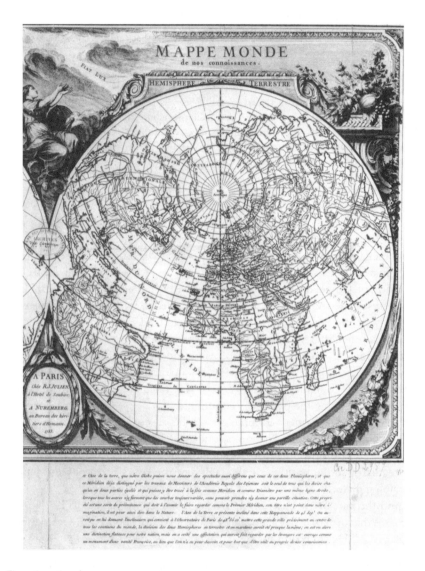

Figure 1 continued.

the continental planisphere is located, by fortunate chance, not far from Paris, at 45 degrees North on the Paris meridian. In his commentary on the map he states:

> The regularity or even the kind of distinction with which the great continents of America and Asia skirt the circle separating them from the seas of the opposing hemisphere, is such that these can hardly be seen as purely the result of chance, but rather they must be viewed as a disposition resulting from some physical law, as among those facts of Nature unknown to us. Thus this world map presents to the natural philosophers of our day a new problem in the Theory of the Earth, the solution of which must be extremely interesting.[14]

The accompanying *Mémoire sur une nouvelle mappemonde* explains a bit more fully Boulanger's assurance that attention devoted to regularities was in the process of yielding unprecedented knowledge of the Earth and its past. This is happening, he thought, through study of "some general or specific phenomena capable of enlightening us about the extraordinary revolutions that apparently have happened in the Earth, but of whose historical details we are ignorant." Evidence is sought specifically "in the general disposition and the correspondence of the seas and the continents, in the nature, form and harmony of mountains, the direction of rivers, and finally in the respective position of the different materials that diversify the surface."[15]

Boulanger uses the same sort of language in one of his writings that is perhaps even more obscure, an article on the "Antiquities and natural curiosities" of Touraine. His brief exposé demonstrates a similar manner of thinking in asserting the basis of a common accord, among scholars of every conviction, on the former "continuance of the seas over the continents": it is a fact, he reports, that the "natural antiquities" constituted by "the treasuries of the sea ... are distributed over the entire globe with so much uniformity and regularity in the position of the beds and the layers of stones and sands containing them."[16]

And the part of the *Encyclopédie* article "Déluge" that is due to Boulanger discusses the form of "alternating angles ... which correspond with such perfect regularity ...." He concludes that

> This admirable disposition in channel width, of valleys and mountains, applies to every place on earth without exception. It is itself one of the most interesting and novel problems that this age's observers have brought into view, the solution to which they are still seeking.[17]

The fact that, in these modest publications, Boulanger draws from these dispositions different conclusions from those taken by such predecessors as Bourguet and Buffon (flexibility of the Earth's crust; "frightful degradations" rather than undersea currents as causes of topography) is in a sense beside the point. What matters is their agreement on the importance of dispositional regularities.

Since Boulanger's unpublished geological writings were more extensive than his short publications, it would be natural to ask whether, in his manuscript *Anecdotes de la nature*, Boulanger shows any different orientation.[18] I hope to be forgiven for not giving a direct answer here. At this point I have done no more than glance again over the manuscript, rather than give it a full re-reading. It does seem, however, that Boulanger was more inclined in this work to emphasize *irregularities* over regularities.

Desmarest

Another figure in whose work we find the thread of this idea of dispositional regularity, from one end to the other of his long career, is Nicolas Desmarest. Desmarest's case, which has been a concern of mine for longer than I sometimes care to admit, is interesting in more than just one respect. Without going too far into details, I think it can be confirmed that Desmarest acquired a viewpoint favorable to regularities quite early, well before beginning to engage in direct geological observation on his own. It also seems clear that he adopted this view at least in part through commitment to a kind of Newtonian phenomenalism. This can be seen best in his lengthy commentaries for the French translation from the English of Francis Hauksbee's *Physico-mechanical Experiments*, in 1754.[19] The main characteristic of this phenomenalism is its manner of praising knowledge of natural relations, established by empirical means, without acknowledging any obligation to be concerned about the causes of those relations. Desmarest's 1757 article on "Géographie physique" in the *Encyclopédie* is filled with expressions of this conception.[20] Just a short time later, upon his introduction to direct geological field investigation, Desmarest immediately seized the opportunity to put to the test several regularities he may have learned by reading Buffon and Boulanger, perhaps also by way of the lectures of Guillaume-François Rouelle and the natural history *promenades* led by Bernard de Jussieu. Traces of this experience are preserved in the manuscripts from some natural historical excursions undertaken by Desmarest, dating to 1761 and 1762.[21]

Desmarest took it upon himself, for example, to take note of what he called the *plans inclinés* and *bords escarpés* of opposing banks in the bends of rivers – the occurrence of gentle slopes paired with steep faces on opposing banks of the curving portions in watercourses. (See Figure 2.) It may be that Desmarest owed his interest in this idea, of matching features in sections across river turnings,

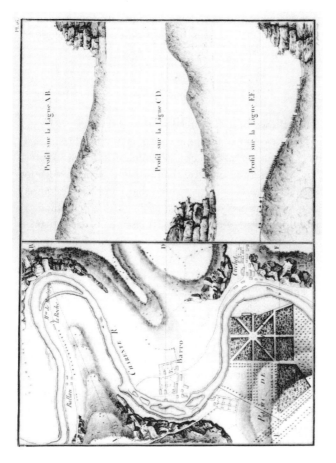

Figure 2. Nicolas Desmarest, "Cours de la Charente avec ses oscillations depuis Ballan jusqu'à Cuchet." Map of river bends (or ox-bows), with profiles. Plate 16 in *Atlas encyclopédique contenant les cartes et les planches relatives à la géographie physique*, by Desmarest and completed by J.-B.-G.-M. Bory de Saint-Vincent (Paris, Agasse, 1827). By permission of Bibliothèque nationale de France. This illustrates Desmarest's interest in the regular occurrence of steep gradients in river banks on the convex sides of river bends (*bords escarpés*), with gentle slopes on the concave sides (*plans inclinés*), as instances of significant regularities. Desmarest saw these features in the transversal form of river beds as evidence that rivers excavate their valleys. The tendency for rivers to run in their beds nearest the steeper banks was one of the dispositional regularities noted by Buffon (Appendix, item 2j). Although this map and accompanying section was published after Desmarest's death, it was prepared in connection with the many pertinent articles published in his *Géographie physique*, whose volumes began to appear in 1794.

Reflections on Natural Laws in Eighteenth-Century Geology   13

to Boulanger. To Desmarest, these features in the transversal form of the beds and banks of rivers seemed to give evidence of slow present causes in valley formation. Desmarest's first field investigations in geology were guided, at least in part, by the hope of confirming and possibly even extending observable relations, phenomena of which he was aware through, I think, a certain aspect of Theories of the Earth. In this fashion it is possible that theories of the earth may well have encouraged empirical field investigations. And it seems to me that Desmarest's transformation from an armchair savant or *scientifique de cabinet* into a field investigator is linked with a pre-existing conviction to the effect that above all else one must keep a sharp eye for tangible regularities.

The same concern is evident in the work of Desmarest's old age, up to five decades later: a determination to see a genuine science of physical geography established upon sure evidence, often in the form of general dispositions. Among Desmarest's works displaying this conviction is his enormous compendium *Géographie Physique*, in the series *Encyclopédie méthodique*.[22] It is true that accounts of Desmarest's most memorable achievements, in Auvergne, had been presented during the 1770s in language less conspicuously marked by this preoccupation. All the same, even here signs of it are evident. Desmarest tells us that it is by way of the *distribution*, the *regular correspondence*, and the *disposition* of volcanic materials that he found himself in a position, in his field investigations during the 1760s, to establish the volcanic origin of columnar basalt, and to discern a distinction among three epochs of volcanism.[23] In my opinion it is not excessive to see Desmarest's participation in a tradition of situational regularities as one major element in the success of his observational geology.

Lavoisier

Also worthy of attention as a seeker of geological regularities is Antoine-Laurent Lavoisier. The famous chemist had been dedicated to geological study during his youth, and returned to this subject in the final years of his shortened life. At an early stage in his geological work he wrote:

> Whatever the disorder that appears to reign in the disposition of beds of earths and rocks that are seen on the surface of the globe we live on, it is nonetheless not difficult to recognize that these irregularities themselves are subject to certain laws, that they follow certain rules.[24]

This passage continues with a distinction among different sorts of *mountains* (evidently meant in a sense similar to the German term *Gebirge*, referring to subsurface rock masses rather than topographic features), in which the dispositions of beds vary discernibly. Generally speaking, the fragments and varied articles in which Lavoisier addressed geological matters emphasize the arrangement and homogeneity of geologically interesting entities. Lavoisier displayed an impulse toward measuring the heights and thicknesses of these units, as well as the limits of their extension, and toward affirming the constancy of their order.

Lavoisier achieved a signal success in reasoning on the regular alternation arising out of different causes of two types of deposits; and the explanation for this that he presented gave him hope that he might "arrive at a determination of the main laws followed by nature in the arrangement of horizontal beds."[25] His detailed examination of *pelagic* and *littoral* beds (a distinction, borrowed from Rouelle, which itself depended on a regular disposition of characters giving evidence of different formative conditions) led him to consider with care the results of action in a sea which, rather than being passive, "leaves its bed only to return, which moves following certain laws, and above all by means of a very slow motion."[26] We see, then, how the effects thus conceived correspond precisely to the dispositions studied by Lavoisier. The spirit of regularities could hardly show itself better vindicated.

De Saussure, Hutton, and Deluc

The case of Horace-Bénédict de Saussure is instructive, with regard to the issue of general dispositions. This prestigious observer of mountains informs us that, in his early alpine journeys, he was looking for general relations, and that he had even published a "discourse on the structure of mountains" founded on such relations.[27] His subsequent investigations on the other hand led him to become more skeptical about the existence of constant and regular relations in the mountains. Despite this conclusion, de Saussure brought forward several general observations that he thought to be valid, among which are some apparent points of resemblance with the assemblage of regularities advanced by Buffon.[28] Disillusioned and probably frustrated in his hopes of seeing his laboriously gathered body of generalized observations yield a solid and convincing theory of the Earth, de Saussure could not escape from this very framework of reasoning. All his life, it seems, he believed that mountains are the best setting for making observations that permit distinctions within the natural divisions of the globe, and fruitful examination of their structures, situations, directions, order, and relationships.[29] De Saussure insisted – and this is not an altogether elementary point – on observation of the normal and the ordinary, not of the unusual.[30] He presented his advice to other observers in the form of an *Agenda*, which is above all an inventory of *things*, of objects to study in order to discover their interconnections or relationships.[31]

The contrast between de Saussure and his Scottish contemporary James Hutton is interesting. Hutton (who, incidentally, quoted de Saussure at great length in his *Theory of the Earth*) was himself a tenacious student of laws.[32] But for Hutton the laws in question related mainly to geological processes, rather than to disposition of material entities of the Earth. The great majority of the regularities to which he calls attention, and which he develops throughout his *Theory*, are addressed to terrestrial actions and operations. It may be said that instead of concentrating on questions about the distribution of the materials of the Earth, for him the key is to ask how terrestrial changes take place. The path Hutton chose, in identifying the objects of inquiry, was clearly different from that developed by de Saussure. The Scot placed his bet, as it were, on the regularities of my fourth class, those of operations. De Saussure's preference, instead, was for the other, dispositional sort of regularity; and in this respect the Genevese savant was in company with many others on the Continent.

Jean-André Deluc, de Saussure's Genevese compatriot (who, however, departed for Great Britain in the middle of his long and prolific career), was perhaps a less conspicuous adherent of the tradition of general dispositions. Nevertheless he was interested in everything that touched the Earth and its history, sometimes explicitly including its regularities. Deluc was suspicious of the idea that had "seduced" Woodward, the contention that the order of strata is related to their specific gravity.[33] But in study of the strata to refute this notion, he felt that he was assisted in discerning "what the real phenomenon is that must be explained."[34] Deluc then manifested some interest in recognizing regularities having to do with, among other things, the order of the strata and the fossil shells found in them.[35]

Toward a Stratigraphic Geology

Our colleagues F. Ellenberger and G. Gohau, who brilliantly demonstrated the influence of Deluc's work on the ideas of Cuvier and Brongniart, made the following observation in their article on this subject:

> Hutton, Dolomieu, de Saussure and their contemporaries lacked for neither ideas nor observations. If they did not make better use of both, this was perhaps because they were unable to find the connecting thread allowing them to place the facts they had gathered in intelligible order. It is doubtless also because they retained the ambition they inherited from their elders, to build a theory of the Earth deducing the globe's formation from comprehensive study of its structure.

Their successors would come to understand how, at once to find unifying concepts and also to limit their ambitions ...."[36]

I think one might add to this judicious statement that a formal aspect of this connecting thread was already in the hands of these forebears of Brongniart and Cuvier, predecessors who bequeathed to them a certain part of their new initiative.

I find it interesting that Brongniart and Cuvier, while omitting mention of Deluc in their first "Essai sur la géographie minéralogique des environs de Paris," make reference to several authors, including Lamanon and Desmarest.[37] It was Lamanon who ascertained the regular presence of remnants of freshwater organisms in certain gypsum deposits, and thus of their "lacustrine formation."[38] And it was Desmarest, addressing himself in his characteristic way to the Montmartre beds so as to distinguish them according to their constitution, disposition, and configuration, who supplied a framework for the division of these beds, to which in turn Cuvier and Brongniart brought the discriminating tool of fossils and the idea of treating them in historical fashion.[39] Once one realizes that they benefitted from previous work done by authors like Lamanon, Desmarest, and Deluc (among others), who were convinced of the capital importance of general dispositions, it is not surprising to find in the Essay by Cuvier and Brongniart expressions that resonate with those very regularities:

> ... but it must not be thought that these varied layers were set in place at random and without any regularity: everywhere they follow the same order of superposition, over the considerable extent of terrain that we have covered ... This constancy in the order of superposition of the thinnest beds, over an extent of at least 12 myriameters [i.e., 12 x 10,000 meters], is, we say, one of the most remarkable facts that we have established in the course of our research. The results that must follow from this fact are consequences, for the arts and for geology, that will be all the more interesting as they acquire more confirmation.[40]

Conclusion

The following general conclusions emerge from the preceding discussion:

1. An important element in the traditions of geological investigations during the eighteenth century was the preoccupation with recognizing regularities or natural laws. This preoccupation was naturally situated mainly within the general enterprise of the Theories of the Earth. It was however capable of being detached from broad theorizing, to be applied to more limited studies. It

reflected the habitual empiricism of the era, and also the high value accorded to the reduction of all knowledge to general terms. It may also have resonated with a Newtonian kind of phenomenalism. In the francophone world, at least, the search for perceptible regularities appears to have been one attractive and cautious way of pursuing empirical knowledge of the Earth, in avoidance of the danger of insufficiently controlled conjecture.

2. Two kinds of laws or putative generalities can be distinguished: one of disposition or situation, the other of process. In the eighteenth century, nothing prevented an investigator from taking an interest in both at the same time. Probably nobody involved neglected one or the other completely. But it is not surprising that often there was an inclination toward one at the expense of the other. With the exception of Hutton (and, I believe, Lamarck, who shared Hutton's taste for processes), a certain preference can be seen, among the geological notables of the period from 1750 to 1800, for the dispositional sorts of regularities.[41]

3. We may of course find some amusement in the apparent naïveté of some of the geological regularities proposed in the eighteenth century. A more constructive reaction, however, is to draw – from what in our sight may seem like credulity – a conclusion about the complicated relations between theory and observation during a period when geological science was in the process of taking its modern form. A simplistic yet still widespread view holds that scientific observation is a simple act that takes place directly and plainly, and that difficulties arise only in the problem of formulating theories to give the observations satisfactory explanation. The historical phenomenon discussed in the present paper, in which we recognize an eighteenth-century taste for regularities, may remind us that the establishment of a sound earth science posed problems that were at once both theoretical and observational. In the absence, in principle, of any limits on the different sorts of potentially relevant observations, it was necessary to prioritize among all conceivably significant observations, to try to define those most worthy of attention, most likely to yield meaningful consequences. In such an effort, separating the empirical and theoretical faculties from each other is not easily done. A great number of savants of good faith, during this period, held the Theory of the Earth to be an unfinished project, and expressed their uncomfortable awareness of the fact that nobody could say with certainty which kinds of geological observations were most important. Regularities responded to the needs of a group of scientific investigators for whom *answers* were difficult, because identification of the proper *questions* had not yet been altogether clarified.

4. In the scientific world during the second half of the eighteenth century, the detection of *dispositional* regularities often provided the occasion for drawing

conclusions about the past condition of the globe (or, at least, of some of its parts). Regularity in the arrangement of the visible effects, traces of changes in a past necessarily inaccessible to our senses, constituted data for repeated efforts to establish knowledge upon past conditions and changes, within a framework that might be limited, or might in other cases be general.[42]

5. Among some French-speaking geologists of the eighteenth century, hope of formulating or discovering laws or generalizations regarding dispositions seems to have promoted field research – it appears to have encouraged the beginnings of the activity of *fieldwork*, so prized a part of geology now for the past two centuries. As our colleague G. Gohau makes clear in his excellent recent book, this may appear at first glance to defy logic. Pioneering geological thinkers might well have asked themselves: if one possesses knowledge of geological laws, of what use would it be to take the trouble to travel widely and observe minutely?[43] But the regularities at issue were not for the most part sufficiently certain to command automatic assent. Rather, they cried out to be confirmed, or refuted, or extended. In any case, in the second half of the eighteenth century, the search for regularities seems to have taken a role in the formation of a healthy tradition of carrying out research in the field. (On this point, I am inclined to take issue with Rachel Laudan's new and very stimulating book, in which the element of field observation in Theories of the Earth is ignored, if not denied: Laudan identifies chemical science as the empirical basis of geology in the late eighteenth century.)[44]

6. More tentatively, I propose a historical link between the preoccupation with general dispositions and the birth of the new stratigraphic geology at the start of the nineteenth century. This is a contention to make with caution: it would be absurd to maintain that the habit of thinking in terms of regularities could directly and by itself have brought about a programme for identifying successions of formations. I am suggesting something far more modest, and less simplistic. The 'game' of regularities could be no more than one element in a complex historical development. Nonetheless it was a game that permitted, indeed encouraged, taking account of different sorts of observed organization and constancy; it may have contributed to the re-conceptualizing of geology in the nineteenth century.

7. If there is any value in this analysis, it might help in clarifying the historical transition from the prevalence of Theories of the Earth to the new geology in the nineteenth century. And it might serve to facilitate a greater appreciation of the methodological richness of geological thinking in the second half of the eighteenth century.[45]

***

I wish to express thanks to my colleagues in COFRHIGÉO for their warm and friendly hospitality, and especially to M. François Ellenberger and M. Jacques Roger, both of whom have been very generous in sharing their knowledge as well as their books and historical materials.

My heartfelt thanks are due also to M. Jean Gaudant for his patient efforts to diminish the number of blunders and awkward expressions in the original French version of this paper. His kind assistance contributed, as well, to strengthening my argument. The errors that remain are, of course, my own responsibility.

VII

APPENDIX

Buffon, *Théorie de la Terre* (1749) with *Preuves*

Regularities and Tendencies

1. Main features in the arrangement of continents and oceans

   (*) a) Global organization of the continents and oceans, from North to South. (The Earth is "divisée d'un pôle à l'autre par deux bandes de terre & deux bandes de mer" ["divided pole-to-pole by two strips of land and two strips of sea"].) (*Preuves*, art. VI).

   b) Angular orientations of the continental masses, according to their greatest lengths; specific obliquities of axes of orientation, relative to the equator (*Preuves*, art. VI).

   c) Notable symmetries of continental masses, considered with regard to their positions (*Preuves*, art. VI):

   – Old World continent [Eastern hemisphere] displaced to the North, New World continent [Western hemisphere] displaced to the South, relative to the equator, each by about 16 to 18 degrees;

   – Equal portions of continental surfaces on each side of axes of maximum length.

   d) A "Mediterranean" interruption or opening in each continental mass (*Preuves*, art. XI).

   e) Principal straits generally oriented East-West (*Preuves*, art. XIX).

   f) Several important continental segments in forms pointed toward South, often with straits and islands (*Preuves*, art. XI).

2. Regularities and tendencies in the disposition of features on land and in the seas

   (*) a) Altitudes of mountains greatest toward the equator, diminishing as

Reflections on Natural Laws in Eighteenth-Century Geology   21

one recedes from the equator (OP, 47, 50; *Preuves*, art. IX).

(*) b) Mountains tend to be situated in continental interiors (OP, 48; *Preuves*, art. X).

c) Altitudes of mountains generally proportional to the number and importance of rivers taking their origin there (i.e., highest mountains located near river sources, "... points de partage ... pour la distribution des eaux") (*Preuves*, art. IX, X).

(*) d) Regular orientations in mountain ranges; East-West tendency in the Old World [Eastern hemisphere], North-South tendency in New World [Western hemisphere] (OP, 47, *Preuves*, IX, X).

e) Networks in groups of mountain ranges; continuity in mountain ranges, especially in continental interiors (OP, 54; *Preuves*, art. IX, XI).

(*) f) Corresponding angles in valleys; rule of salients and re-entrants (OP, 47–48; *Preuves*, art. V, VII, IX, XIII; Concl.).

g) Opposing hills have the same altitudes, on each side of valleys with corresponding angles (OP, 48; *Preuves*, art. VII, IX).

h) Tendencies in rivers to run parallel to mountains, near their sources; but perpendicular to the coasts near their mouths (OP, 48; *Preuves*, art. X).

i) Tendency for river courses to run straight near their sources, and sinuously near their mouths (*Preuves*, art. X).

(*) j) Positions of streams in their beds, in relation to gradients of banks: Near the center when the slopes are equal, otherwise close to the steeper slope, in proportion to the difference in gradient (*Preuves*, art. X).

k) Depth of seas along coasts, in proportion to the altitudes of adjacent lands; the greatest differences being situated near the equator (*Preuves*, art. IX, XIII).

l) Volcanoes placed in mountains, never in valleys or plains; volcanic *foyers* [hearths, heat sources] near summits, above level of surrounding plains; possible subterranean connections among volcanoes (OP, 48, 60).

3. Regularities in the structures and materials comprising the Earth's surface and crust

   a) Parallelism and constant thickness of beds (OP, 48, 49, 50, 53, 54; *Preuves*, art. IV, VII, IX, XIII).

   b) Correspondence of mineralogical characters of materials, across valleys between mountains, or across straits between land masses (OP, 48; *Preuves*, art. VII, IX, XIII).

   c) Presence of fossil vestiges of marine organisms in sedimentary rocks, in every part of the world (OP, 48; *Preuves*, art. IV, VIII).

   d) Fossil shells lying flat, parallel to beds (*Preuves*, art. VIII).

   (*) e) Similarity between rock matrix and matter filling fossils (i.e., matter of which fossils are composed) (*Preuves*, art. VIII).

   f) Certain (weak) correlations between rocks and kinds of fossils present in them (*Preuves*, art. VIII).

   g) Perpendicular cracks (joints) in rocks, notably in stratified rocks (OP, 48, 58; *Preuves*, art. VIII, IX, XVII; Concl.).

   h) Exact correspondence of the two sides of such cracks (joints) (OP, 59).

   i) Variable distances between cracks (joints), depending on the nature of the rocks; variable orientation of cracks depending on character of the rock (OP, 58; *Preuves*, art. IX, XVII).

   j) General correlation between topographic form and stratigraphic structure: surface shapes of mountains tend to follow the slopes of the beds (*Preuves*, art. VII).

   k) No direct relation between order of beds and their specific gravities (contradicting the opinions of Whiston, Woodward) (*Preuves*, art. II, IV, VII, VIII).

4. Regularities of processes, in perceptible terrestrial operations

   (*) a) General movement of ocean waters, by the flux and reflux (tides), from East toward the West (OP, 51, 55; *Preuves*, art. XII).

b) Growing effect of this movement as one approaches the equator; movement of waters toward the equator (OP, 50).

c) Similar behavior of the winds: from East to West, stronger toward the equator – augmenting the effects of movement of the waters (OP, 52, 54; *Preuves*, art. XIV, XIX).

d) Movements of the ocean water are not merely superficial: transmission of surface agitation to considerable depths (OP, 52; *Preuves*, art. XIX).

e) Erosion of continents by fluvial waters (OP, 55, 61).

f) Regularities in the movements of river waters:

– The speed of a river depends more on the amount of water upstream than on the gradient;

– The speed of a current varies depending on depth: the greatest speeds occur in middling depths;

– The height of the surface of running waters varies depending on speed (which reduces the effect of weight); in a rapidly-moving current, the water near the center flows faster and at a higher level than that flowing near the edges (*Preuves*, art. X).

OP = *Oeuvres philosophiques de Buffon*, edited by Jean Piveteau (Paris, Presses Universitaires de France, 1954) [this edition does not include all of the *preuves*].

(*) = this regularity, or a variant form of it, is found in Bourguet.

NOTES

All translations from French to English, in this essay, are mine unless otherwise indicated. Quotations in French are rendered in their original form, without modernization or correction in spelling, including accents.

Where quotations in the paper's text are presented in English, the corresponding French passages appear in the notes. Quotations presented only in the notes are given both in the original French and in English translation.

1   Peter Simon Pallas, for instance, refers in his *Observations sur la formation des montagnes et des changements arrivés au globe* (St. Pétersbourg: De l'imprimerie de l'Académie Impériale des Sciences, [1777]) to the assumed connections of mountains across the Arctic polar region, a "continuation qui devient probable par sa conformité aux loix, que la Nature semble observer dans la continuité des chaines montagneuses du globe ...." ["continuation which becomes probable by its conformity to the laws that Nature appears to observe in the continuity of the globe's mountain chains"] (p. 12). According to Pallas, who inquires into the orientation and organization of mountains for insight into the changes the Earth has undergone: "On entrevoit de certaines loix, à l'égard de l'arrangement respectif de cet ordre secondaire d'anciennes roches, par tout les systèmes de montagnes qui appartiennent à l'Empire russe" ["Glimpses are seen of certain laws regarding the respective arrangement of this secondary order of ancient rocks in all the mountain systems belonging to the Russian Empire"] (p. 27).

Lavoisier, too, speaks of *laws* in the arrangement of beds (see below, notes 24 to 26).

Buffon, on the other hand, makes use of this word in the sense of rules governing the actions that fashion the globe: " ... la mer a des limites & des loix, ses mouvemens y sont assujétis, l'air a ses courans réglés ...." [" ... the sea has limits and laws, to which its motions are subject, the air has its regulated currents ...."] (p. 69 of *Second Discours, Histoire & Théorie de la Terre*, in *Histoire naturelle, générale et particulière*, t. I, Paris, De l'Imprimerie Royale, 1749). The correspondence of alternatingly opposed angles, Buffon says, is a consequence "fondée sur les loix du mouvement des eaux & l'égalité de l'action des fluides ...." ["based on the laws of motion of water and the equality of action in fluids"] (p. 451 of *Preuves de la théorie de la Terre*, same volume). The sea's currents should be regarded as "comme de grands fleuves ou des eaux courantes, sujettes aux mêmes loix que les fleuves de la terre ...." ["like great rivers or flows of water, subject to the same laws as rivers on the surface of land"] (*Preuves*, p. 452).

Similar language is taken up, three decades later, by the Historian of the Academy of Sciences (at this point in time, Condorcet), reporting on a new discovery, near Paris, of beds filled with fossil shells: "Il est important de multiplier les observations de ce genre; ce n'est que par une connoissance exacte & détaillée du Globe, qu'on peut parvenir à se former une idée des révolutions qu'il a éprouvées, des causes des changemens qu'on y observe, & des loix suivant lesquelles ces causes ont agi" ["It is important to multiply observations of this kind; it is only by exact and detailed knowledge of the Globe, that one can arrive at a proper idea of the revolutions it has undergone, of the causes of the changes observed in it, and of the laws under which these causes have acted"] (*Histoire de l'Académie Royale des Sciences*, 1779 [1782], p. 12).

2   Buffon, *Théorie de la Terre*, pp. 68–69:

Commençons donc par nous représenter ce que l'expérience de tous les temps & ce que nos propres observations nous apprennent au sujet de la Terre. Ce globe immense nous offre à la surface, des hauteurs, des profondeurs, des plaines, des mers, des marais, des fleuves, des cavernes, des gouffres, des volcans, & à la première inspection nous ne découvrons en tout cela aucune régularité, aucun ordre. Si nous pénétrons dans son intérieur, nous y trouvons des métaux, des minéraux, des pierres, des bitumes, des sables, des terres, des eaux & des matières de toute espèce, placées comme au hasard & sans aucune règle apparente; en examinant avec plus d'attention, nous voyons des montagnes affaissées, des rochers fendus & brisés, des contrées englouties, des isles nouvelles, des terreins submergés, des cavernes comblées; nous trouvons des matières pesantes souvent posées sur des matières légères, des corps durs environnés de substances molles, des choses sèches, humides, chaudes, froides, solides, friables, toutes mêlées & dans une espèce de confusion qui ne nous présente d'autre image que celle d'un amas de débris & d'un monde en ruine.

[Let us therefore begin by representing to ourselves what common experience and our own observations teach us on the subject of the Earth. At its surface this immense globe shows us uplands, depths, plains, seas, swamps, rivers, caverns, abysses, volcanoes – and at first inspection we find in all of this no regularity, no order. If we penetrate into the interior, we find there metals, minerals, stones, bitumens, sands, earths, waters and materials of all kinds, placed as if by chance and without any apparent regulation. By looking with greater attention, we see collapsed mountains, split and broken rocks, engulfed regions, new islands, submerged lands, caverns filled in. We find heavy materials often resting upon lighter materials, hard bodies surrounded by soft substances, things that are dry, damp, hot, cold, solid, friable – all mixed together and in a sort of confusion that strikes us with no image other than that of a heap of debris and a world in ruins.]

The next paragraph, by contrast, evokes the stable and harmonious conditions of life amidst these ruins, " ... où tout est animé & conduit avec une puissance & une intelligence qui nous remplissent d'admiration & nous élèvent jusqu'au Créateur:"

Ne nous pressons donc pas de prononcer sur l'irrégularité que nous voyons à la surface de la Terre, & sur le désordre apparent qui se trouve dans son intérieur, car nous en reconnoîtrons bien-tôt l'utilité & même la nécessité; & en y faisant plus d'attention nous y trouverons peut-être un ordre que nous ne soupçonnions pas, & des rapports généraux que nous n'apercevions pas au premier coup d'oeil.

[... where everything is animated and conducted with a power and an intelligence that fills us with admiration and lifts us up toward the Creator.

Let us not hasten therefore to pronounce on the irregularity that we see on the Earth's surface, and on the apparent disorder that is found in its interior, for we shall soon recognize its utility and even its necessity; and by giving it greater attention we will perhaps find there an order that we had not suspected, and general relationships that we had not perceived at first glance (pp. 69–70).]

    3 Louis Bourguet, *Lettres philosophiques sur la formation des sels et des crystaux* ... *Avec un mémoire sur la théorie de la Terre* (Amsterdam: Chez François L'Honoré, 1729). The classes of regularity found in Bourguet are three in number:

    1) Phenomena concerning the globe's surface (pp. 192–200);
    2) Phenomena concerning the interior structure of the solid part of the globe (pp. 201–208);
    3) Phenomena concerning the destruction of the earth (pp. 209–211).

    4 Buffon states, in particular: " ... mais cette espèce d'organisation de la Terre que nous découvrons partout, cette situation horizontale & parallèle des couches, ne

peuvent venir que d'une cause constante & d'un mouvement réglé & toujours dirigé de la même façon." ["... but this sort of organization in the Earth that we discover everywhere, this horizontal and parallel situation of the beds, can come only from a constant cause and a motion that is regulated and always directed in the same manner"] (p. 81).

And, a bit further on, he continues: "Il est aisé d'apercevoir que cette uniformité de la Nature, cette espèce d'organisation de la terre, cette jonction des différentes matières par couches parallèles & par lits, sans égard à leur pesanteur, n'ont pû être produites que par une cause aussi puissante & aussi constante que celle de l'agitation des eaux de la mer, soit par le mouvement réglé des vents, soit par celui du flux & du reflux, &c." ["It is easy to perceive that this uniformity in Nature, this sort of organization in the Earth, this joining of different materials in parallel layers and beds, regardless of their differing weight, can only have been produced by a cause as powerful and constant as that of the agitation of the sea's waters, whether by the regulated motion of the winds, or that of the tidal flux and reflux, &c"] (p. 93).

5   "Ces rapports peuvent tenir à quelque chose de général que l'on découvrira peut-être, & que nous ignorons" (*Preuves*, article VI, p. 209). This declaration is preceded by these sentences:

Voilà ce que l'inspection attentive du globe peut nous fournir de plus général sur la division de la Terre. Nous nous abstiendrons de faire sur cela des hypothèses & de hasarder des raisonnemens qui pourroient nous conduire à de fausses conséquences, mais comme personne n'avoit considéré sous ce point de vûe la division du globe, j'ai cru devoir communiquer ces remarques. Il est assez singulier que la ligne qui fait la plus grande largeur des continens terrestres, les partage en deux parties égales; il ne l'est pas moins que ces deux lignes commencent & finissent aux mêmes degrés de latitude, & qu'elles soient toutes deux inclinées de même à l'équateur.

[Here we have what a careful inspection of the globe can provide us by way of general understanding about the Earth's divisions. We will refrain from making up hypotheses and guessing at reasons that might lead us to false consequences, but since nobody had previously considered the division of the globe from this viewpoint, I thought I should pass on these observations. It is quite odd [singular] that the lines of greatest extent through the terrestrial continents divides them in two equal parts. It is no less odd that these two lines begin and end at the same degrees of latitude, and that they both are inclined to the equator to the same degree.]

6   As one of the era's most eloquent defenders of empiricism, Buffon nonetheless lacked a reputation as a keen geological observer. Some of his contemporaries regarded him as a victim of one of the classic weaknesses of the Theory of the Earth tradition, namely the inclination to judge prematurely the meaning of phenomena.

7   Bourguet, "Mémoire sur la théorie de la terre," pp. 175–220 of his *Lettres philosophiques* (see note 3). See the biographical article on Bourguet by François Ellenberger, *Dictionary of Scientific Biography*, vol. 15 (Supplement I), pp. 52–59 (New York, Charles Scribner's Sons, 1978). Buffon criticizes Bourguet's general theory, but at the same time praises him for his discovery of the correspondence of salient and re-entrant angles (notably in the *Preuves*, pp. 193, 321, 324). Buffon even states that the

correspondence of angles and of the heights of opposing hills provided him with a conceptual key: "... c'est par des observations réitérées sur cette régularité surprenante & sur cette ressemblance frappante, que mes premières idées sur la théorie de la Terre me sont venues...." [" ... it was through repeated observations of this surprising regularity and this striking resemblance that my first ideas on the theory of the Earth came to me ...."] (*Preuves*, p. 457).

8 Kenneth L. Taylor, "Natural Law in Eighteenth-Century Geology: The Case of Louis Bourguet," XIIIth International Congress of the History of Science, *Proceedings*, vol. VIII (Moscow, Editions 'Nauka', 1974), pp. 72–80. See also the recent article by Marguerite Carozzi, "From the Concept of Salient and Reentrant Angles by Louis Bourguet to Nicolas Desmarest's Description of Meandering Rivers," *Archives des Sciences*, vol. 39 (1986), pp. 25–51.

9 Bourguet, "Mémoire sur la théorie de la terre," pp. 189–190: "sur les hauteurs des Montagnes, sur la disposition de leurs différens *Strata*, sur l'inclinaison de ces *Strata* ou Lits à l'horison, sur la pésanteur [sic] spécifique des matériaux dont ils sont composés, ..." etc.

10 John Woodward, *An Essay toward a Natural History of the Earth* (London, Printed for Ric. Wilkin, 1695), especially the first part, pp. 75 ff.

11 Jacques Roger, "La théorie de la terre au XVIIème siècle," *Revue d'histoire des sciences*, vol. 26 (1973), pp. 232–48. See also, by the same author, "The Cartesian Model and Its Role in Eighteenth-Century 'Theory of the Earth'," in *Problems of Cartesianism*, ed. Thomas M. Lennon, John M. Nicholas, and John W. Davis, McGill-Queen's Studies in the History of Ideas, no. 1 (Kingston & Montreal, McGill-Queen's University Press, 1982), pp. 95–125. According to Roger, the Theory of the Earth represents a way of thinking that (1) depends in part on the post-Copernican vision of the Earth as an autonomous object, not as a cosmic region; (2) responds also to the growing prestige of the physical sciences and especially the new mechanical philosophy, encompassing explanatory principles to which the Earth is subject in its entirety and through its history; and (3) tends to incorporate an emerging historical sensibility, or rather an interest in questions about the generation of things. This insightful analysis provides us with three fundamental criteria in identifying a Theory of the Earth: such a Theory should treat the Earth as an entity or a coherent system; it should take account of, or be consistent with, the prevailing principles, laws, or basic processes to which the Earth is understood to be subject, and which constitute its essential nature; and it should give grounds for a rational accounting of the development of the Earth or of its changes.

12 Rachel Laudan, *From Mineralogy to Geology: The Foundations of a Science, 1650–1830* (Chicago & London, University of Chicago Press, 1987), p. 8.

13 [Nicolas-Antoine Boulanger], *Nouvelle mappemonde, dédiée au progrès de nos connoissances* (Paris, Chés R.J. Julien; Nuremberg, Au Bureau des Héritiers d'Homann, 1753). See also the study by Paul Sadrin, *Nicolas-Antoine Boulanger (1722–1759), ou avant nous le déluge*, vol. 240 of *Studies on Voltaire and the Eighteenth Century* (Oxford, The Voltaire Foundation, 1986).

14 Boulanger, *Nouvelle mappemonde*, text on the map. The entire text is reproduced by Paul Sadrin in the notes of his edition of Boulanger's *L'Antiquité dévoilée par ses usages*,

2 vols., *Annales littéraires de l'Université de Besançon*, no. 215 (Paris, Les Belles Lettres, 1978), vol. 2, p. 43. ["La regularité où (sic) même l'espece d'affectation avec laquelle les grands continens de l'Amérique et de l'Asie cotoyent le cercle qui les sépare des mers de l'Hémisphere opposé, est telle que l'on ne peut gueres la regarder comme un pur effet du hazard, mais plus-tot comme une disposition conséquente de quelque loy Physique, et de quelques uns de ces faits de la Nature que nous ignorons. C'est ainsi un nouveau Problême dans la Théorie de la Terre que cette Mappemonde offre aux Physiciens de nos jours, dont la solution ne peut être que fort interessante." (As throughout this paper, spelling and accented characters are unaltered from the original document.)]

15 [Boulanger], *Mémoire sur une nouvelle mappemonde* ([no place], 1753), pp. 3–4. Study of "quelques Phénomènes généraux ou particuliers capables de nous éclairer sur les révolutions extraordinaires qui semblent y être arrivées, mais dont nous ignorons les détails historiques." Evidence sought "dans la disposition générale & la correspondance des Mers & des Continens, dans la nature, la forme & l'ensemble des Montagnes, dans la direction des Fleuves, & enfin dans la position respective des diverses matieres qui en varient la superficie."

16 Boulanger, "Antiquités et curiosités naturelles," *Almanach historique de Touraine pour l'année 1755* (Tours, Chez François Lambert, 1755), unpaginated (4 pp.). [Concernant le] "séjour des mers sur les continens": ... les "antiquités naturelles" [que sont] "les trésors de la mer ... sont distribuées par toute la terre avec une uniformité & une régularité si grande dans la position des lits & des couches de pierres & de sables qui les contiennent." I thank Rhoda Rappaport for having suggested that I look at these works by Boulanger.

17 "Déluge," *Encyclopédie*, vol. IV (1754), p. 797b. This is an "article in which all that is placed within quotation marks is by M. Boulanger" (p. 802a). ["Article où tout ce qui est en guillemets est de M. Boulanger".]

["angles alternatifs ... qui se correspondent avec une si parfaite régularité ...."]
"Cette admirable disposition des détroits, des vallées & des montagnes, est propre à tous les lieux de la terre sans aucune exception. C'est même un problème des plus intéressans & des plus nouveaux que les observateurs de ce siècle se soient proposés, & dont ils cherchent encore la solution."

18 Boulanger, *Anecdotes physiques de l'histoire de la nature*, MS 869 of the Bibliothèque du Muséum National d'Histoire Naturelle.

19 Francis Hauksbee, *Expériences physico-méchaniques sur différens sujets, et principalement sur la lumière et l'électricité, produites par le frottement des corps. Traduites de l'anglois ... par feu M. de Brémond ... Revûes & mises au jour, avec un discours préliminaire, des remarques & des notes, par M. Desmarest*. 2 vols. (Paris, Chez la Veuve Cavelier, & Fils, 1754). See for instance in his "Discours historique et raisonné sur les principes & sur les expériences de M. Hauksbée" in vol. I, Desmarest's remarks on pages xxi–xl; and his comments in vol. II, pp. 80–89, 134–136. Desmarest disparaged the Cartesian impulsionist doctrine, and admired the Newtonian attractionists' clear renuniciation of any knowledge of the gravitational force's basic cause:

"If I am fortunate enough to find a representational standard [*module général*] by which to understand the effects of interest that I perceive, what does it matter whether

I know the essence and cause of such a standard?" ["Si j'ai eu le bonheur de trouver un *module général*, auquel je puisse rapporter exactement les effets qui me frappent & qui m'intéressent, qu'importe que je connoisse l'essence & la cause de ce *module.*"] (Vol. II, p. 89). Desmarest provides, incidentally, an extended critique of the claim "that different strata are arranged in order of the specific gravity of their components" ["que les différentes couches sont disposées suivant la pesanteur spécifique des substances"] (vol. II, pp. 548–555). But this does not prevent him from accepting as true certain other regularities, the main one being the existence in stratigraphic disposition of a "relation so decided, as regards their differing sinuosities, that this must be the effect of a uniform cause" ["rapport si décidé, par rapport à leurs différentes sinuosités, qu'il doit être l'effet d'une cause uniforme"] (vol. II, p. 554).

20 Desmarest, "Géographie physique," *Encyclopédie*, vol. VII (1757), pp. 613–626. He declares (p. 616) that "true philosophy consists in discovering relations hidden to those with little vision and inattentive minds" ["la vraie Philosphie consiste à découvrir les rapports cachés aux vûes courtes & aux esprits inattentifs"].

21 "Voyage dans une partie du Bordelois et du Périgord," Bibliothèque Municipale de Bordeaux, MS 721; "Remarques de Mr Desmarest (de l'Académie des Sciences) sur la géographie physique, les productions & les manufactures de la généralité de Bordeaux, lors de ses tournées depuis 1761 jusqu'en 1764," Archives Départementales de la Dordogne, MS 26. The title of this latter manuscript is in error, as the travels in question ended in 1762. The geological content of the second manuscript is of greater interest than that of the first.

22 Desmarest, *Géographie-physique*, 5 vols. (Paris, H. Agasse, an III [1794] - 1828). The fifth and posthumous volume was not due to Desmarest; his last volume (IV) is dated 1811. It is hardly practical to assemble a complete list of the articles in the first four volumes in which Desmarest appeals to dispositional regularities. Skimming over the volumes' contents, one finds this feature everywhere. Choosing only from subjects relating to the phenomena of running water, here are some of the articles in which Desmarest's insistence on certain general dispositions is evident:

– Angles in the confluence of rivers, in relation to the volume, speed, and slope or gradient of the waters' flow ("Affluence," vol. II, pp. 182–184; "Angle de confluence," vol. II, p. 593; "Bec," vol. III, pp. 99–100; "Confluences," vol. III, pp. 450–451; "Confluens," vol. III, p. 451).

– Rules governing changes in the slope or gradient of streams, in relation to the distance from the river source or mouth ("Bassins des rivières de France," vol. III, pp. 68–73; "Fleuves," vol. IV, p. 176).

– Systematic distribution of fluvial debris, of varying size ("Crémens du Rhone," vol. III, pp. 544–546; "Eau," vol. IV, p. 5).

For Desmarest the special interest in such general effects is justified through a methodical principle, through which one proceeds backwards in time, starting with the most recent changes. Thus,

It is thus quite certain ... that study of the globe's phenomena should be begun through the latest degradation by flowing waters, whether fluvial or torrential. Although these effects often cause confusion, and it is difficult without a long preliminary study to analyse and place them in order, nonetheless the means of linking the facts together are easily found if one has followed the great masses, if these are given systematic study, and if these facts are assigned to their general categories. ("Analyse du globe de la terre," vol. II, pp. 494–495.)

[Il est donc bien certain, … qu'on doit commencer l'étude des phénomènes du globe par les dernières dégradations des eaux courantes, soit fluviales, soit torrentielles. Quoique souvent ces effets fassent confusion, & qu'il soit difficile, sans une longue étude préliminaire, de les analyser & de les ranger par ordre, cependant on trouvera facilement les moyens de lier les faits si l'on a suivi les grandes masses, qu'on en ait fait une étude raisonnée, & qu'on ait rapporté tous ces faits à des classes générales.]

23 Desmarest, "Mémoire sur l'origine & la nature du basalte à grandes colonnes polygones, déterminées par l'histoire naturelle de cette pierre, observée en Auvergne," *Mémoires de l'Académie Royale des Sciences,* 1771 (1774), pp. 705–775; "Mémoire sur le basalte. Troisième partie, où l'on traite du basalte des anciens; & où l'on expose l'histoire naturelle des différentes espèces de pierres auxquelles on a donné, en différens temps, le nom de basalte," ibid., 1773 (1777), pp. 599–670; "Extrait d'un mémoire sur la détermination de quelques époques de la nature par les produits des volcans, & sur l'usage de ces époques dans l'étude des volcans," *Observations sur la physique,* vol. XIII (1779), pp. 115–126.

In his first memoir, a vocabulary of position is especially visible in the opening section (pp. 705–725). Then, in this article's second part, Desmarest develops more fully his conception of the former integrity of physiographic features preceding their degradation, also expressed similarly in terms of regularity in arrangement. For example:

On the facing slopes of gaps between hills, I first saw lava beds and horizontal rows of prisms which presented for me a very striking correspondence, by the stone's similar grain, by the mixtures of extraneous matter they contained, by the form and dimension [*module*] of the prisms; finally, by the level [*niveau*] and number of layers or horizontal rows of these prisms (p. 737). [Sur les faces des coupures qui séparoient les collines, j'aperçus d'abord des lits de laves & des rangées horizontales de prismes, qui m'annoncèrent une correspondance très-frappante, par le grain semblable de la pierre, par les mélanges étrangers qu'elle renfermoit, par la forme & le module des prismes; enfin, par le niveau & le nombre des étages ou des rangées horizontales de ces prismes.]

In the third part, Desmarest pushes even further the relations between regularity of distribution and knowledge of the past. Thus,

… it is easy to see in the general and regular disposition of all these beds that they are as it were a subsequent superfluity to the lavas, and that they did not exist when the lavas were melted liquids (p. 645). [… il est aisé de voir dans la disposition générale & régulière de toutes ces couches, qu'elles sont pour ainsi dire, une superfétation postérieure aux laves, & qu'elles n'existoient pas lorsque les laves ont été fondues ….]

In the vocabulary of the 1779 "Extrait" – speaking of *disposition, distribution,* and *limits* of volcanic materials – Desmarest especially considers what he calls the *circumstances* of the rocks. This is a somewhat ambiguous word which seems to encompass at once the *position*, and the *state* or *degree of alteration*, of the pertinent rocks.

It is interesting to see, in Desmarest's very first printed statement about the volcanic origin of basalt, that he insists on the orientation of basaltic prisms as a regularity indicative of the original proportions of the melted material:

From thoughtful reflection on all these combined dispositions, M. Desmarest believes he is entitled to conclude that the prisms' axis is always subject to the smallest dimension of a mass consisting of an assemblage of prisms, such that the bases are part of [i.e., they rest upon] the greatest surface of these masses ("Basalte d'Auvergne," Règne Minéral, Planche VII, in *Encyclopédie, Recueil de planches,* vol. VI, 1768). [De l'examen réfléchi de toutes ces dispositions combinées, M. Desmarest croit être autorisé à conclure que l'axe des prismes est toujours assujeti à la plus petite dimension d'une massse composée d'un assemblage de prismes, de sorte que les bases font partie des plus grandes surfaces de ces masses.]

24 *Oeuvres de Lavoisier*, 6 vols., ed. J.-B. Dumas, E. Grimaux, and F.-A. Fouqué (Paris, Imprimerie Impériale / Imprimerie Nationale, 1862–1893), vol. V (1892), "Note de géologie," p. 12 (an undated manuscript note, identified by the editors as one of Lavoisier's youthful writings). ["Quel que soit le désordre qui règne en apparence dans la disposition des couches de terres et de pierres qui se présentent à la surface du globe que nous habitons, il n'est pas néanmoins difficile de reconnaître que ces irrégularités mêmes son assujetties à de certaines lois, qu'elles suivent de certaines règles."]

On Lavoisier's geological work see the articles by Rhoda Rappaport, especially "Lavoisier's Geologic Activities, 1763–1792," *Isis*, vol. 58 (1967), pp. 375–384; and "Lavoisier's Theory of the Earth," *British Journal for the History of Science*, vol. 6 (1973), pp. 247–260; also "The Geological Atlas of Guettard, Lavoisier, and Monnet: Conflicting Views of the Nature of Geology," in *Toward a History of Geology*, ed. Cecil J. Schneer (Cambridge, Mass., M.I.T. Press, 1969), pp. 272–287; and "The Early Disputes between Lavoisier and Monnet, 1777–1781," *British Journal for the History of Science*, vol. 4 (1969), pp. 233–244.

25 Lavoisier, "Observations générales, sur les couches modernes horizontales, qui ont été déposées par la mer, et sur les conséquences qu'on peut tirer de leurs dispositions, relativement à l'ancienneté du globe terrestre," *Mémoires de l'Académie Royale des Sciences*, 1789 (an II), p. 352. Lavoisier's hope to ["parvenir à déterminer les principales loix qu'a suivi la nature dans l'arrangement des couches horizontales."] This sentence follows several lines after another invocation of geological laws:

Study of horizontal beds presents still another very remarkable singularity: the sand and calcareous materials are not usually mixed together, or at least they are not except in the vicinity of the point of contact, in certain cases, and following certain laws. Most of the sands, those called fine sand [*sablons*], contain no calcareous earth, and conversely chalk and most calcareous rocks, taken from within the bed, contain no sand nor siliceous earth. [L'examen des couches horizontales présente encore une autre singularité très-remarquable: le sable et les matières calcaires ne sont point communément mêlés ensemble, ou au moins ils ne le sont que dans les environs du point de contact, dans certains cas, et suivant de certaines loix. La plupart des sables, ceux qu'on nomme sablons, ne contiennent point de terre calcaire, et réciproquement la craie et la plupart des pierres calcaires, prises en plein banc, ne contiennent point de sable ni de terre silicieuse.]

26 Ibid., p. 359. [Lavoisier considers the action of an active sea, which "sort de son lit pour y rentrer, qui se déplace suivant de certaines loix, et sur-tout en vertu d'un mouvement très-lent."] Lavoisier conceives the possibility of grasping rules for the formation of littoral beds in these terms: "But what might escape a first glance, but will be easily understood from a few moments of reflection, is that the materials of which littoral beds are formed cannot be indiscriminately mixed, but rather to the contrary that they must be arranged and disposed in accord with certain laws" (p. 356). ["Mais ce qui pourroit échapper au premier coup-d'oeil, et ce que l'on concevra facilement, cependant, par quelques instants de réflexions, c'est que les matières dont sont formés les bancs littoraux, ne doivent point être indistinctement mélangées, qu'elles doivent être au contraire arrangées et disposées suivant de certaines loix."]

In 1766 Lavoisier gave thought to application of a chemical principle discovered by Rouelle, to the interpretation of changes in the globe. In analyzing different types of gypsum according to distinctions established by Rouelle – different amounts of acid in their compositions – Lavoisier referred to a series of geological masses that corresponded to the *schisty and calcareous belts (bande schisteuse, bande calcaire)* demarcated by Guettard, differing from one another by systematic difference in chemical conditions. He wrote, "We will try to show that this variety observed in gypsum is not the result of chance, that on the contrary it follows constant and invariable laws, and that this very system – so bizarre in appearance – is in fact connected with the physical system of the earth" (p. 129 of "Sur le gypse," in *Oeuvres de Lavoisier*, vol. III (1865), pp. 128–144; see also "Extrait de deux mémoires sur le gypse," pp. 106–110, and "Analyse du gypse," pp. 111–127). ["Nous tâcherons de faire voir que cette variété qu'on observe dans les gypses n'est point l'effet du hasard, qu'elle suit au contraire des lois constantes et invariavbles, et que cet arrangement même, si bizarre en apparence, tient au système physique de la terre."]

For Lavoisier, chemical study of waters was also a source of hope for geological understanding, through a presumed correspondence between waters and the prevalent materials at their places of origin. The idea was that if there exists a constant law, an invariable relationship between the character of a locality and the waters derived from it, what surer guide for the mineralogist could there be than an analysis of the waters? Would this not be one more method for probing nature? Although nature works in secret and hides carefully from our inquiry, would not the waters springing ceaselessly from its laboratory sometimes betray nature and reveal its secrets? ("De la nature des eaux d'une partie de la Franche-Comté, de l'Alsace, de la Lorraine, de la Champagne, de la Brie et du Valois," *Oeuvres de Lavoisier*, vol. III, p. 146.)

27 Horace-Bénédict de Saussure, *Voyages dans les Alpes, précédés d'un essai sur l'histoire naturelle de Genève*, 4 vols. (Neuchâtel, Chez Louis Fauche-Borel, 1779–1796), vol. IV, p. 464.

28 Ibid., p. 465.

29 See, in *Voyages dans les Alpes*, the "Discours préliminaire" (vol. I, pp. i–xxiv) and the "Agenda" of vol. IV (see note 31 below). On de Saussure, see Albert V. Carozzi's article in *Dictionary of Scientific Biography*, vol. XII (New York, Charles Scribner's Sons, 1975), pp. 119–123.

30 "Agenda," in *Voyages dans les Alpes*, vol. IV, p. 533.

31 "Agenda, ou tableau général des observations & des recherches dont les résultats doivent servir de base à la théorie de la terre," in *Voyages dans les Alpes*, vol. IV, pp. 467–539. The "Agenda" was also published separately in de Saussure's time.

32 James Hutton, *Theory of the Earth, with Proofs and Illustrations*, 2 vols. (London & Edinburgh, Printed for Messrs Cadell, Junior and Davies; and William Creech, 1795). Hutton was given to quoting Pallas, Dolomieu, and especially de Saussure, and to praising their observations, while at the same time taking issue with their theoretical views. In his equally extended critiques of Deluc, a tone of approval of his observations is lacking.

33 Jean-André Deluc, *Lettres physiques et morales sur l'histoire de la terre et de l'homme adressées à la reine de la Grande Bretagne*, 5 vols., in 6 (La Haye, Chez De Tune, 1779), vol. I, pp. 257, 276, 304.

34 Ibid., vol. I, pp. 307–308.

["quel est le vrai phénomène qu'il faut expliquer"]

35 Ibid., vol. I, pp. 308–310:

Now here is what I have frequently noticed. One hill [rock mass] contains shells; but not through its entire extent indifferently. They are in particular beds. ... I have noticed in general that when the beds are composed in this way of a single species of shellfish in great abundance, these are most often bivalves. These shellfish move much more slowly than the univalves, and some even hardly move at all. Thus they remain much more together, and multiply prodigiously. ... [These] beds alternately of shells and of sand are also of stones mixed in, that is often recognized as having belonged to known mountains [masses], composed of materials easy to distinguish, and still existing in the neighborhood. Here is a phenomenon of capital importance for the Theory of the Earth, which merits being explained fundamentally. [Or voici ce que j'ai remarqué fréquemment. Une colline renferme des *Coquillages*; mais ce n'est point dans toute sa hauteur indifféremment: c'est dans des couches particulières." ... "J'ai remarqué en général, que quand les *couches* sont ainsi composées d'une seule espèce de *coquilles* en très-grande abondance, ce sont le plus souvent des *bivalves*. Ces *coquillages* se meuvent avec beaucoup plus de lenteur que les *univalves*, quelques uns même ne se meuvent pas du tout; ils restent ainsi beaucoup plus ensemble, & peuplent prodigieiusement."

... Ces "*couches* alternatives de *coquillages* & de *sable*, sont aussi mêlées de *pierres*, que l'on reconnoit souvent pour avoir appartenu à des montagnes connues, composées de matières faciles à distinguer, & qui subsistent encore dans le voisinage. C'est ici un phénomène capital dans la Théorie de la Terre et qui mérite d'être expliqué à fond.]

From Deluc's viewpoint the organization of continents and rivers shows significant order and arrangement: "For us it suffices to see that the regularity of the dry surface of the Earth excludes collapse as a feasible cause of its present arrangement" (vol. I, p. 360). ["Il nous suffit de voir, que la régularité de la surface sèche de la Terre, exclut tout éboulement comme cause de son arrangement actuel."]

36 François Ellenberger and Gabriel Gohau, "A l'aurore de la stratigraphie paléontologique: Jean-André Deluc, son influence sur Cuvier," *Revue d'histoire des sciences*, vol. 34 (1981), p. 218. ["Ce ne sont ni les idées ni les observations qui manquent à Hutton, Dolomieu, de Saussure et à leurs contemporains. S'ils n'en font pas un meilleur usage c'est peut-être faute d'avoir trouvé le fil conducteur permettant la mise en ordre des faits recueillis. C'est sans doute aussi qu'ils gardent l'ambition, héritée de leurs aînés, de bâtir une théorie de la Terre déduisant la formation du globe de l'étude d'ensemble de sa structure.

Leurs successeurs sauront, à la fois, trouver des concepts unificateurs et limiter leurs ambitions ...."]

37 Georges Cuvier and Alexandre Brongniart, "Essai sur la géographie minéralogique des environs de Paris," *Annales du Muséum d'Histoire Naturelle*, vol. XI (1808), p. 294.

38 Robert de Paul de Lamanon, "Description de divers fossiles trouvés dans les carrières de Montmartre près Paris, & vues générales sur la formation des pierres gypseuses," *Observations sur la physique*, vol. 19 (1782), pp. 173–194.

39 Desmarest, "Mémoire sur les prismes qui se trouvent dans les couches horizontales de plâtre et de marnes des environs de Paris, et sur leur analogie avec les prismes du basalte," *Mémoires de l'Institut National des Sciences et Arts. Sciences mathématiques et physiques*, vol. IV (an XI, 1802), pp. 219–231; "Second mémoire sur la constitution physique des couches de la colline de Montmartre et autres collines correspondantes," *Ibid.*, vol. V (an XII, 1803), pp. 16–54. Cuvier and Brongniart allude also to the "information provided by this same savant on the Seine River basin, in the Encyclopédie méthodique." It is indeed true that one finds, dispersed through Desmarest's *Géographie physique*, descriptions and discussions of the flow of the Seine and its tributaries, and of the terrain of the Paris region. But, as the third volume of this compendium is dated 1809, and volume IV is dated 1811, presumably Cuvier and Brongniart are referring only to articles published in the first two volumes (1794, 1803), unless they had access to articles (or maps) in advance of their publication. Volume I includes "Notes et observations tirées de l'ouvrage de Boulanger sur les cours de la Marne" (pp. 8–25), and in the article on Buache some commentary on a "Carte physique du bassin terrestre de la Seine" (pp. 65–66). And in volume II there are, on the Paris region, the articles "Argenteuil" (pp. 780–782) and "Auteuil" (pp. 873–876), in addition to pertinent information in the articles "Amas" (especially p. 348), "Anecdotes de la nature et de l'histoire de la terre" (particularly pp. 543, 556, 557, 571), and "Arêtes" (pp. 765–767); perhaps not to be overlooked is the article "Aube" (pp. 857–858).

40 Cuvier and Brongniart, "Essai," pp. 307–308. ["... mais il ne faut pas croire que ces divers bancs y soient placés au hasard et sans règles: ils suivent toujours le même ordre de superposition dans l'étendue considérable de terrain que nous avons parcourue ...

Cette constance dans l'ordre de superposition des couches les plus minces, et sur une étendue de 12 myriamètres au moins, est, selon nous, un des faits les plus remarquables que nous ayons constatés dans la suite de nos recherches. Il doit en résulter pour les arts et pour la géologie des conséquences d'autant plus intéressantes, qu'elles sont plus sûres."]

41 Shortly after this paper's oral presentation [in February 1988], I realized that one might add a third sort of regularity, that of condition or state. Such a third regularity would take into account cases where the observer alleges the presence of certain features or characteristics in a phenomenon, not belonging under either the category of position or that of operation. It seems to me that this third class might apply, for example, to the following case, from the "Notice d'un voyage au Mont-Rose" by de Saussure (*Observations sur la physique*, vol. 37 [1790], p. 8, no. 1:

The *secondary* calcareous rocks, or those that were formed since the revolution following which the seas were populated by fish and shellfish, are almost always overlain by sandstones, breccias, and conglomerates, that is to say by debris of the rocks that were broken and pulverized during that revolution. It is the debris interposed between the beds of primitive rocks and secondary rocks that forms the transitions I have frequently observed, especially at the foot of the Buet (*Voyages*, paragraph 594). The *primitive* calcareous rocks, by contrast, or those that existed before this revolution, present no transitional features, or else transitions of a totally different kind. This distinction, which I believe is new, appears to me as important for the theory of the earth.

[Les pierre calcaires *secondaires*, ou celles qui ont été formées depuis la révolution à la suite de laquelle les mers ont été peuplées de poissons & de coquillages, sont presque toujours recouvertes de grès, de brêches, de poudingues, c'est-à-dire, des débris des rochers qui ont été rompus & broyés dans cette révolution. Ce sont les débris interposés entre les couches de roches primitives & celles de pierres secondaires qui forment les transitions que j'ai fréquemment observées, & spécialement au pied du Buet (Voyages, paragraphe 594). Les calcaires *primitives*, au contraire, ou celles qui ont existé avant cette révolution, ne présentent aucune transition, ou ce sont des transitions d'un tout autre genre. Cette distinction que je crois nouvelle me paroît importante pour la théorie de la terre.]

42 An example of a limited sort of historical inference would be seen in Desmarest's "Détermination de quelques époques de la nature" (1779), where the research's results were confined to just parts of the temporal transformation of just one region. For contrast, such restraint is far less evident in the following passage from Pallas's *Observations sur la formation des montagnes* (1777):

We could speak more decisively about the *secondary and tertiary mountains* of the [Russian] Empire, and it is [indeed] from those – the nature, arrangement, and contents of their strata, the great inequalities and form of the European and Asian continents – that one can draw more confidently some insight into the changes the habitable earth has undergone. These two orders of mountains present the oldest chronicle of our globe, the least susceptible to correction, and at the same time more easily interpreted than the features of the primitive chains. These are nature's archives, older than the oldest human letters and traditions, that it was reserved to our observation-minded century to excavate, analyze and interpret, yet which centuries to come will not be able to exhaust (p. 29).

[Nous pourrions parler plus décisivement sur les *montagnes secondaires & tertiaires* de l'Empire, & c'est de celles-là, de la nature, de l'arrangement, & du contenû de leurs couches, des grandes inégalités & de la forme du continent d'Europe & d'Asie, que l'on peut tirer avec plus de confiance quelques lumières sur les changemens arrivés aux terres habitables. Ces deux ordres de montagnes présentent la chronique de notre globe la plus ancienne, la moins sujette aux falsifications & en même tems plus lisible que le caractère des chaines primitives; ce sont les archives de la Nature, antérieures aux lettres & aux traditions les plus reculées, qu'il étoit réservé à notre siècle observateur de fouiller, de commenter & de mettre au jour, mais que plusieurs siècles après le notre n'épuiseront pas.]

43 Gabriel Gohau, *Histoire de la géologie* (Paris, Éditions La Découverte, 1987), p. 137.

44 Rachel Laudan, *From Mineralogy to Geology: The Foundations of a Science, 1650–1830* (Chicago & London, University of Chicago Press, 1987). While in my judgement Rachel Laudan's treatment assigns too little importance to the fieldwork done by 18th-century geologists, still I should not fail to acknowledge the real merits of her arguments emphasizing chemical traditions, which histories of geology's early development have tended to neglect.

45 François Ellenberger, "Trois aspects de la notion de lois dans les sciences de la terre du XVIIème au XIXème siècles," in Centre Interdisciplinaire d'Étude de l'Évolution des Idées, des Sciences et Techniques, *Séminaires et tables rondes de l'année universitaire 1978–1979* (Orsay, C.I.E.E.I.S.T., 1980), pp. 83–87.

VIII

The Historical Rehabilitation of Theories of the Earth

Lyell on Speculation Versus Geology

In the highly influential historical chapters with which Charles Lyell opened his *Principles of Geology* (1830), he warned that a review of the history of geology during the century or so before 1800 requires one "to dwell on futile reasoning and visionary hypothesis, because the most extravagant systems were often invented or controverted by men of acknowledged talent". This historical inquiry, he said, "although highly interesting to one who studies the philosophy of the human mind, is singularly barren of instruction to him who searches for truths in physical science" (p. 30).

Lyell was an excellent writer, as well as a brilliant scientist (Fig. 1). Reading his *Principles* today remains interesting and instructive. In my opinion this is true not only for purposes of studying the philosophy of the mind, but also for insight into the nature of science. The availability of a facsimile edition in paperback makes it easy enough for anyone to try (Lyell, 1990).

To learn from Lyell, obviously, it is not necessary to agree with him. And I do quarrel with Lyell's way of drawing the early history of geology. I am far from alone in this; a generation of scholarly work has done much to establish a reformed historiography of geology. Historians of science learned some time ago that Lyell's history cannot be taken at face value, but must be understood as part of a strategy both to present his own doctrines as the latest stage in the scientific progress of geology, and to stigmatize his opponents by tying them with retrograde parts of geology's development (Rudwick, 1969; Porter, 1976; Ospovat, 1976). For a time, it would seem, Lyell's admirers forget something that his contemporary opponents could not, namely that Lyell had a legal training and showed a tendency in his geological writing to argue in adversarial fashion. However, it is one thing to see the tendentiousness of a prestigious historical account, and another to replace it with a better one. My comments here concern the way historians of geology have been reconstituting our view of the seventeenth- and eighteenth-century theories of the Earth.

VIII

The Historical Rehabilitation of Theories of the Earth

Figure 1. Charles Lyell (1797–1875).
The historical account of geology in the opening chapters of Lyell's *Principles of Geology* (Vol. 1, 1830) had a sustained influence on later histories of geology. From Katherine M. Lyell, ed., *Life, Letters and Journals of Sir Charles Lyell*, Bart., 2 vols., John Murray, London, 1881 (v. 1, frontispiece).

Actually, Lyell hardly recognized the theories of the Earth. And that is central to the problem. Lyell castigated as altogether speculative the "physico-theological cosmogonies" and certain other systems of which he disapproved (especially Werner's), whereas the geological activities and schemes it suited his purposes to present favorably (notably Hutton's) were treated as the results of unprejudiced observation and shrewd reasoning. In Lyell's history, the geological arena before his own lifetime was rather clearly divided between advocates of fanciful cosmogonical doctrines and hypothetical systems on one hand, and properly induced principles on the other.

In history as in every discipline, decisions about classification of the objects of interpretation go a long way toward shaping the result. Among the choices Lyell made in fixing his historical taxonomy, one of the most significant was to deny the basic cognitive integrity and coherence among theories of the Earth.

Instead he carefully distinguished cosmogonies and systems he considered disreputably speculative, from that sort of geology which was to be portrayed in a favorably empirical light. Although one might not realize it from reading Lyell's history, the latter group often included geological ideas developed within the framework of theories of the Earth. Lyell's historical classification broke apart what had been, for many scientists from geology's formative period, a more continuous theoretical tradition extending even into the first decades of the nineteenth century.

Thus, for example, Lyell characterized Whiston's theory, with its hypothesis of the Deluge's causation by a comet's passage near the Earth, as having "retarded the progress of truth, diverting men from the investigation of the laws of sublunary nature, and inducing them to waste time in speculations...." (p. 39). Werner is portrayed as an unoriginal contriver of "cosmological inventions" (p. 69), a dogmatic purveyor of an empirically unfounded theory that was "one of the most unphilosophical ever advanced in any science" (p. 59). The Huttonian doctrine, by contrast (notwithstanding the inconvenient baggage of its title: *Theory of the Earth*), is delineated as the product of the patient empirical labors of a naturalist making discoveries "by fair induction from an independent class of facts" (p. 62), to arrive "at grand and comprehensive views in geology" (p. 61). Figures like Pallas and Saussure are presented as observers working from essentially inductive procedures, with the implication that their theoretical views were irrelevant to their good empirical work (p. 54).

There are at least three seriously misleading consequences of Lyell's determination to disregard a continuing eighteenth-century tradition of theories of the Earth. First, it falsifies a view widely accepted during the eighteenth century, when theories of the Earth usually were taken as sharing common membership of a certain style or genre of science. In addition, it artificially separates theory from empirical investigation in geological science up to the end of the eighteenth century. Finally, many writers who adopted these Lyellian notions about geology's early history drew out of them the inference (intended by Lyell's strategy) that the theories of the Earth differed from genuine geology not only in being incorrect, but also by virtue of being unscientific. Historical scholarship has lately done much to rectify these three points.

The Historiographic Legacy of Lyell

It is not surprising that historians of science of my generation (my education in the subject dates from the 1960s) found Lyell's historiographic posture unappetizing. Many of us were brought up on historical precepts exemplified by scholars such as Edwin Arthur Burtt, Alexandre Koyré, and Herbert Butterfield.

According to their "conceptualist" perspective, which then retained an aura of novelty, most scientific change of consequence is best understood as arising in conditions encouraging intellectual reconceptualization of phenomena (Burtt, 1932; Koyré, 1939, 1957; Butterfield, 1957). Rather than expecting new scientific understanding to follow directly from unbiased empirical discovery, we learned to see adjustments in theory as enabling – perhaps requiring – the comprehension of facts in a new light. We were taught to regard the doctrine of scientifically neutral empiricism as philosophically primitive and historically unilluminating.

This stance contradicted the conventional scientific inductivism that most of us had learned in school, and that had served as a predominant explanatory matrix in much of the older literature in the history of science. We saw historical revision based on the conceptualist approach – and grounded in thorough, scholarly reconsideration of primary source materials – as one of our most important tasks. Many of us relished Agassi's wicked attack (1963) on the historiographic consequences of naïve inductivist epistemology. Kuhn's articulation of a general socio-philosophical model for scientific change (1962), with its emphasis on the incommensurability of different theoretical ways of ordering data, for a time provided important elements in the vocabulary of historians of science. I think one of the main reasons Kuhn's scheme appealed so much to historians of science was that in the theory-fact relationship it denied an especially privileged place to empirical data; in any case Kuhn's was only the most widely celebrated among a number of fresh approaches to scientific knowledge that satisfied a prevailing appetite among historians for noninductivist visions of science.

Meanwhile, although Collier (1934) had published a large study of cosmogonies based on admirably close study of original texts, her book had done little to foster a new historical orientation. Possibly this owed something to the fact that Collier's narrative emphasized description of her material, and was cautious as regards interpretation. So with a few notable exceptions (e.g., Greene, 1959), by the 1960s the prevailing historiography of geology had changed relatively little since Lyell's time. Clearly there was work to be done.

It should not be assumed, by the way, that the historical taxonomy by which Lyell separated theory from observation appeared in the *Principles* out of nowhere. Like so much else used by great scientists in great scientific syntheses, Lyell's scheme of classification was adapted from some of his predecessors; but that is another story. It was mainly through Lyell that the thesis of a dualism of scientifically sterile cosmogony and geologically fruitful observation became a historical convention lasting well over a century.

The Historical Rehabilitation of Theories of the Earth

Consider, for example, how K.A. von Zittel's classic *History of Geology and Palaeontology* (1901) organizes his comments about geology before 1800, in terms of "two methods of research, the empirical and the speculative":

> The one had for its immediate aim the determination of facts, and in its further outlook, the possible construction of some suitable theory; the other contented itself with a minimum of observations, accepted the risks of error, and set about explaining the past and the present from the subjective viewpoint. This latter method naturally attained no higher results than the geogenetic fantasies of classical antiquity. And it certainly could never have gathered sufficient energy to roll aside the mass of philosophical and doctrinal tradition that blocked the path of progress (p. 23).

By Zittel's account, the positive achievements that were managed by eighteenth-century geologists came out of opposition to pernicious speculative tendencies. Zittel's discussion of the theorists is separated from the section entitled "Beginnings of Geological Observation" (p. 34). Moro is among those who are said to have "sought to counteract the tendency of their time toward the theoretical construction of an earth history" (p. 31). And "the true spirit of research was still kept alive by men who confined themselves to special subjects of investigation, or described the stratigraphy of particular localities" (p. 34). In a conflicted discussion of Buffon (whose place in eighteenth-century geology is always a problematic question if approached through the conviction that speculation and synthetic systematizing can lead to no good), Zittel concludes that it was the failures of theorizing that finally brought men to the clear perception that they must, after all, observe geologically (p. 44).

> The characteristic features of this age [the "Heroic Age" of 1790–1820, in which true geology emerged], and that which gave it a rejuvenating significance in the development of geology, was the determined spirit that prevailed to discountenance speculation, and to seek untiringly in the field and in the laboratories after new observations, new truths (p. 46).

Probably the single most influential and well-known history of geology in the English language during our century has been Archibald Geikie's *The Founders of Geology* (1905; Oldroyd, 1980). It is marked throughout with a similar thematic differentiation between fact and speculation.

According to Geikie (Fig. 2) the seventeenth-century theories, including particularly the "grotesque speculations" (p. 61) of Burnet and Woodward, "obstructed the progress of inquiry, inasmuch as they diverted attention from the observation of Nature into barren controversy about speculations" (p. 65). Compared with Lyell, Geikie was rather kinder to what he called the "scientific cosmogonists" (distinguished from the "physico-theological" type), including Descartes (Figs. 3 and 4), Leibniz, de Maillet, and Buffon.

Figure 2. Archibald Geikie (1835-1924), Scottish geologist and historian of geology.
Geikie's *The Founders of Geology*, first published in 1897 and reissued in an enlarged 1905 edition, has probably been this century's most widely read book on the subject in English. One of its main themes, as in Lyell's historical chapters, was the opposition between geological observation and theorizing. From Geikie, *A Long Life's Work: An Autobiography*, Macmillan, London, 1924 (frontispiece).

Although speculative, the work of these men was connected with some "keen and shrewd" observing and reasoning (p. 84, in reference to Maillet), and was scientifically worthy in part because distinguishable from Biblical historical interpretations. Nonetheless according to Geikie these cosmogonical writings were to be viewed differently, as they also had been in Lyell's history, from the more purely observational (and thus more excellent) activities of scientists like Arduino, Lehmann, Füchsel, Pallas, and Saussure. Where Geikie acknowledges that such investigators had theoretical commitments (as in the case of Saussure, p. 185–87), the scientist's observations are understood as emerging only in spite of theory, not through it. Hooke is treated as untainted by speculation, but as a man of "remarkable powers of acute observation and sagacious

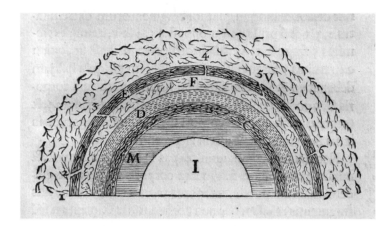

Figure 3. A stage in the Earth's formation, from René Descartes' theory of the Earth in his *Principia Philosophiae*, Elsevier, Amsterdam, 1644, p. 215.

According to Descartes' theory the Earth formed as a cooling star. In the process of differentiation of the moving matter near the surface, above the "central fire" (I), solid layers of differing density were formed (M, C, E). The highest and lightest of these (E) separated, with zones of water (D) and air (F) beneath. Movement of air through pores in the upper crustal layer produced progressively larger cracks (1, 2, 3, 4, 5, 6).

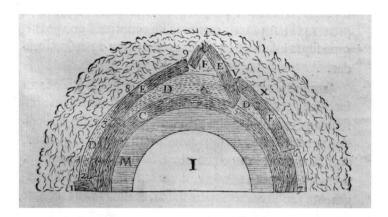

Figure 4. A later stage in the Earth's formation, from Descartes' theory, p. 215. The upper crustal sections have collapsed where the cracks had developed. The Earth's surface now shows major irregular features in accord with experience, including oceans (left and right), continental plains (8 to 9, and V to X), and mountains (4). Considerable bodies of water remain beneath the continents, and large cavities (F) are assumed to exist beneath the mountains.

reflection" (p. 68). Guettard's geological work is presented, in admirable detail, as "shunning any approach to theory" (p. 104). Hutton's achievement is drawn as the antithesis of a theory of the Earth (despite the title), mainly on the grounds of its supposedly inductive character and its deliberate obliviousness to the issue of "the origin of things." Geikie's book ends with a warning against the effects of dogmatism in geology (p. 472–473); the book as a whole is suffused with a message strongly associating dogma with speculation, whereas empiricism is dogma's enemy.

It must be said that in Geikie's book, far more than in Zittel's, one gets a picture of geological activity of real significance during the eighteenth century. According to Geikie, though, the real geological accomplishments of the eighteenth century were gained only by overcoming or disregarding theory. While Geikie provided ample narrative evidence of the intellectual virtuosity of several eighteenth-century geologists, he would have us see their achievements as the product of skilled observation, free of theoretical contamination.

The other classic history of geology in English, that of Frank Dawson Adams (1938), has less to say about the eighteenth century than Geikie's, and little indeed about the theories of the Earth. The few words he has, however, are not such as to encourage serious consideration of these theories as science. The cosmogonists "presented theories spun out of their own vivid and brilliant imaginations" (p. 209). The results were "fables" fit to be "read by all who are in need of mental recreation and who possess the required leisure and a certain sense of humor" (p. 210). Buffon's *Théorie* and *Epoques* represented "the close of this long period of imaginative effort..," and "the time had now come when those who looked for an explanation of the origin of the earth should turn from mere speculation to a study of earth itself...." (p. 209–210). The theory of the Earth was displaced by Field Geology.

During the 1960s, when I first began to read the history of geology, the books by Zittel, Geikie, and Adams were still practically standard. Most of the historical scholars who gave students of my age the apparatus for reconnecting theory and observation in early geology had not yet applied those tools to this subject to very great advantage. For instance, Charles Gillispie's stimulating study of geology in the first half of the nineteenth century dismissed geological literature before the end of the eighteenth century as "fanciful in content and probably in intent" (Gillispie, 1951, p. 41). Gillispie's summary comments on geology before Hutton implied a polar opposition between speculative theory of little value, and good empirical work by investigators like Pallas, Saussure, Dolomieu, and Michell (p. 42). In a widely read set of lectures that he published a decade later, comprehensively addressing the development of modern science, Gillispie's views on this topic had not changed discernibly; he wrote that "the science of geology represents a coming together of lore from the ancient practice of mineralogy with speculations about the origin of the earth, seventeenth- and eighteenth-century cosmogonies which have in them more of science fiction than of science" (Gillispie, 1960, p. 291–292).

As I mentioned, things have changed much in a generation's time. Even so, the theme of the pre-nineteenth-century prevalence of reckless theorizing as opposed to sober observation is not dead yet. In the authoritative *Dictionary of Scientific Biography*, itself a production of the self-conscious new professionalism in the history of science, one may still find the eighteenth-century theories of the Earth characterized as unscientific (Eyles, 1972, p. 580). Even in a recent and informative study of early nineteenth-century geology, one encounters the assertion that in the eighteenth century geological thinking had been dominated by "the evil spirit of idle speculation and of deductive cosmogony" (Rupke, 1983, p. 180).

Geological Observations Within Geological Theory: the Example of Desmarest

The reassessment of theories of the Earth and their place in early geology is only one part of the historical revisionism that has been in progress since I was a student. I did not think for some time, in fact, that it was one of the parts that mattered most. Perhaps a brief sketch of my own path toward rethinking the roles of theories of the Earth may help put some of these changes into perspective.

I had decided by early in 1965 to write my doctoral thesis on Nicolas Desmarest, already celebrated by Geikie as a pioneer observer of the volcanic phenomena of Auvergne. I am sure my choice of Desmarest was inspired in part by the remarkably lively and well-informed discussion that I read in Geikie's book. And I am pretty certain (it is difficult to recall clearly how I was thinking more than 25 years ago about a subject that has been a major preoccupation ever since) that while I appreciated Geikie's awareness of the richness of eighteenth-century geological accomplishments, particularly those by Desmarest, the historical shortcomings of Geikie's stark speculation-versus-observation framework of interpretation were apparent to me. I was at least dimly aware that other scholars were forming a picture of a more complicated and ambiguous connection between theory and factual knowledge among eighteenth-century geologists. Notably, in a broad treatment of the development of evolutionary thinking, John Greene (1959) had discussed a number of earth-theorists with evident historical sympathy for the empirical problems they faced. And Rhoda Rappaport, who had made a close study of

Guettard, had remarked that this paragon of atheoretical science had notions of a theory of the Earth (1964, p. 66).

My teachers had trained me to focus especially on efforts, through close examination of published texts and other historical documents, to determine the underlying premises and presuppositions of interesting expressions of scientific thought. In the process of establishing a more detailed and thorough account of Desmarest's life in science, I hoped to be able to enlarge a little on our knowledge of the ideas that guided his work.

In retrospect it does seem to me, though, that I went about my research in the expectation of locating information tending to show an unconstructive role for theories of the Earth, in the work of Desmarest and his scientifically progressive contemporaries. In accepting (from Geikie and other authorities) the persuasive thesis that the reformulation of geological ideas near the end of the eighteenth century ushered in the foundations of modern geological thinking, I believe that I also absorbed a more dubious element of Lyell's interpretive schema: That the comprehensive geological theorizing still so abundant in the second half of the eighteenth century was an obstacle to scientific change, and perhaps was even of marginal scientific standing. That I remained to some extent under the influence of the theory-versus-observation syndrome seems clearer to me now, certainly, than it did at the time. I now believe I was looking then mainly for evidence to help explain why Desmarest was different from the theorists of the Earth, not for signs of his affinity with them.

Incidentally, I had help from Desmarest himself in forming such an attitude. Desmarest was one of those whose remarks on theories of the Earth, especially from late in his career, were ultimately adapted by Lyell in his articulation of the speculation-versus-geology theme. To cite just one example, Lyell's often-noted comment that "geology differs as widely from cosmogony, as speculations concerning the creation of man differ from history" (Lyell, 1990, p. 4) seems probably to have been inspired by a very similar formulation of Desmarest's (1794/95, p. 1).

Once I got deeply enough into Desmarest's writings, it was fairly easy to see that, whatever his public utterances about the necessity of making observations independently of prevailing doctrines, he was in fact subject to belief in some fairly traditional theories, notably "Neptunistic" ones. There could be no question that Desmarest was a scientific observer for whom traditional theories were a major consideration, playing a large role in shaping his sense of what his volcanological observations meant (Taylor, 1969). What was more difficult to learn, though – and took me more time to realize – was that many of the observational preoccupations that became characteristic of Desmarest's geological work through his long scientific career were derived actually from theories of the Earth written by people such as Louis Bourguet, Nicolas-Antoine Boulanger, Georges-Louis Leclerc de Buffon, and Guillaume-François Rouelle. Desmarest's observational interests were influenced greatly by these and other authors, through the emphasis placed in their theories on natural "regularities," particularly in what they often termed "dispositional" regularities in the configurations of geological features of a topographic, structural, or compositional nature (Taylor, 1974, 1988).

That is, I began to see that it was not only Desmarest's interpretations of his geological observations that were informed by theories of the Earth; so also were the very sorts of things he thought it worth observing. This in turn made it apparent that the distinctions between the traditional theories of the Earth and the new ideas out of which geology was being formed were far more blurred than I had expected. In time I came to the view that the theories of the Earth, far from being sterile notions which had to be jettisoned before any constructive scientific changes could occur, were apparently among the sources out of which those changes came. Theories of the Earth, Lyell to the contrary, did not divert attention from study of the laws of nature; they actually encouraged active pursuit of natural laws suitable to an understanding of the Earth. Now I think of Desmarest as one of many geological characters in the second half of the eighteenth century whose contributions toward creation of a modern geological outlook were made as much within the context of theories of the Earth as in a posture of rebellion against them, possibly more so.

Theories of the Earth Historically Rehabilitated

The viewpoint to which I all too slowly came, regarding the wide scientific context of Desmarest's geological work, evidently entails a greater continuity in the traditions of theories of the Earth than I had once appreciated. This perspective of course requires that one regard the underlying unity within those traditions as belonging more to their form than their content. It was notorious even in the eighteenth century that the various theories of the Earth were not compatible with one another, and any hope of reaching consensus about many major issues seemed frustratingly distant.

As one example of how one might see formal coherence among competing systems, the French historian Jacques Roger, who has made important contributions to our understanding of eighteenth-century geology, has proposed (1973, 1982) that the defining features of the theories of the Earth can be distilled into three points: (1) in the post-Copernican era, the Earth comes to be seen as an entity rather than a cosmic region, and in being redefined now as a thing rather than a place, is subject to examination and explanation in accord

VIII

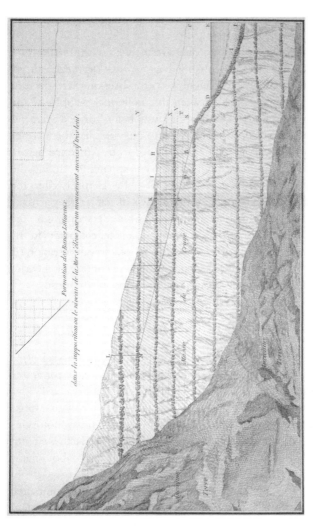

Figure 5. An ideal section of a coastal cliff of chalk with layers of flint, originally published with a paper (1789) by Antoine-Laurent Lavoisier (1743–1794).

Lavoisier's paper, based on research done in the 1760s, made a distinction between littoral and pelagic deposits, and presented evidence of repeated incursions and retreats of the sea. This diagram addresses the effect of the sea's shoreline action on the cliff, depending on the tides' maximum height, and the resulting littoral deposits. The observations Lavoisier was concerned to summarize in diagrams such as these were directed toward establishment of a theory of the Earth. Lavoisier's observations were carried out in part under the influence of ideas drawn from existing theories of the Earth. From *Oeuvres de Lavoisier*, Vol. 5, Imprimerie Nationale, Paris, 1892, pl. 2. (Plate dimensions: 177 x 291 mm.)

with the procedures of natural philosophy (physics) and natural history; (2) there is convincing evidence that the Earth's present state is different from its former or original condition, so an important new task is to provide an account of its alteration; and (3) the account so provided, addressing as comprehensively as possible the Earth's surface features and internal structure, is presumed to be consistent with prevailing understanding of basic scientific principles.

If these basic tenets regulated theories of the Earth, there was a lot of latitude in how the theories might accomplish their goals. For instance, the format outlined by Roger can embrace both those theories which aimed at consistency with Scripture and those which sought to erect resolutely materialist systems; the degree to which a particular theory involved a moral response to geological phenomena need not affect its membership in the genre (Herries Davies, 1989). Roger's analysis of theories of the Earth also permits inclusion of theories which addressed the generation of the Earth, as well as those in which the question of the Earth's origins is conspicuously left untouched. This is not to say that these sorts of differences are unimportant, only that such differences can have been delineated within an ongoing framework of thinking sufficiently sturdy to furnish some direction to the authors of each theory.

As I have indicated, my own view is that much of the geological observation done during the eighteenth century not only was carried out in the expectation that it would be suitably fitted into the framework of one or another theory of the Earth; it also was actually inspired in many instances by the intellectual orientations provided by theories of the Earth. I should point out that the second part of this contention is not yet clearly established. I hope soon to present a more thorough case for it. If an empirically directive role for theories of the Earth is currently unproven, however, I think most of my historical colleagues have come to take the first point as nearly axiomatic: Eighteenth-century geological observers did not (probably could not) seal away their fact-gathering in a compartment separate from their theoretical concerns.

The interconnection between geological observation and comprehensive theorizing among eighteenth-century geological scientists is supported in a growing volume of historical research, only part of which can be cited here (Roger, 1962; Herries Davies, 1969; Eyles, 1969; Ospovat, 1971, 1980; Ellengberger, 1975-1977; Porter, 1977, 1979, 1980; Rappaport, 1978; Ashworth, 1984; M. Carozzi, 1986; M. and A.V. Carozzi, 1987; Laudan, 1987; Gohau, 1990). Just a few specific examples may suffice to illustrate. Rappaport's research (1968, 1973) makes clear that the considerable interest in geological investigation shown by the chemist A.L. Lavoisier, whose initiation into natural science came partly through geological fieldwork, was oriented toward a planned theory of the Earth (Fig. 5). Neve and Porter (1977) have demonstrated that the geological fieldwork of Alexander Catcott during the 1740s through the 1760s was undertaken out of conviction in a Scripturally revealed cosmogonical theory. And a series of publications by A.V. Carozzi and M. Carozzi show that the extensive geological observations made by Saussure and Pallas were closely linked with their respective interests in elaborating a theory of the Earth (A.V. Carozzi, 1989; A.V. Carozzi and M. Carozzi, 1991; M. Carozzi and A.V. Carozzi, 1987; Carozzi and Newman, 1990).

I daresay that most scholars now working on the early history of geology would agree with the remarks on this subject by Rachel Laudan, in her valuable discussion of geology's foundation (1987, p. 8). She says that by now we should have gotten past a problem that has tended to obfuscate much previous consideration of the historical relationship between theory and evidence,

> ...namely, too sharp a contrast between 'speculators' and 'empiricists.' All the geologists in the eighteenth and nineteenth centuries expected theory or 'system' to play a role in geology. And they all demanded that theory should be warranted by evidence. The point at issue was not whether to opt for theory or for fact gathering. Rather, it was the relationship between theory and facts.

It might be added, though, that if all geologists from about 1700 forward took this attitude toward their *own* work, it does not follow that they always thought the same was true of *others* among their contemporaries or predecessors. The example of Lyell's historiography is a reminder of that.

These are interesting times for historians of early geology, which has in a sense been effectively reconnected with the history of modern geology. We have much to learn about how geological theory and observation were related to each other before 1800. There are significant historical questions to answer about the evolution of the genre of theories of the Earth. I am confident that in the coming years historical research will yield some new and interesting insights regarding the foundation of observational procedures and ideas in geology. And I expect that the traditions of theories of the Earth will be seen more often in roles that facilitated observational innovation, and will be cast less often as obstructions to it.

✱✱✱

This essay is based in part on research supported by a grant from the National Science Foundation. This support is gratefully acknowledged. The illustrations were provided courtesy of the History of Science Collections, University of Oklahoma Libraries.

REFERENCES

Adams, Frank Dawson, 1938, *The Birth and Development of the Geological Sciences*: Williams and Wilkins, Baltimore, v, 506 p.

Agassi, Joseph, 1963, Towards an Historiography of Science: *History and Theory, Studies in the Philosophy of History*, Beiheft 2; Mouton & Co., 'sGravenhage, viii, 117 p.

Ashworth, William, B., Jr., 1984, *Theories of the Earth, 1644–1830: The History of a Genre – An Exhibition*; with bibliographic descriptions by Bruce Bradley: Linda Hall Library, Kansas City, Missouri, 68 p.

Burtt, Edwin Arthur, 1932, *The Metaphysical Foundations of Modern Physical Science* (2nd ed., rev.): Routledge & Kegan Paul, London, xi, 343 p.

Butterfield, H., 1957, *The Origins of Modern Science, 1300–1800* (rev. ed.): Free Press, New York, 250 p.

Carozzi, Albert V., 1989, Forty Years of Thinking in Front of the Alps: Saussure's (1796) Unpublished Theory of the Earth: *Earth Sciences History*, v. 8, no. 2, p. 123–140.

Carozzi, Albert V., and Carozzi, Marguerite, 1991, Pallas' Theory of the Earth in German (1778). Translation and Reevaluation by a Contemporary: H.-B. de Saussure: *Archives des sciences*, v. 44, fasc. 1, p. 1–105.

Carozzi, Albert V., and Newman, John K., 1990, Dialogic Irony: An Unusual Manuscript of Horace-Bénédict de Saussure on Mountain Building: *De Montium Origine* (1774): *Archives des sciences*, v. 43, fasc. 2, p. 235–263.

Carozzi, Marguerite, 1986, From the Concept of Salient and Reentrant Angles by Louis Bourguet to Nicolas Desmarest's Description of Meandering Rivers: *Archives des sciences*, v. 39, fasc. 1, p. 25–51.

Carozzi, Marguerite, and Carozzi, Albert V., 1987, Sulzer's Antidiluvialist and Catastrophist Theories on the Origin of Mountains: *Archives des sciences*, v. 40, fasc. 2, p. 107–143.

Collier, Katharine Brownell, 1934, *Cosmogonies of Our Fathers: Some Theories of the Seventeenth and the Eighteenth Centuries*: Columbia University Press, New York, 500 p. Reprinted 1968 by Octagon Books, New York.

[Davies, Gordon L.: See Herries Davies, G.L.]

Desmarest, Nicolas, 1794/95 [an III], *Encyclopédie méthodique. Géographie-physique*; Vol. I: H. Agasse, Paris, 858p.

Ellenberger, François, 1975-1977, A l'aube de la géologie moderne: Henri Gautier (1660-1737): *Histoire et nature*, no. 7, p. 3–58, & nos. 9–10, p. 3–145.

Eyles, V.A., 1969, The Extent of Geological Knowledge in the Eighteenth Century, and the Methods by Which It Was Diffused, *in* Cecil J. Schneer, ed., *Toward a History of Geology*: M.I.T. Press, Cambridge, Mass., and London, p. 159–183.

Eyles, V.A., 1972, Hutton: In *Dictionary of Scientific Biography*: Charles Scribner's Sons, New York, v. 6, p. 577–589.

Geikie, Sir Archibald, 1905, *The Founders of Geology* (2nd ed.): Macmillan and Co., London, x, 486 p.

Gillispie, Charles C., 1951, *Genesis and Geology: A Study in the Relations of Scientific Thought, Natural Theology, and Social Opinion in Great Britain, 1790–1850*: Harvard University Press, Cambridge, Mass., xiii, 315 p.

Gillispie, Charles C., 1960, *The Edge of Objectivity: An Essay in the History of Scientific Ideas*: Princeton University Press, Princeton, 562 p.

Gohau, Gabriel, 1990, *Les Sciences de la terre aux XVIIe et XVIIIe siècles: Naissance de la géologie*: Albin Michel, Paris, 420 p.

Greene, John C., 1959, *The Death of Adam: Evolution and Its Impact on Western Thought*: Iowa State University Press, Ames, Iowa, 388 p.

Herries Davies, G.L., 1969, *The Earth in Decay: A History of British Geomorphology, 1578–1878*: American Elsevier, New York, xvi, 390 p.

Herries Davies, G.L., 1989, A Science Receives Its Character, *in* G.L. Herries Davies and Antony R. Orme, *Two Centuries of Earth Science, 1650–1850*: William Andrews Clark Library, University of California, Los Angeles, p. 1-28.

Koyré, Alexandre, 1939, *Etudes galiléennes*: Hermann, Paris, 335 p.

Koyré, Alexandre, 1957, *From the Closed World to the Infinite Universe*: Johns Hopkins University Press, Baltimore and London, xii, 313 p.

Kuhn, Thomas S., 1962, *The Structure of Scientific Revolutions*: The University of Chicago Press, Chicago and London, xv, 172 p.

Laudan, Rachel, 1987, *From Mineralogy to Geology: The Foundations of a Science, 1650–1830*: University of Chicago Press, Chicago and London, xii, 278 p.

Lyell, Charles, 1830, *Principles of Geology*, Vol. I: John Murray, London, xv, 511 p.

Lyell, Charles, 1990, *Principles of Geology*, Facsimile repr. of 1st ed., Vol. I, with introduction by Martin J.S. Rudwick: University of Chicago Press, Chicago and London [lvii], xv, 511 p.

Neve, Michael, and Porter, Roy, 1977, Alexander Catcott: Glory and Geology: *British Journal for the History of Science*, v. 10, no. 34, p. 37–60.

Oldroyd, D.R., 1980, Sir Archibald Geikie (1835-1924), Geologist, Romantic Aesthete, and Historian of Geology: The Problem of Whig Historiography of Science: *Annals of Science*, v. 37, p. 441–462.

Ospovat, Alexander M., 1976, The Distortion of Werner in Lyell's *Principles of Geology*: *British Journal for the History of Science*, v. 9, no. 32, p. 190–198.

Ospovat, Alexander, M., 1971, translation with introduction and notes, *Short Classification and Description of the Various Rocks*, by Abraham Gottlob Werner: Hafner, New York, xii, 194 p.

Ospovat, Alexander M., 1980, The Importance of Regional Geology in the Geological Theories of Abraham Gottlob Werner – A Contrary Opinion: *Annals of Science*, v. 37, p. 433–480.

Porter, Roy, 1976, Charles Lyell and the Principles of the History of Geology: *British Journal for the History of Science*, v. 9, no. 32, p. 91-103.

Porter, Roy, 1977, *The Making of Geology: Earth Science in Britain, 1660–1815*: Cambridge University Press, Cambridge, xi, 288 p.

Porter, Roy, 1979, Creation and Credence: The Career of Theories of the Earth in Britain, 1660-1820, *in* Barry Barnes and Steven Shapin, eds., *Natural Order: Historical Studies of Scientific Culture*: Sage Publications, Beverly Hills & London, p. 97–123.

Porter, Roy, 1980, The Terraqueous Globe, *in* G.S. Rousseau and Roy Porter, eds., *The Ferment of Knowledge, Studies in the Historiography of Eighteenth-Century Science*: Cambridge University Press, Cambridge, p. 285–324.

Rappaport, Rhoda, 1964, Problems and Sources in the History of Geology, 1749–1810: *History of Science*, v. 3, p. 60–77.

Rappaport, Rhoda, 1968, Lavoisier's Geologic Activities, 1763-1792: *Isis*, v. 58, no. 193, p. 375–384.

Rappaport, Rhoda, 1973, Lavoisier's Theory of the Earth: *British Journal for the History of Science*, v. 6, no. 23, p. 247–260.

Rappaport, Rhoda, 1978, Geology and Orthodoxy: The Case of Noah's Flood in Eighteenth-Century Thought: *British Journal for the History of Science*, v. 11, no. 37, p. 1–18.

Roger, Jacques, 1962, introduction and notes, *Les Epoques de la Nature*, by G.L. Leclerc de Buffon: Mémoires du Muséum National d'Histoire Naturelle, ser. C (Sciences de la Terre), v. 10.

Roger, Jacques, 1973, La Théorie de la terre au XVIIème siècle: *Revue d'histoire des sciences*, v. 26, p. 23–48.

Roger, Jacques, 1982, The Cartesian Model and Its Role in Eighteenth-Century 'Theory of the Earth,' *in* Thomas M. Lennon, John M. Nicholas, and John W. Davis, eds., *Problems of Cartesianism*: McGill-Queen's University Press, Kingston and Montreal, p. 95–125.

Rudwick, Martin J.S., 1969, The Strategy of Lyell's *Principles of Geology*: *Isis*, v. 61, no. 206, p. 5–33.

Rupke, Nicolaas A., 1983, *The Great Chain of History: William Buckland and the English School of Geology (1814–1849)*: Clarendon Press, Oxford, xii, 322 p.

Taylor, Kenneth L., 1969, Nicolas Desmarest and Geology in the Eighteenth Century, *in* Cecil J. Schneer, ed., *Toward a History of Geology*: M.I.T. Press, Cambridge, Mass., and London, p. 339–356.

Taylor, Kenneth L. 1974, Natural Law in Eighteenth-Century Geology: The Case of Louis Bourguet: XIIIth International Congress of the History of Science, *Proceedings*, v. 8, Editions Nauka, Moscow, p. 72–80.

Taylor, Kenneth L., 1988, Les Lois naturelles dans la géologie du XVIIIème siècle – Recherches préliminaires: *Travaux du Comité Français d'Histoire de la Géologie*, ser. 3, v. 2, no. 1, p. 1–28.

Zittel, Karl Alfred von, 1901, *History of Geology and Palaeontology to the End of the Nineteenth Century*; Translated by Maria M. Ogilvie-Gordon: Walter Scott, London, xiii, 562 p.

IX

*Volcanoes as Accidents:
How 'Natural' Were Volcanoes
to 18th-Century Naturalists?*

Historical examination of the ways volcanoes were discussed during the 18th century reveals significant differences of view about their status within the framework of natural phenomena. At one end of a spectrum of opinion, volcanoes were often treated as 'accidents.' At the other end, some investigators of the later part of the century came to regard volcanoes as – to use James Hutton's phrase – "a natural ingredient in the constitution of the globe." This essay explores some of the conceptual underpinnings of the idea that volcanoes must be considered as accidental rather than ordinary, and thus not altogether natural. Such attitudes are shown to be linked with traditional ideas about knowledge of nature being established properly upon recognition of natural regularities. This helps to account for prevalent resistance, during much of the 18th century, to acceptance of volcanic heat as a formative agent on a level comparable to aqueous agency. Reluctance of geologists to adjust to thinking of volcanoes as fully part of nature's regular operations is illustrated in the work of several figures who can be thought of as volcanological 'progressives,' including Buffon, Desmarest, and Strange.

> Having thus found a distinguishing character for those fused substances called, in general, lavas, and having the most visible marks for that which had been actually a volcano, naturalists, in examining different countries, have discovered the most undoubted proofs of many ancient volcanos, which had not been before suspected. Thus, *volcanos will appear to be not a matter of accident, or as only happening in a particular*

IX

> *place, they are general to the globe*, so far as there is no place upon the earth that may not have an eruption of this kind; ...
> *Volcanos are natural to the globe, as general operations*; ... .
>
> – James Hutton, *Theory of the Earth* (1788)[1]

What is Natural is Ordinary

Hutton insisted on the status of volcanoes not as accidents or features of merely local interest, but as phenomena fully integrated within a global system of natural processes. A geology embracing the generality of volcanic operations did eventually receive widespread assent, in the decades following the Scottish philosopher's death. Yet Hutton's expression of this view betrays his awareness that he was in some noteworthy manner contradicting prevailing ideas. In setting forth his opinion through a contrast between a vision of the naturalness and generality of volcanoes, on one hand, and notions of their local and accidental character on the other, Hutton implicitly acknowledged a need to overcome current views standing in the way of volcanoes' full membership in the natural economy of the globe. I wish here to comment on the historical underpinnings of contemporary resistance to volcanic 'naturalness', a conception that Hutton was by no means alone in advocating during the later part of the 18th century.[2] Perhaps we can gain some useful historical perspective by considering elements of the outlook against which Hutton's declaration was directed.

What stands out as peculiar, and therefore historically intriguing, in Hutton's affirmation of the naturalness of volcanoes, is the opposition it invokes between the *accidental* and the *general*. Such a locution is no doubt a bit strange to modern ears – with the probable exception of those attuned keenly to the traditions of philosophy. It reflects an idea that by Hutton's day had been part of European intellectual baggage for many centuries. An apt comparison can be made with a passage from a work of science dating back

---

[1] HUTTON (1788), p. 274. Emphasis added.

[2] Alberto Fortis, William Hamilton, and Rudolf Erich Raspe, to name just three, were among those giving support to the new conception of volcanic generality and naturalness.

seventeen centuries before Hutton – the *Natural Questions* of Lucius Annaeus Seneca. At a point in Book II of Seneca's dialogue there is discussion of the possible causes of the phenomenon of thunder; the question is raised whether thunder might not be caused, sometimes, by shooting stars falling into clouds. To this question the answer suggested is to the effect that this may happen sometimes, "but we are now seeking the natural and usual cause, not the rare and accidental" (or, in another translation, "it is not the occasional chance cause but the natural normal one that we are in search of").[3] This way of thinking, linking 'nature' with ordinariness and normality while distinguishing it from ephemerality and irregularity, is readily seen to be associated with the emphasis prevalent in ancient philosophical doctrines upon explanation as a matter of understanding the universal rather than the particular case, and upon a distinction between the essential nature of a thing on one hand and its accidental or circumstantial properties on the other.

So we encounter here a mental framework in which not everything in nature is equally natural: What is truly natural, in some sense, is ordinary. This habit of separating different levels or degrees of naturalness in things is disclosed in scientific writings across the ages from antiquity to the relatively recent past. It is one of the persisting hierarchical features in early modern thought which, when we take the trouble to recognize it, marks it as foreign to the homogeneity of 'modernity'. Within this traditional framework of thought, for a proper understanding of nature one sought explanations in terms of what is usually or normally ('naturally') found to be the case, not through rare or uncharacteristic ('accidental') occurrences. According to such a viewpoint, events or objects that stood outside what was thought to be the ordinary course of nature, while worthy of interest and quite possibly even of serious study, could not be expected to serve the purposes of the seeker after genuine knowledge, which consisted centrally of an understanding of the general case. By this way of thinking, *extraordinary* things, which in a literal sense stand outside rule, could not represent the fundamental order of nature.[4]

---

[3] "Nunc naturalem causam quaerimus et assiduam, non raram fortuitamque." SENECA (1971), Vol. 1 (containing Seneca's Books I - III), Book II, Chapter 55: Latin text p. 186, translation p. 187. The alternate translation given here is from SENECA (1910), p. 99.

[4] It is obvious, I trust, that we are dealing here with only one cluster of meanings assigned at various times to the multivalent terms *natural* and *nature*. The ideas of the natural as normal or ordinary, and of the accidental as exceptional, are evidently related to (although perhaps not

IX

With these considerations in mind, it is not difficult to see that part of Hutton's objective (as stated in the epigraph) was to refute the idea of the extraordinariness of volcanoes. For Hutton, bringing volcanoes within the domain of the ordinary – domesticating them, situating them within a natural economy – was vitally important for his geological system. To accomplish this act of domestication, however, Hutton did not present the argument we might expect of a modern geological counterpart, to the effect that there are no such things as accidents in the sense of their being less than entirely natural; he maintained instead that volcanoes are natural and *not accidental*. Hutton thus preserved a traditional distinction between the natural (ordinary) and the accidental (extraordinary), while undertaking to see volcanoes transferred out of the accidental category. That Hutton here chose to speak in this way is, I think, significant, and reflective of mental habits among naturalists of the time which we should not ignore. And that domesticating measures were called for – in other words, that there is testimony from Hutton's predecessors and contemporaries in support of a common view of volcanoes as extraordinary and in some way not fully natural – is a point to which we must return momentarily.

Terrestrial Order and Change: Knowledge Through Regularity

In a paper for the IXth INHIGEO Symposium of 1980, in Paris, Rhoda Rappaport presented an outstanding demonstration of how close attention to the vocabulary used in geology during the 18th century can provide important

---

necessarily entailed by) a conception of nature as essence. A conception of the natural as opposed to the supernatural is beside the point here. It may be pertinent to add that, while it seems to me that the peculiarity to which I point, in the view of nature revealed in Seneca (and Hutton), has largely faded away in the way modern people think when engaged in the sciences, it is far from unknown in ordinary talk about ourselves: we speak as if someone's *nature* is identified with the person's *usual* conduct and demeanor, and one is commonly said to act *out of character* in doing something inconsistent with that nature. The durability of outmoded scientific conceptions for ordinary-language use is also illustrated in the persistence of the Hippocratic-Galenic doctrine of humours: the practice of distinguishing someone's temperament as sanguine or phlegmatic, while medically extinct, continues in the common culture. Modern Western habits of thought in the routines of human relations, it would seem, sometimes retain features which have been suppressed in more formally disciplined modes of thinking.

clues to its characteristic ways of thinking.⁵ Two of the three terms she chose for examination were 'revolution' and 'accident'. Regarding the former, she commented on the concern of certain geologists, in the 18th and early 19th centuries, to dissociate the course of geological change from disorder; she also noted the identification by at least one leading geologist (Leopold von Buch) of fire or heat with disorderly change. As for 'accidents', Rappaport discussed these primarily as "the legacy of rationalism." That is to say, she situated the conception of accident with respect to early modern philosophy and historiography, in which features of time, place, and circumstance were distinguished from the fundamental order of natural and human reality. She called attention to the highly variable ways different geological commentators of the 18th century demarcated accidents from the normal and natural; and she drew attention to complications brought on by the contemporary tendency to confuse accidents, as circumstantial and local aspects of some thing or event, with revolutions, as changes which while local and circumstantial might be part of general and recurrent patterns.⁶

---

⁵ RAPPAPORT (1982). Examining uses of the terms *monument, revolution,* and *accident* in 18th-century geological literature, Rappaport showed that the concept of *accidents* persisted in geology into the early part of the 19th century, as she illustrated through the resistance offered to the finding, by Cuvier and Brongniart, that freshwater or non-marine sediments must be seen as integral parts of the earth's historical record, rather than as accidental exceptions to the norm of marine sedimentation. Rappaport noted the fact that relegation of accidents to a subordinate cognitive status came to an end more or less simultaneously in both geology and history, through the early 19th-century reforms of those branches of knowledge; in effect, the category of 'accident' disappeared as agreement emerged that everything that happens counts in reconstructing the past. Within her broader examination of geological vocabulary's import for understanding deep conceptual changes, she explicitly noted the general 18th-century treatment of volcanoes and earthquakes as deviations from the normal or natural.

⁶ Such confusion was easy; the term *accident* had a range of meanings and was used in a variety of contexts. A traditional meaning somewhat removed from this essay's 18th-century locus, but perhaps not completely antiquated by that time, concerned the *accidental properties* of mineral bodies: a medieval mineralogist such as Albertus Magnus used this kind of criterion in differentiating stones by characters such as color, hardness, and so on, as has been discussed recently by ANGEL (1992). While 18th-century naturalists may not have lost completely the habit of discussing accidental properties of objects in the mineral realm as characters affording distinctions within static ensembles for purposes of identification or classification, as far as I can tell it had then become common to use the term 'accident' in a dynamic sense: accidents had comparatively little to do with a mineral object's intrinsic or permanent features, and much to do with *what happened* to the object during its production and especially its subsequent

IX

Through her discussion of the tangled ramifications of the terms 'revolution' and 'accident', Rappaport in effect pointed to prolonged efforts within 18th-century geology to reconcile the increasingly evident reality of enormous past terrestrial changes with abiding conviction in both the natural world's fundamental stability and (no less important) the proper expression of truth in terms of permanence. For us to appreciate fully how hard such an accommodation must have been is difficult, from our standpoint two centuries and more after the fact. We are, after all, heirs to generations of historicist ways of thinking which were then only in the course of being devised. Strengthened confidence in an invariable natural order, and in rational access to at least the superficially discernible forms of that order, was part of the legacies 18th-century thinkers derived from Cartesian and Newtonian science. The effects of this confidence on the minds of the varied lot of geological observers and thinkers during this period of 'proto-geology' should not be underestimated. Most of them – members of overlapping groups identifiable with such endeavors as mineralogy, physical geography, or the Theory of the Earth – responded to the encouragements of a logic of experience; they supposed that success in building scientific understanding of the earth and its changes could be accomplished only by an empirical elaboration of generalizations about terrestrial features or processes. Hardly any of them would have disputed the tenet that scrupulous attention to facts was indispensable for such success. But by the empiricist credo of the new philosophy, which was perceived as triumphing not only in attainment of abstract laws but in enlarged grasp of

---

preservation and alteration. Thus in Jean-Étienne GUETTARD'S "Mémoire sur les accidens des coquilles fossiles" (1759), written to refute Élie Bertrand's contention that some 'figured stones' are "primitive and essential to the earth," the French naturalist presented his proofs that figured stones represent the remains of marine bodies like those now living in the seas, and did so by dividing the fossils' accidents in four categories: their attachments [*attaches* – principally, other creatures fastened to them], their conservation, their destruction, and their deformation. These categories correspond to events and circumstances bearing on the natural 'experience' of a fossil body. In WERNER'S *Short Classification and Description of the Various Rocks* (1971, p. 54) there is a perhaps somewhat more equivocal example: "The feldspar in gneiss is in a few rare cases somewhat weathered or even completely dissolved into porcelain earth and the mica in part changed into steatite; but since this seldom occurs and is brought about by special causes, it cannot be listed as an essential part of the description of this rock but must be considered and noted as an accidental occurrence [*eine zufällige Beschaffenheit*] in it." Werner's book, to be sure, was aimed at an understanding of what *kind* of object each mineral body is, but the *accidental status* of feldspar as spoken of here is a function of special causes operating in its production.

natural history knowledge as well, not all facts were necessarily of equal value; that credo tended to elevate usual, recurrent facts above strange and unusual ones. The most highly respected models of advancement in natural knowledge drew attention to what was regular and general, not irregular and merely local. This is an important part of the reason, I think, why so many reputable naturalists and natural philosophers of the 18th century continued to speak of local and circumstantial aspects of phenomena as belonging on a different and lower plane than those aspects that appeared to be universal and essential.[7]

The power of the impulse toward generalizing about stable and constant configurations and operations (as distinct from the possibilities, gradually coming to be seen as offering comparable promise, of ordering events in historical sequence) is widely displayed throughout the century's geological writings.[8] Even when the efforts to organize geological information assumed a developmental or historical form, the 'historicity' they tended to exhibit was one distinctly limited to the patterns of unfolding events ordained (or programmed, we might say) in the nature of things, as several scholars have observed.[9] In contrast to the contingent historicism which only took distinct shape in the 19th century, this determinist or universalist history emerging among a number of 18th-century naturalists had little affinity with phenomena bearing marks of uniqueness, chance, or local circumstance. Thus even those geologists who were prepared to find historical patterns in the data they assembled were inclined to multiply observations of common elements in what was seen, rather than look to singularities or exceptions for significant guid-

---

[7] I have tried in several previous essays to address some of the themes summarized in this paragraph: TAYLOR (1974, 1979, 1981-82, 1988, 1992a, 1992b). The patterns of thinking out of which modern geology emerged are treated by, among others, PORTER (1977), LAUDAN (1987), GOHAU (1990), ELLENBERGER (1988-1994) and OLDROYD (1996).

[8] BUFFON's 1749 *Theory of the Earth* famously began by asserting at length the fundamental order and regularity to be discerned behind the superficial appearance of disorder in the Earth's features. From later in the century, an elegant and highly sympathetic expression of "the aim of mineralogical science," written by a *savant* who can be considered a well-informed outsider in geological matters, and summing up the goals of this science in terms of order and regularity superceding misleadingly apparent difference and confusion, is found in CONDORCET's historical eulogy of Guettard (1786), at pp. 52-53. Going back to near the beginning of the century, we may take note of Fontenelle's more equivocal epigram: "On ne sçauroit guere attribuer à la Nature trop d'uniformité dans les Regles generales, & trop de diversité dans les applications particulieres" – as quoted by RAPPAPORT (1991), p. 286.

[9] For example, OLDROYD (1979) and GOHAU (1990). See also TAYLOR (1992a).

IX

ance. During the middle decades of the 18th century, when the closest thing to a dominant geological model was coming to be organized around the differentiation of distinct rock assemblages identified principally by their physical and lithological qualities but conceived as representing a universal sequence (primitive or primary, secondary, tertiary, and alluvial, sometimes with added groups and variants), it was usual to look on most evident deformations of these assemblages as deviations from a normal arrangement; in these conditions it was perhaps not strange that seemingly local and exceptional phenomena would tend to be regarded as no real help to the geologist, but more nearly as distractions from the important task of building general knowledge.[10] How often are the texts of even the most determinedly anti-speculative of these observers punctuated by allusions to *comparing and combining facts* as ever-necessary steps along the way to genuine comprehension! It scarcely needs to be added, that exceptional facts are the last to lend themselves to comparison and combination.

An especially resonant expression of the priority that was assigned, in geological explanation, to normality and recurrence, and of the irrelevance of whatever is merely occasional or abrupt, is found in Buffon's 1749 *Theory of the Earth.*

> Furthermore, causes whose effects are rare, violent, and sudden, should not concern us; they are not found in the ordinary conduct of nature; but effects that occur every day, motions that recur and are renewed without interruption, constant and ever-repeated operations, those are our causes and explanations.[11]

According to Buffon, only general and uniform causes (which he linked with water, not fire – a point to which we will return below) can have "acted to construct and shape the present surface of the earth"; accidents can produce only particular effects, not the general ones which are the proper object of a naturalist's attention.[12]

---

[10] A point noted in different ways by, among others, GOHAU (1983), p. 9; and CAROZZI (1987), p. 204.

[11] BUFFON (1749), p. 99.

[12] Jacques Roger discussed Buffon's belief, as expressed in the 1749 *Théorie de la Terre*, in fire's irrevocably destructive character: BUFFON (1962), p. xxi. Roger remarked that for Buffon, fire's destructive nature permits it to be primitive only, thus separating earth's formation from its subsequent revolutions; only water is eternal. However, Roger went on to say that the new

Buffon's point that naturalists must turn to general processes for understanding of the Earth's features was not his invention, of course. He had predecessors who emphasized the idea, such as Louis Bourguet, the "Pliny of Neuchâtel".[13] And the same view is echoed by a number of geological figures of the later part of the century. Thus Jean-Louis Giraud Soulavie wrote, in his natural history of southern France, that the naturalist must "follow Nature in its most general and widest operations."[14] In the geological lecture notes of John Walker, professor at the University of Edinburgh, one sees his clear orientation toward the properly generalized form of sound geological observation, where evidences of whatever is rare, episodic, or violent warrants its relegation to a status of subordinate geological importance.[15] The tirelessly dedicated Alpine traveller-naturalist H.-B. de Saussure, in his "Agenda," stressed the need for the naturalist-collector to select ordinary and common specimens, avoiding the temptation to choose instead special or unusual ones.[16] In another manual intended to aid geological observation, Besson argued similarly that not accidents but general and universally prevalent aspects of nature provide insight into the age, formation, and revolutions of a land.[17]

This centralization of the regular, and a concomitant marginalization of the extraordinary and irregular, sometimes found expression in the association of exceptional phenomena with disturbance, disruption, or even disfigurement of the natural. Thus Wallerius admitted that the much-discussed rule of salients and re-entrants, or 'correponding angles' in mountainous topography (which he believed to be accounted for by the regular action of running water), is not always observed in nature, but in those instances where it is not "the effect must be attributed to a different accidental cause which has

---

Buffon of the *Époques de la Nature* now (1778) accepted successions and history, and saw heat as a continuous factor in earth's history.

[13] BOURGUET (1729), esp. his "Mémoire sur la théorie de la terre," pp. 175-220; and BOURGUET (1742), e.g., pp. 32-33, 66, 84. In one passage here, for instance, Bourguet spoke of the action of a violent *bouleversement* as distinct from the "ordinary or natural" (p. 74).

[14] GIRAUD SOULAVIE (1780-1784), Vol. 1, p. 150.

[15] WALKER (1966), pp. 21-23, 170-176, 180-185.

[16] DE SAUSSURE (1779-1796), Vol. 4 (1796), "Agenda", pp. 467-[539]. At p. 533.

[17] BESSON (1794), at pp. 302-303. For a particularly nice statement of the priority sound naturalists were supposed to place on observation of general facts, see the "Avertissement" to the second edition of Philippe BERTRAND (1782), p. vi.

IX

disturbed" the symmetrical arrangement.[18] Wallerius' way of speaking here is perhaps representative of a fairly common practice in 18th-century scientific language, according to which the 'natural' is what was original, and 'accidental' refers to what is subsequent; in other words, accidents alter the first or natural state.[19] This corresponded (or perhaps in some quarters still corresponds), especially in the understanding of living things, with speaking of what is innate as natural, and of acquired characters as accidental.[20] Modern usage preserves the process of "denaturing" certain materials, in the sense of alteration which brings significantly changed properties. It was not unusual for mineralogists of the late 18th and early 19th centuries to say that an altered mineral was denatured. However, in at least some cases from 18th-century geology there is discussion of transformations from the natural described in terms which appear to signify something other than neutral change: a damaging kind of alteration.[21]

We have seen enough, I think, to ascertain that there persisted in the discourse of a number of 18th-century naturalists an old notion that extraordinary phenomena are in a sense excluded from the fully natural, and do not really figure in what counts as a proper understanding of the natural. I have indicated the ancient lineage and venerability of this notion by introducing it

---

[18] WALLERIUS (1780), p. 258. On the concept of corresponding angles in 18th-century geology, see TAYLOR (1974), and CAROZZI (1986).

[19] The Edinburgh lectures of WALKER (1966) commonly imply that accidental effects are the same as subsequent or recent ones; e.g., pp. 175-176 (I thank Kerry Magruder for calling this to my attention).] MARSILI wrote in 1710, in *Brieve ristretto del saggio fisico intorno alla storia del mare*, of the parts of the Earth's crust other than those dating from the Creation (which he called "essential") as "accidental" (according to RAPPAPORT, 1997, chap. 7). See also CONDILLAC (1951), where "nature" means "le premier état d'une chose", and "naturel" applies to what is innate, not acquired (pp. 398-399); so "accident" had strong associations with alteration, disturbance of the original.

[20] On this point as seen in Lamarck, for example, see JORDANOVA (1984), p. 40.

[21] That 'nature' means preservation and excludes that which does not endure belongs to venerable traditions of scientific thought. This can be seen in the appeal made in the 16th century by Copernicus to long-respected principles in defending the 'naturalness' of the Earth's rotation: "Yet if anyone believes that the earth rotates, surely he will hold that its motion is natural, not violent. But what is in accordance with nature produces effects contrary to those resulting from violence, since things to which force or violence is applied must disintegrate and cannot long endure. On the other hand, that which is brought into existence by nature is well-ordered and preserved in its best state." COPERNICUS (1978), Bk. I, Ch. 8, p. 15.

through Seneca. But that Roman author's connection with some of the 18th-century bearers of the idea in geological science is actually closer than I have mentioned. Scholars have recognized Seneca's high standing in some Enlightenment quarters; what deserves notice here is that at least two significant geological figures of the period went out of their way to signal their affinity with Seneca. I do not believe this is just a coincidence. One of the two was Nicolas Desmarest, whose habit of distinguishing between regular and accidental geological phenomena will be noted below. Desmarest's interest in Seneca was marked in part by his participation in La Grange's 1778 translation of the Roman philosopher's works, for which he provided commentary on the *Natural Questions* in several dozens of notes. In addition to these notes, there is much additional evidence within Desmarest's writings over a half-century span to indicate that there were both moral and physical elements of Seneca's thought that appealed to him. Among the recurrent features of Desmarest's resort to Seneca's authority, one notes a relentless naturalism, and a sense of nature conceived as a system of continuous cycles of destruction and renovation.[22]

The second geological figure I have in mind was the diplomat and virtuoso John Strange. So decided was Strange's admiration for Seneca's emphasis on "seeking the natural and usual cause, not the rare and accidental" that he chose this very quotation from the *Natural Questions* for the epigraph of his 1774 report on the volcanic phenomena of the Veneto district.[23]

---

[22] SENECA (1778). Desmarest's notes commonly exploited opportunities provided in SENECA's text to approve (and in some cases to disapprove or correct) the Roman author's ideas about natural phenomena, and to expand on fundamental questions regarding nature and human understanding of it. Desmarest had earlier shown his affinity with moral aspects of Seneca's Stoicism (e.g., DESMAREST, 1756, p. 59), and cited him also in his *Encyclopédie* article on springs (DESMAREST, 1757, pp. 82a, 98a). He later included a lengthy article on Seneca in his compendium of physical geography (DESMAREST, 1794-1828, I: 436-489); here Desmarest called him the most enlightened of the ancient scientists. Many parts of Desmarest's notes from the La Grange edition of Seneca reappeared in the encyclopedic *Géographie physique*, not only in the article devoted to Seneca but in various other articles as well. On the 18th-century fashion for ancient knowledge and taste, a starting point is the recent study by GRELL (1995).

[23] STRANGE (1775). This was a letter sent to the Royal Society in 1774. The epigraph is rendered: "Naturalem causam quærimus et assiduam, non raram et fortuitam." The same quotation appears (in slightly different form) on the title page of the volume Strange published in Italian (1778). On Strange see CIANCIO (1995); also DE BEER (1951), and STOKES (1971).

IX

John Strange's Two Types of Volcanic Productions: Ordered, and Confused

The case of John Strange brings us back to the natural-accidental distinction in the particular case of volcanic phenomena. Was the aphorism from Seneca really indicative of something fundamental for Strange? We can see that it was, by considering his remarks on two sets of columnar basalts (or "Giant's Causeways"), in the Euganean Hills near Padua, and near San Giovanni Ilarione in the Val d'Alpone in the Vicentine pre-Alps. Convinced of the igneous origins of these "groups of prismatic basaltine columns," Strange gave a lengthy discussion of the significance of his observations. He argued that the ordered and regular basalts represented a "vulcanic" production quite different from the confusion seen in "real" or recently active volcanoes, such as Vesuvius or Etna or Santorini. The order and regularity of disposition seen in the basaltic structures, similar to the displays of columnar basalts he had seen in Auvergne, Velay, Vivarais and other parts of France, reflected their generation in a manner and condition quite different from the "violent convulsions" yielding "tumultuary and inordinate aggregates" thrown out "from the bowels of the earth, by subterraneous explosions."

> For in fact, what does Vesuvius, Ætna, the Monte di Cenere, and such like eructed piles, present to us but a heap of ruins, which evidently manifest the casual and extraordinary cause, to which they avowedly owe their origin? But this origin seems irreconcileable with the regular structures before mentioned, as may perhaps satisfactorily appear, from my considerations on the particular phaenomena, that characterize them. And though it is very possible, that such organizations may sometimes take place upon the concretion of liquified matter thrown up in vulcanic eruptions; yet, however similar they may be, from the nature of their origin, I can hardly imagine they can form other than imperfect and irregular masses. For however wonderful the rivers of *lava* of Vesuvius or Ætna may appear to us; they, in reality, are but partial and tumultuary efforts of nature, that by no means seem adequate to the production of a Giants Causeway, or the basaltine organizations of Auvergne and Velay, several of which continue, almost uninterruptedly, for many miles. [24]

Strange thought that these regular basaltic masses displayed evidences of their origin "upon a more steady and uniform principle" [25] than unruly explo-

---

[24] STRANGE (1775), pp. 19-21.
[25] STRANGE (1775), p. 21.

sions, and of their formation from pre-existing structures which "have suffered fire *in statu quo*, or locally, without the least appearance of subversion, or change of place."

> From the preceding observations it appears, I think, evident, that subterraneous explosions and eruptions are merely accidental phaenomena, that are by no means essential to the production of all vulcanic mountains, as has been commonly imagined.[26]

Such accidents, Strange proposed, have the effect of concealing from the observer of active volcanoes their original structure and order, still extant beneath the ejecta scattered out upon it.

> Nor have we any foundation, from the external appearance of such mountains, to conclude, that all others, that have suffered fire, are of the same character. We see nothing but a heap of ruins, cast up from their bowels, and we are apt to imagine, that such inordinate materials compose the intire mass; and analogy, too often seducive [sic] in similar matters, leads us to conclude the same of other vulcanic mountains in general. But I am much inclined to think, that the materials thrown up by burning mountains, are only lodged superficially, as it were, on their sides; and though they may considerably increase their bulk, as well as alter their form, yet they do not seem to constitute the intire mass of those mountains, as might be reasonably imagined from their external appearance. For it has been observed, both by PADRE DE LA TORRE, M. DE LA LANDE[27], and others, that the inner sides even of the funnel of mount Vesuvius preserve manifest vestiges of its primary organization, in regular, parallel, and nearly horizontal *strata*, like those of other common mountains. And does it not appear more than probable from hence, that an original mountain lies under the *lava* of Vesuvius, serving in a manner, as its base, and which, whatever local alteration it may have received intrinsically, from the subtle element, that wastes its bowels, still maintains its primary undisturbed structure, like the vulcanic mountains of the Veronese territory described.[28]

For Strange, as for a good many of his contemporaries, it was evident that volcanoes *happen in* certain mountains, into which channels are opened for

---

[26] STRANGE (1775), p. 27.
[27] Here Strange cited Lalande: "Voyages d'Italie, tom. VII. p. 169, 176."
[28] STRANGE (1775), pp. 28-29.

IX

the venting of the action of fire. From this incidental quality of volcanoes, as well as their ruinous disorder, it appeared clear to Strange that doctrines according to which entire mountain ranges are uplifted by subterranean fire stood exposed and unsupportable. (He referred to writings by Hooke, Ray, and Raspe, as well as what he considered the highly unsuccessful theory proposed by Moro.) Extensive and integrated surface structures such as the Andes could only derive from some more controlled cause than fire.

> If they form integral parts of a continued chain, as it is natural to suppose, is it not even absurd to imagine, that they can have had such an origin? Is it not, on the contrary, rather to be presumed, that channels only have been opened along this chain, by different explosions, where these volcanos respectively exist; and that the sides of these channels form integral parts of its original structure, as in the case of mount Vesuvius before remarked, and which here seems to receive the strongest confirmation. For however the eruptions of the volcanos of the Andes may have loaded their sides and summits in particular parts; yet surely inferior masses exist of a much prior origin, and whole continuity sufficiently seems to prove, that such eruptions are, relatively, only accidental phaenomena.[29]

It followed from the accidental status of erupting volcanoes, Strange said, that their study really provides rather little geological insight.

> It also plainly appears, if I mistake not, from what has been before said, that the phaenomena of recent volcanos are very little calculated to give us much instruction about the more curious igneous concretions, and the origin of vulcanic mountains in general; and that a few days tour in such countries as Auvergne, Velay, and the Venetian state are worth a seven years apprenticeship at the foot of mount Vesuvius or Ætna; where nothing but a heap of uninstructive ruins, and a sameness of phaenomena appear. And since our ideas, concerning vulcanic effects, have been almost exclusively drawn from recent volcanos, we cannot much wonder if they yet remain so imperfect.[30]

The uninstructively "constant sameness of phaenomena" that Strange believed was seen at an active volcano like Vesuvius or Etna[31] might at first confuse us, if we think of sameness as equivalent to uniformity. Actually, he meant just the opposite; the order and regularity noted in assemblages of columnar

---

[29] STRANGE (1775), p. 31.
[30] STRANGE (1775), pp. 32-33.
[31] STRANGE (1775), pp. 28-32.

basalt work as a basis for scientific instruction, as the chaos of cinder-cones does not, because geological uniformity implies the possibility of making distinctions. Strange meant to contrast the haphazard homogeneity of volcanic ejecta with order. He appears to have assumed, with many of his contemporaries (as I emphasized in the previous section), that scientific intelligibility (theory) arises out of patterns of coherence among which there is the possibility of discrimination; neither coherence nor the opportunity for instruction could be found, he thought, in the undifferentiated volcanic products of ruinous explosion.

Volcanoes As (Partly) Accidental Phenomena

Strange's opinion that the organized basaltic remnants of past volcanism represent an order of volcanicity entirely different from that seen in the destructively explosive eruptions of modern experience was not shared by all of his contemporaries. But neither was it an idea peculiar to him alone. And many of his views – that volcanoes are mountains in which pre-existing materials are in varying degrees altered in place, that regular productions of igneous agency such as prismatic basalts are difficult to reconcile with the uncontrolled force apparent in active volcanoes, and that the theories of mountain-elevation by subterranean heat are far-fetched and unsustainable – appear to have been more the norm than the exception among geological observers during the second half of the 18th century, a period which saw the beginnings of the systematic study of the productions of extinct volcanoes, and significant advances in close examination of active volcanoes.

As I noted at the outset, James Hutton appears to have assumed, in advocating the general involvement of volcanoes in the terrestrial system, that some of his contemporaries would need to be convinced that volcanoes are natural and general, not accidental. Strange's report on some of the basalts of northern Italy in the *Philosophical Transactions* helps us see how this was so. A few further illustrations of scientists' doubts about the full membership of volcanic phenomena within the category of the earth's general operations support the impression that Hutton had good reason for thinking that his conception could be met with suspicion. Some of these examples may also widen our sense that Strange was not alone in having complicated and ambivalent notions about the relations of volcanicity with order and disorder.

In his theory of the earth, Louis Bourguet had found a place for discussion of volcanoes only under the heading of "phenomena concerning the des-

IX

truction of the earth," not in the corresponding sections on the globe's "surface phenomena" or "interior structure".[32] Bourguet shared with many 18th-century naturalists a presumption that *structure* was a geological category difficult to reconcile with *destruction*, as the first represented a tranquil order while the second connoted violent disorder. Buffon, more deeply influenced by Bourguet than he cared to acknowledge, expanded on this theme. In his 1778 *Époques de la nature*, for example, Buffon elaborated on how earthquakes and volcanoes formerly "ravaged" the Earth by "commotions" of far greater intensity than in our own later days. While acknowledging the considerable effects of past earthquakes and volcanic action, particularly on land (submarine volcanoes being subject to the extinguishing capacity of overbearing waters), Buffon described these in terms of violence and explosion yielding terrible disorder, in contrast to the tranquil uniformity deriving from the regularity of aqueous action:

> Thus after the water, by uniform and constant movements, had accomplished the horizontal construction of the Earth's beds, the fire of volcanoes, by sudden explosions, overthrew, cut into and covered over several of these beds; ... [Well before the start of volcanic eruptions, earthquakes exerted themselves by] violent shocks which produced effects every bit as violent and far more extensive than those of volcanoes.[33]

We should cease to be surprised, said Buffon, by destruction and disordering (which is "particular") of the general organization in aqueous products, brought on by accidental earthquakes and volcanoes:

> This disorder ... nonetheless merely conceals Nature from the eyes of those who only see it on a small scale, and who fabricate a general and constant cause from an accidental and particular effect. It is water alone which, as a general cause and one subsequent to that of the primordial fire, has acted to construct and shape the present surface of the Earth. And whatever is lacking in uniformity in this universal construction is only the particular effect of the accidental cause of earthquakes and volcanic action.[34]

Were we to probe Buffon's geology more deeply, we would find that the clear division of aqueous and igneous processes proclaimed here, along lines

---

[32] BOURGUET (1729), pp. 175-220: "Phenomènes concernant la destruction de la terre" at pp. 209-211.
[33] BUFFON (1962), pp. 128-9.
[34] BUFFON (1962), pp. 129-130.

of general construction and particular defacement, soon becomes more problematic. The same can be seen through the field-investigations performed by some of Buffon's compatriots, on the remains of extinct volcanoes. In the prolonged disputes over the ability of fire or heat to generate structured rocks, Jean-Étienne Guettard spoke for many contemporaries in rejecting the volcanic origins of columnar basalt, at first, by referring to volcanoes as agents of confusion, antithetical to order and regularity.

> ... this opinion seems to me all the less probable, [in proportion] as in the eruption of volcanoes everything happens in confusion, and as they confuse and mix all the materials, far from arranging them with order and regularity.[35]

In time, however, Guettard was brought to a substantial change of mind. And even in his earliest volcanic observations, Guettard by no means dealt with volcanic phenomena exclusively in terms of irregularity. For instance, in recognizing some of the Auvergne *puys* as formed by volcanic accumulation, he explained the differences in degrees of symmetry in volcanic cones through varied but on the whole regular accumulations of volcanic material: the highly conical peaks had experienced frequent and repeated eruptions, and so amassed sufficient matter to even out surface irregularities.[36]

Nicolas Desmarest, no less than his older contemporary Guettard, was guided in his observations by an ideal of regularity. Desmarest's meticulous and imaginative studies contain indications of concern that anything produced

---

[35] GUETTARD (1768-1786), Vol. 2 (1770), Préface, p. xxvi. See also pp. 251-2. In his famous report on the volcanic character of the Auvergne *puys*, Guettard repeatedly linked the words "fire" and "violent", and similarly coupled "water" with "tranquil" (GUETTARD, 1752, e.g. pp. 51, 55). Werner's fundamental conviction in the irreconcilability between volcanic action and crustal organization was expressed with some qualifications: "These [true volcanic] rocks lie in their formations in a very disorderly manner, but in a position which comes fairly close to being layered, depending upon whether they were eruptively heaped one upon the other or whether they flowed out . . . ." (WERNER, 1971, p. 80). But for Werner "even the exterior appearance of true volcanic rock formations is distinctly disorderly . . . ." (p. 82). The conception opposing volcanic fire's action to consequences of water's law-like production of ordered materials was repeated later in Leopold von Buch's remarks about the controversy between Kirwan and Hall over the igneous formation of granite. Von Buch said that nature's revolutions, however terrible in appearance, must always have followed the same fixed laws – and this eliminates the possibility of so regular a form as columnar basalt having been formed by fire. (Cited by RAPPAPORT, 1982, p. 36).

[36] GUETTARD (1752), p. 44.

IX

by igneous causes might resist incorporation within ordered science: In his pioneering cartographic studies of the extinct volcanic districts of Auvergne, he tended to speak of volcanoes as accidental and intermittent, and was hesitant to expect that anything about them would disclose sufficient organization to permit sensible interpretation.

> At first, certain productions of heat displayed a correspondence as regular as it was instructive: ... But then, he encountered such apparent disorder in the lavas' arrangement, so little uniformity [*ensemble*] in their distribution, that he was tempted to attribute this state to the tumultuous attacks of fire, and to the irregularities of its effects; ... but several considerations undeceived him.
>
> He understood from the start that the conflagrations of volcanoes being accidents in the order of the phenomena of Nature, their recurrences had not been subject to any invariable period.[37]

It appears that the main thing which undeceived Desmarest, aside from the basaltic structures whose organization so impressed others like Strange, was his success in combining a keen topographic vision of the arrangements of Auvergne landforms (their 'dispositions' and 'distributions') with a perception of the progressively destructive effects of moving water and time on their integrity.

A worthy French successor to Desmarest in the interpretation of volcanic landforms, Jean-Louis Giraud Soulavie also used language revealing of a divided mind on the issue of volcanic susceptibility to order. Soulavie confirmed the disorderly quality of volcanic districts ("Confusion reigns there on all sides"), and identified volcanic action with "recent irregularity" in surface features whereas visible regularities could be attributed to such causes as universal gravitational attraction or the solvent action of water.[38] Yet Soulavie's treatment of the volcanic phenomena he studied in Vivarais tended to emphasize basic orderliness barely concealed by disorder. Thus, in imagining the extrusion of basalts and their solidification on the surface: "... notwithstanding the disorder of the eruptions, this all took place in a proportionate and consequential manner [*avec poids et mesure*]; this melted basalt followed laws."[39]

---

[37] DESMAREST (1779), p. 116. In this digest of a 1775 presentation to the Royal Academy of Sciences, Desmarest refers to himself using the third person.

[38] GIRAUD SOULAVIE (1780-1784), "Confusion" at Vol. 1, p. 141; "irregularity" at Vol. 1, p. 90.

[39] GIRAUD SOULAVIE (1780-1784), Vol. 2 (1780), p. 26.

Toward the Ordinariness of Volcanoes

Desmarest and Soulavie, like many other geological writers in the later part of the 18th century, showed a growing amenability to organize phenomena in order of time; their examination of volcanicity had significant connections to the growing historical sensibility in geology. If the end of the notion of 'accidents' in geology, as in history, accompanied the establishment of the historicist outlook, who would not suppose it useful to search the grounds out of which historicism grew for better understanding of so curious an adjustment in geological thinking? Yet this will almost certainly not be enough, as the example of the notoriously unhistorical-minded Hutton attests.

It is worth asking questions about the origins, career and fate of these ideas in geology – distinctions which for a time marked off the natural, uniform and general from the accidental, irregular, and local. Without asking such questions, how will we understand the differences in outlook on display, for example, between two British diplomats posted contemporaneously in Italy, John Strange and William Hamilton? Each combined a fascination with observing geological phenomena of acknowledged volcanicity, with strong artistic and antiquarian interests – but what a difference in geological attitude! The one saw at least as much difference as similarity between igneous productions of past and present, amd disparaged the notion of gaining worthwhile knowledge from "uninstructive ruins" such as Vesuvius or Etna. The other seems not to have doubted the widest applicability of what he saw in his fervent study of the behavior of these volcanoes; in an uncharacteristically expansive moment of theorizing broadly from his observations, Hamilton declared, "Upon the whole, if I was to establish a system, it would be, that *mountains are produced by volcanos, and not volcanos by mountains.*" [40] Hamilton's embrace of the generative capacities of volcanoes stood in sharp contrast to Strange's assumption that the action of volcanoes is mainly disfigurative; yet it also deserves notice that Hamilton's formulation, like Hutton's later assertion of the generality of volcanic dynamics, implies an entrenched contrary view to which it is opposed.

Let me close with a few related words from another British visitor, the admirably observant traveller Patrick Brydone, from his 1770 journey in the Neapolitan region. Brydone reflected on how contrary to expectation it is

---

[40] HAMILTON (1769), p. 21. Emphasis in the original.

IX

when science informs us that volcanic power is a creative as well as a destructive force.

> Now, should I tell you that this immense coast, ... covered over with an everlasting verdure, and loaded with the richest fruits, is all the produce of subterranean fire; it would require, I am afraid, too great a stretch of faith to believe me; ... It is strange, you will say, that nature should make use of the same agent to create as to destroy; and that what has only been looked upon as the consumer of countries, is in fact the very power that produces them.[41]

It would not be too long a time, in the course of changing scientific ideas, before what appeared strange to Hutton's younger contemporary Brydone – nature's use of the same agent to create as to destroy – would come to be for most geologists all but self-evident. What would then become strange was what Brydone had taken as obvious: that a volcanic agency's capacity for productive action was in the nature of things incompatible with its clear role in ruin.

---

[41] BRYDONE (1773), Vol. 1, pp. 22-23.

References

ANGEL M. (1992). *Propriétés accidentelles des pierres: couleur, dureté, fissilité, porosité et densité selon Albert le Grand*. Travaux du Comité Français d'Histoire de la Géologie, sér. 3, t. 5, pp. 87-92.

BERTRAND P. (1782). *Lettre à M. le Comte de Buffon, ... ou Critique, et nouvel essai sur la Théorie Générale de la Terre*. 2nd edition. Besançon, Chez Charmet; Paris, Chez Esprit.

BESSON [no first name known]. (1794). *Sur les moyens de rendre utiles les voyages des naturalistes*. L'Esprit des Journaux, françois et étrangers, 23, April, pp. 292-308.

BOURGUET L. (1729). *Lettres philosophiques sur la formation des sels et des crystaux, et sur la génération et le mechanisme organique des plantes et des animaux, avec un mémoire sur la théorie de la terre*. Amsterdam, Chez François l'Honoré.

BOURGUET L. (1742). *Traité des pétrifications*. Paris, Briasson.

BRYDONE P. (1773). *A Tour Through Sicily and Malta*. 2 vols. London, Printed for A. Strahan and T. Cadell.

BUFFON G.-L. LECLERC, Comte de (1749). *Histoire & théorie de la terre*. In Vol. 1 of *Histoire naturelle, générale et particulière, avec la description du cabinet du roi*. Paris, De l'Imprimerie Royale.

BUFFON G.-L. LECLERC, Comte de (1783-1788). *Histoire Naturelle des Minéraux*. 5 vols. Paris, De l'Imprimerie Royale.

BUFFON G.-L. LECLERC, Comte de (1962). *Des Époques de la Nature*. Édition critique avec le manuscrit, une introduction et des notes par J. ROGER. In: *Mémoires du Muséum National d'Histoire Naturelle*, Sér. C, Sciences de la Terre, tome 10. Paris, Éditions du Muséum (Réimpression 1988).

CAROZZI A.V. (1987). *La Géologie*. In: J. TREMBLEY (ed.), *Les Savants genevois dans l'Europe intellectuelle du XVIIe au milieu du XIXe siècle*. Geneva, Éditions du Journal de Genève. Pp. 203-265.

CAROZZI M. (1986). *From the Concept of Salient and Reentrant Angles by Louis Bourguet to Nicolas Desmarest's Description of Meandering Rivers*. Archives des sciences, 39, pp. 25-51.

CIANCIO L. (1995a). *The Correspondence of a "Virtuoso" of the Late Enlightenment: John Strange and the Relationship Between British and Italian Naturalists*. Archives of Natural History, 22, pp. 119-129.

CIANCIO L. (1995b). *Autopsie della terra: Illuminismo e geologia in Alberto Fortis (1741-1803)*. Biblioteca di "Nuncius": Studi e testi, 18. Firenze, Leo S. Olschki.

IX

CIANCIO L. (1995c). *A Calendar of the Correspondence of John Strange, FRS (1732-1799)*. London, The Wellcome Institute for the History of Medicine. (Occasional Publication 2).

CONDILLAC É. BONNOT, abbé de. (1951). *Dictionnaire des synonymes*. In Oeuvres philosophiques de Condillac. Texte établi et présenté par G. Le Roy. Vol. 3. Paris, Presses Universitaires de France.

CONDORCET J.A.N. DE CARITAT, Marquis de. (1786). *Éloge de M. Guettard*. Histoire de l'Académie Royale des Sciences, Paris [published 1788], pp. 47-62.

COPERNICUS N. (1978). *On the Revolutions*. Edited by Jerzy Dobrzycki, translation and commentary by Edward Rosen. Baltimore, Johns Hopkins University Press.

DE BEER G.R. (1951). *John Strange, F.R.S., 1732-1799*. Notes and Records of the Royal Society of London, 9, pp. 96-108.

DESMAREST N. (1756). *Conjectures physico-méchaniques sur la propagation des secousses dans les tremblemens de terre, et sur la disposition des lieux qui en ont ressenti les effets*. N.p., no publ.

DESMAREST N. (1757). *Fontaine*. In: Encyclopédie, ou Dictionnaire raisonné des sciences, des arts et des métiers. Vol. 7, pp. 80-101.

DESMAREST N. (1779). *Extrait d'un Mémoire sur la détermination de quelques époques de la nature par les produits des volcans, & sur 'usage qu'on peut faire de ces époques dans l'étude des volcans*. Observations sur la physique, sur l'histoire naturelle et sur les arts, 13, pp. 115-126.

DESMAREST N. (1794-1828). *Géographie-physique*. In Encyclopédie méthodique. 5 vols., 1 atlas. Paris, H. Agasse.

ELLENBERGER F. (1988-1994). *Histoire de la Géologie*. 2 vols. Paris, Technique et Documentation (Lavoisier).

GOHAU G. (1983). *Idées anciennes sur la formation des montagnes*. In: Cahiers d'histoire et de philosophie des sciences. Nouvelle sér., no. 7.

GOHAU G. (1990). *Les Sciences de la terre aux XVIIe et XVIIIe siècles: Naissance de la géologie*. Paris, Albin Michel.

GRELL C. (1995). *Le Dix-huitième siècle et l'antiquité en France, 1680-1789*. In: Studies on Voltaire and the Eighteenth Century, Vols. 330-331. Oxford, Voltaire Foundation.

GUETTARD J.-É. (1752). *Mémoire sur quelques montagnes de la France qui ont été des volcans*. Mémoires de l'Académie Royale des Sciences, Paris [published 1756]. Pp. 27-59.

GUETTARD J.-É. (1759). *Mémoire sur les accidens des coquilles fossiles, comparés à ceux qui arrivent aux coquilles qu'on trouve maintenant dans la mer.* Mémoires de l'Académie Royale des Sciences, Paris [published 1765]. Pp. 189-226, 329-357, 399-419.

GUETTARD J.-É. (1768-1786). *Mémoires sur différentes parties des sciences et des arts.* 5 vols. in 6. Paris, Chez Laurent Prault.

HAMILTON W. (1769). *A Letter ... Containing Some Farther Particulars on Mount Vesuvius, and Other Volcanos in the Neighbourhood.* Philosophical Transactions, 59, pp. 18-22.

HUTTON J. (1788). *Theory of the Earth; Or, an Investigation of the Laws observable in the Composition, Dissolution, and Restoration of Land upon the Globe.* Transactions of the Royal Society of Edinburgh, Vol. I, Part II (read 1785), pp. 209-304.

JORDANOVA L. J. (1984). *Lamarck.* Oxford & New York, Oxford University Press.

LAUDAN R. (1987). *From Mineralogy to Geology: The Foundations of a Science, 1650-1830.* Chicago and London, University of Chicago Press.

OLDROYD D. R. (1979). *Historicism and the Rise of Historical Geology.* History of Science, 17, pp. 191-213, 227-257.

OLDROYD D. R. (1996). *Thinking About the Earth: A History of Ideas in Geology.* London, Athlone Press.

PORTER R. (1977). *The Making of Geology: Earth Science in Britain, 1660-1815.* Cambridge, Cambridge University Press.

RAPPAPORT R. (1982). *Borrowed Words: Problems of Vocabulary in Eighteenth-Century Geology.* The British Journal for the History of Science, 15, pp. 27-44. [A summary was published with other IXth INHIGEO Symposium papers in Histoire et Nature, 1981-82, no. 19-20, pp. 57-58.]

RAPPAPORT R. (1991). *Fontenelle Interprets the Earth's History.* Revue d'histoire des sciences, 44, pp. 281-300.

RAPPAPORT R. (1997). *When Geologists Were Historians, 1665-1750.* Ithaca, Cornell University Press.

ROGER J. (1974). *Le feu et l'histoire: James Hutton et la naissance de la géologie.* In: *Approches des Lumières: Mélanges offerts à Jean Fabre.* Paris, Klincksieck. Pp. 415-429.

SAUSSURE H.-B. de (1779-1796). *Voyages dans les Alpes, précédés d'un essai sur l'histoire naturelle des environs de Genève.* 4 vols. Neuchâtel, Chez Samuel Fauche.

SENECA L.A. (1778). *Questions naturelles.* In: *Les Oeuvres de Sénèque le Philosophe.* Vol. 6. Translation by La Grange. Paris, Chez les frères De Bure. [Notes by Desmarest and others.]

IX

SENECA L.A. (1910). *Physical Science in the Time of Nero. Being a Translation of the Quaestiones Naturales of Seneca*. By John Clarke. London, Macmillan.

SENECA L.A. (1971). *Naturales Quaestiones*. English translation by Thomas H. Corcoran. Loeb Classical Library. 2 vols. London, William Heinemann; Cambridge, Massachusetts, Harvard University Press.

SOULAVIE J.-L. GIRAUD (1780-1784). *Histoire naturelle de la France méridionale*. 7 vols. Nîmes, Chez Belle.

STOKES E. (1971). *Volcanic Studies by Members of the Royal Society of London, 1665-1780*. Earth Sciences Journal, 5, pp. 46-70.

STRANGE, J. (1775). *An Account of Two Giants Causeways, or Groups of prismatic basaltine Columns, and other curious vulcanic Concretions, in the Venetian State in Italy; with some Remarks on the Characters of these and other similar Bodies, and on the physical Geography of the Countries in which they are found*. Philosophical Transactions, 65, pp. 5-47.

STRANGE J. (1778). *De' monti colonnari e d'altri fenomeni vulcanici dello stato veneto*. Milano, Marelli.

TAYLOR K.L. (1974). *Natural Law in Eighteenth-Century Geology: The Case of Louis Bourguet*. International Congress of the History of Science, 1971, Proceedings, 8, pp. 72-80. Moscow, Editions 'Naouka'.

TAYLOR K.L. (1979). *Geology in 1776: Some Notes on the Character of an Incipient Science*. In: SCHNEER, C. J. (ed.), *Two Hundred Years of Geology in America*. Hanover, University Press of New England. Pp. 75-90.

TAYLOR K.L. (1981-82). *The Beginnings of a French Geological Identity*. Histoire et nature, no. 19-20, pp. 65-82.

TAYLOR K.L. (1988). *Les Lois naturelles dans la géologie du XVIIIème siècle: Recherches préliminaires*. Travaux du Comité Français d'Histoire de la Géologie, sér. 3, t. 2, no. 1, pp. 1-28.

TAYLOR K.L. (1992a). *The Époques de la Nature and Geology During Buffon's Later Years*. In GAYON, J. (ed.), *Buffon 88: Actes du Colloque international pour le bicentenaire de la mort de Buffon*. Paris, J. Vrin. Pp. 371-385.

TAYLOR, K.L. (1992b). *The Historical Rehabilitation of Theories of the Earth*. The Compass, 69, pp. 334-345.

WALKER J. (1966). *Lectures on Geology*. Edited with notes and an introduction by Harold W. Scott. Chicago, London, University of Chicago Press.

WALLERIUS J.G. (1780). *De l'origine du Monde, et de la Terre en particulier*. Translation by Jean-Baptiste Dubois de Jancigny. Varsovie, et se trouve à Paris, Chez J. Fr. Bastien.

WERNER A.G. (1971). *Short Classification and Description of the Various Rocks*. Translated with an introduction and notes by Alexander M. Ospovat. New York, Hafner.

X

Two Ways of Imagining the Earth at the Close of the 18th Century: Descriptive and Theoretical Traditions in Early Geology

„At the time when the Kurze Klassifikation was first published, there was great interest in the study of rocks and in the history of the earth. In fact, this was an unusually active period geologically, as a perusal of the literature clearly indicates. But in all the outpouring of geological literature, little concern was shown for a geological synthesis that would tie together in a meaningful way the many descriptions of investigations of particular and isolated areas. In addition, no fewer than twenty-seven systems of mineralogy had been put forward in the relatively short period between 1647 and 1775, most of them including classifications of rocks. There was, therefore, a great need not only for a geological synthesis but also for a general agreement on nomenclature and descriptive techniques."[1]

Alexander M. Ospovat, 1971

---

[1] Werner, Abraham Gottlob (Alexander M. Ospovat, translator and editor): Short Classification and Description of the Various Rocks. New York, Hafner Pub. Co., 1971, p. 1 of Ospovat`s Introduction.

1. Introduction

My remarks today are an attempt historically to analyze some of the elements of geological thinking around 200 years ago or more, in the later parts of the 18th century. I intend to dwell more on some aspects of the form of evolving geological ideas than on their content.

Let me warn you that this talk is a reflection piece, and some will think it consists of too much abstract discussion and too little history. I suppose my defense is that our historical understanding profits from general as well as specific discussions of the problems we want to understand. Also, my experience generally in the history of science suggests how much the nature of ideas is governed by the forms in which they are exercised.

I begin by invoking some comments about Abraham Gottlob Werner, whose achievements and influence we are here in Freiberg to explore and celebrate. The remarks in the epigraph above are those of my friend and colleague Alexander Ospovat, from the Introduction to his 1971 translation of Werner's Kurze Klassifikation, first published in 1786. Allow me to call to your attention Alex's reference to the abundant evidence of great interest, during the 1780s, in geological information in the form of descriptions of investigations of particular and isolated areas. In addition, Alex made the interesting assertion that while around this time there was a great need for a geological synthesis, there had not been evidence of substantial concern to find or create such a synthesis.

In the same Introduction, Alex went on to say that Werner's Kurze Klassifikation helped to further subsequent efforts toward more standardized and uniform nomenclature, that it served to establish petrography as an independent branch of the geological sciences by separating rock classification from mineral classification, and that it was important in offering a history of the earth's crust. In Werner's system, Alex stated, the doctrine of geological succession is a cardinal principle of earth history.[2]

With due allowance for the fact that Alex Ospovat was discussing the larger context of the work of a particular Freiberg academic, it is apparent that he meant his remarks to apply broadly. These passages from Alex's discussion of Werner's Kurze Klassifikation illustrate what I think is a significant point about the condition of geological writings in the closing decades of the 18th century, and about our historical analysis of them. Alex's discussion recognized three distinguishable aspects of the cognitive or scientific consideration of the earth and its accessible constituents, namely the descriptive, historical, and theoretical modes. These modes of treatment, intertwined though they often are in science, are recognizable and familiar to

---

[2] Ospovat, in Werner (note 1), pp. 2-3.

us, as I believe they were also to most of those whose efforts during the later 18th century were directed at advancing geological science.

2. Descriptive, Historical, and Theoretical Modes of Comprehension

Let me begin my general comments on three distinct ways of conceptualizing the earth, with remarks on the second one as shown in Table 1 – the historical mode. This is because historical ways of thinking about the earth figure least in what I have to say. As the subtitle of this essay indicates, I intend to focus mainly on the other two.

Table 1.: *Three cognitive modes for treatment of the earth, and their characters.*

Cognitive modes in natural science	Corresponding purposes, methods, approaches
Descriptive mode	Identifying and distinguishing objects; inventorying; naming, classifying; nature's specificity; nature's plan
Historical mode	Placing objects and events in temporal sequence; ordering in time; nature's evolution; nature becoming
Theoretical mode	Relating objects, events, processes to timeless principles; ascertaining permanent underlying order; recognizing atemporal causal basis; nature's generality, universality; nature being

To dispense quickly with the historical mode may seem a bit odd, since geology is widely seen as the first of the natural sciences to have appropriated historical forms of thinking as central to its cognitive structure. I do not argue against such a point of view. In fact, I generally agree with it; I even consider that the incorporation of historical conceptions into efforts to comprehend nature, during the last three centuries, constitutes possibly the single most momentous intellectual reorientation in the modern transformation of science.

I do think, however, that in the 18th century this change had not yet been really accomplished. Its ingredients were only beginning to be assembled. In our retrospective view on the early development of geology, we may sometimes overestimate the significance of apparently historical ways early geological scientists wrote. This may happen in part because we are wise after the event: we know how the story later turned out – we know how geology in fact became quite historical in the 19th century.

In any case, without denying there were some extremely interesting test runs of a semi-historical sort during the 18th century, indeed even in the 17th, I will very nearly ignore these today. And you will see that I imply that the authors of these historical trials might not have conceived them in as fully historical a way as we sometimes imagine. I have become convinced that geological mentalities before the end of the 18th century tended to remain dominated by the other two cognitive modes – the descriptive and the theoretical. Even those who floated quasi-historical geological ideas could not foresee how this might result in a 19th-century reconfiguration of geological knowledge and geological research.

As for the other two modes, descriptive and theoretical, it seems to me that together they represent the classic impulses of natural science as it was practiced in the Western tradition for over two millennia. Their functions correspond closely to those of natural history on one hand and natural philosophy on the other. The one asks „What is it?" while the other asks „How does it work?" Descriptive science identifies what sorts of things there are in the world, and provides knowledge of their similarities and differences. Theoretical science offers explanatory principles, seeking ultimately to find a causal account for what happens – what was frequently called in the 18th century the „operations of nature."

Distinctions such as these in fundamental approaches to comprehension of nature are not merely artificial. The historical actors we try to understand, in the remote past as well as more recently, generally understood and respected the differences. This is not to say, of course, that a person simply did one or the other. In practice, there was nearly always a lot of interchange and overlap, in both the activities and the identities of naturalists and natural philosophers - or physicists [physiciens], as the 18th-century French subjects of my research often called the latter group. In fact, I intend here to try to point out an interesting kind of interplay between these two cognitive styles among 18th-century geological writers.

One other point about the descriptive and theoretical modes before we move beyond their general definitions: Like it or not, over the long term, a position of privilege has generally been reserved for theoretical science. Some have not liked it. But that theoretical or universal knowledge has generally been seen as a form of understanding superior to knowledge of the particular has been true for a long time, going back at least as far as Aristotle's insistence that the highest knowledge is of the general case. I make this assertion not just about opinions prevalent among historians of science. What matters is that for ages it has tended to be true about attitudes among considerable numbers of scientific thinkers, reflecting hierarchical distinctions within the ranks of science. One should not underestimate the strength of this point of view among scientists of the 18th century, including many geological scientists. It was commonplace for geological writers of the later 18th century to express commitment to the objective of making out of geology a science of a high level of generality (a condition it was widely agreed not yet to have attained), thus reducing its identity as a science of specificity.

In the late 18th century, as Alex Ospovat suggested in the passage quoted at the head of this essay, there was a general perception that the growing body of descriptive knowledge yielded, in an era of very active geological interest, was not matched with a comparably satisfactory synthesis. It is one thing, however, to note that in the last decades of the 18th century, geologists thought that the information they increasingly possessed lacked a proportionately adequate unifying framework. It is quite another to say that, at least until the later 1780s, this state of affairs reflected a lack of concern. This is the facet of Alex's claims with which I take issue here.

I will sketch an argument that the extent of concern during the late 18th century about building a geological synthesis has been easy to underestimate because much of it was expressed in apparently descriptive rather than clearly recognizable theoretical terms. I have come to think that a retrospective reading of the geological literature leading up to the century's turn too easily yields a misleading sense of contemporary resignation, if not complacency, about theoretical resolution of data, on account of a sort of descriptive disguise worn by part of the theoretical impulse.

3. Bourguet's Rule of Corresponding Angles: An instance of dispositional regularity

Allow me to offer a rather autobiographical explanation of what I mean. Like so much else of my acquaintance with 18th-century geology, my encounter with this point started years ago with reading in the works of Nicolas Desmarest (1725–1815). More specifically, it happened when I was trying to plow my way through the bulkiest of Desmarest's works, and in some ways his most baffling. This is the multi-volume set on physical geography that he prepared for the Encyclopedie méthodique.[3] „Géographie physique" was Desmarest's preferred term for systematic earth science. These volumes began to appear as Desmarest approached 70 years of age; they were the production of his golden years.

It was while trying to acquaint myself with Desmarest that I first noticed in this, his most wide-ranging and comprehensive work, that Desmarest evidently had an acute interest in what I have come to call dispositional regularities. Actually, my attention was first drawn to one specific kind of regularity by which Desmarest seemed inordinately intrigued. It took a while for me gradually to realize how his interest in this one type of regularity was symptomatic of a commitment he had made to investigation over a wide range

---

[3] Desmarest, Nicolas: Géographie physique. Paris, Agasse, 1794–1811, 4 vols. [a 5th volume by other hands appeared in 1828].

of dispositional regularities, as the methodology proper to an earth scientist. I was slow, also, in realizing that this commitment was spread over Desmarest's lifetime of work in science, going back to the beginnings around mid-century. And certainly, it was only by stages that I arrived at the point I am asserting now: that this habit I first saw in Desmarest only, of insisting on empirically described dispositional regularities as the basis for a generalized geological science, was actually rather widely acknowledged, even advocated rather aggressively, by geological authors of varied descriptions in the later decades of the 18th century.[4]

The specific dispositional regularity that caught my eye – through Desmarest's strikingly recurrent treatment of it – was sometimes called the rule of corresponding angles, maybe even more often that of salients and re-entrants. Through Desmarest I traced the lineage of this topographical generalization back as far as Louis Bourguet (1678–1742), the Neuchâtelois naturalist and earth-theorist from earlier in the century. Its more direct access to Desmarest and others of his generation was possibly through Buffon, who made much of this generalization in his Théorie de la terre published in 1749. Bourguet had emphasized the configurational regularity of promontories opposed to corresponding indentations, in the topography of mountains and valleys, as the main empirical clue to a true theory of the earth.

It became apparent to me that among his geological successors in the second half of the century, Bourguet's theory of the earth was not accorded extraordinary respect, certainly no more than many others among the theories that proliferated as the century advanced. What is more, Bourguet's rule of corresponding angles was very often challenged or contradicted. Desmarest, through whom I had first met the idea, was himself not convinced of its universal applicability. But despite this, Desmarest – and, I gradually realized, many others, especially in the francophone geological world whose preoccupations I was trying to gauge – seemingly could not stop discussing salients and re-entrants, even when it was to raise objections to the relation's general validity. A dialogue on the empirical legitimacy of corresponding angles pervades a significant portion of the geological literature up to 1800, and even beyond. An abbreviated list of the authors in whom one sees this in some part of their work includes, besides Buffon and Desmarest (a

---

[4] I first discussed these perceptions in Taylor, Kenneth L.: Natural Law in Eighteenth-Century Geology: The Case of Louis Bourguet, in: XIIIth International Congress of the History of Science (Moscow, 1971), Proceedings 7 (1974), pp. 72–80. See also Carozzi, Marguerite: From the Concept of Salient and Reentrant Angles by Louis Bourguet to Nicolas Desmarest's Description of Meandering Rivers, in: Archives des Sciences (Société de Physique et d'Histoire Naturelle de Genève) 39 (1986), pp. 25–51. For an account of the early formation of Desmarest's outlook see Taylor, Kenneth L.: La Genèse d'un naturaliste: Desmarest, la lecture et la nature, in: Gohau, Gabriel (Ed.): De la Géologie àson histoire. Paris, CTHS, 1997, pp. 61–74. My views on the historical role of "dispositional regularities" are partly developed in Taylor, Kenneth L.: Les Lois naturelles dans la géologie du XVIIIème siècle: Recherches préliminaires, in: Travaux du Comité Français d'Histoire de la Géologie, 3ème série, 2 (1988), no. 1, pp. 1-28.

francophone bias is noticeable): Bergman, Pallas, Lamanon, Soulavie, Collini, Hamilton, Dolomieu, Saussure, Palassou, Haidinger, Launay, Walker, and Lamétherie.

Why this persistent life for a generalization about topographical relations, even in the discourse of some who thought it was erroneous? The answer, I think, has to do with the idea's form rather than its substance.

The rule of corresponding angles offered the merits of foundation in observational descriptions, and of susceptibility to extensive empirical testing. These aspects of the exercise were praised, explicitly or implicitly, if the rule itself often was not. Empirical grounding and observational derivation were attractive features for ideas then contending for scientific consideration – in an era when there was increasingly strong agreement about the defects of speculation as an avenue to resolution of general geological issues, and about the correspondingly powerful appeal of the „epistemological modesty" found in the triumphant natural philosophy of the Newtonians.

4. Dispositional regularities

So, shortly after becoming aware of the surprisingly dynamic career of the rule of salients and re-entrants, I came to believe that a leading reason for the vitality of debate over its validity lay in a favorable contemporary judgment about its formal qualities. Empirical-minded geological scientists like Desmarest, skeptical of theories that overreached, but committed nevertheless to the objective of finding a true theory of the earth, admired the style and formal quality of the kind of inquiry exemplified by corresponding angles, even if they were not so enthralled by this particular result. On the view I thus attribute to Desmarest and like-minded geologists, the search for a satisfactory theory of the earth ought to emulate this effort; it should seek out legitimate descriptively-founded generalizations about the relations of geological data. They believed that a proper theory, the contours of which might for the present remain obscure, could be anticipated as the eventual outcome of patient descriptive effort.

The school of historical study of science in which I was brought up emphasizes identification and analysis of guiding ideas manifested in the work of one's historical characters. My effort to comprehend Desmarest's fascination with Bourguet's law of corresponding angles led me to think that attitudes underlying the business of theorizing among 18th-century geologists were at least as consequential as any specific theories they entertained. Here, it seemed to me, was an attitude of more than trivial interest: Some leading geologists of the later 18th century appear to have adopted a policy of treating descriptive generalization as something like the equivalent of theorizing. Describing in ways that transcend the particular – by trying to recognize generally valid patterns which describe terrestrial features – was apparently considered a part of the theoretical enterprise. There was abroad, among at least a certain group of Enlightenment geological writers, a conviction that expressing general relations among things through exact description actually constituted a sensible way of theorizing. If this is so, it evidently means that these authors were blurring the taxonomy of scientific approaches I have proposed in this essay.

As you may imagine, at a certain point I began jotting out lists of types of descriptive generalizations that seemed to fascinate Desmarest. As the lists grew, so did the number of my encounters with these same descriptive generalizations in the writings of other geologists, especially from the last third of the 18th century. Table 2 summarizes a small selection from my lists.

Table 2.: *Some types of dispositional regularity discussed by 18th-century geologists*

Type of dispositional regularity	Instances
Land-sea configurations	Geographic relations of continents and oceans
Symmetries, patterns in mountain configurations	Distributions, linear arrangements of ranges; Directional orientations of steep & gentle slopes
Configurations of river/stream junctions	Concordant junctions: common levels of confluence; Angular orientations of conjunctions (e.g., relative to volume or swiftness of stream flow)
Patterns, structures within rocks	Fentes perpendiculaires: cracks (joints)
Configurations of rock structure	„Parallelism of beds"; Lithostratigraphic arrangements

I will cite just one specific text where certain dispositional regularities (many of which are taken up by later writers) are discussed with special intensity: Buffon's Théorie de la terre (1749).[5] Here one discovers, for example, assertions about the significance of macrogeographical arrangements of the continents and oceans on the globe -- that is, the presumably non-accidental configuration of land and sea. Putative regularities in this broad category included proposed geometric patterns in the orientation of main axes of symmetry, displayed in large continental masses. Also, one meets up with discussions of a suspected significance in the fact that major continents narrow to the south rather than the north (Africa, South America, the Indian subcontinent).

Another class of regularities consisted of supposed symmetries in mountain configurations. A long-standing conception of the world-wide continuity of mountain ranges persisted in some quarters through the 18th century. Variants

---

[5] Buffon, Georges-Louis Leclerc, Comte de: Second discours. Histoire & Théorie de la Terre, in: Buffon, Histoire naturelle, générale et particulière, Vol. 1. Paris, De l'Imprimerie Royale, 1749. Amplification on some of the dispositional regularities considered by Buffon and others is found in Taylor: Les Lois naturelles .... (cited in note 4).

of this idea included attention to geometric rationalization of the distribution and location of mountain ranges within continents, and of their angular relations with one another. (So, the loxodromist project of Alexander von Humboldt, or the geometric elegances of Léonce Elie de Beaumont's réseau pentagonal, of the 19th century, had their 18th-century precursors.) Not uncommonly, claims were made also about the orientations of the steep faces and gentle slopes of mountains – as expected to submit to a general rule.

Still another family of regularities sometimes encountered in the late-18th-century geological literature has to do with the junctions of streams and rivers. Members of this group include general observations about the usual common elevation of stream junctions; and proposed correlation of the angles at which stream tributaries meet, with variables such as gradient, stream volume, or rate of flow of water. One also runs across related generalizations about the sorting of transported debris in streams, in relation to distances from presumed sources of the rock, and varying declivity over the range of distribution.

Other clusters of regularities receiving persistent attention, that I will mention in passing, include patterns in joints, or swarms of cracks in rocks. My files on this topic are labelled fentes perpendiculaires (perpendicular cracks), on account of the currency of that term in the late 18th century – again, with direct or implied expectation that some regular configurational patterns among these cracks will come into focus with enough study, and shed light on the theory of the earth.

The type of regularity from my lists that I will mention last is one I suspect many attending this symposium might by now have expected to see coming: namely, the abundantly-remarked (in the 18th century) „parallelism of beds," and the well-documented undertaking, especially from mid-century onward, to identify and situate distinct groupings of lithological units. This was perhaps implied already, under a category I mentioned before: regular patterns in mountain configurations. Readers of works by 18th-century rock and mountain classifiers are familiar with the strong association, for example, between early lithostratigraphical efforts and an appetite to discover symmetries in rock structures along axes of mountain ranges.

Historical accounts of early lithostratigraphic geology tend to approach these early efforts as harbingers of historical geology. That is all very well, but it is one of the risks of diachronic history that with the historian's gaze fixed on an ultimate destination, the outcome readily tends to take over and obscure the origins. It seems to me that it would be profitable to consider how much early lithostratrigraphic study seems to have been framed in a spirit of recognizing static dispositional regularities – and with correspondingly less anticipation of organizing these regularities as historical sequences.

To consider this sort of thing – the attitude and outlook of the geologists of the late 18th century who displayed the preoccupation with dispositional regularities I have been insisting upon – it helps, of course, to forget momentarily about the later scientific fruitfulness of one regularity or another. Making an effort at such judicious historical suspension of our own memory of subsequent results serves our appreciation of what these geological ancestors thought they were doing. They cannot really have known which among the range of conceived dispositional regularities to bet on. I get the feeling, reading the literature of their day, that some were betting on all of them more or less indiscriminately.

Conclusion

I have tried here to call attention to a form of geologial thinking in the later part of the 18th century whose prevalence in geological rhetoric I believe has not been sufficiently recognized. A strong current of the era's scientific ethos, emphasizing inductive caution among some geologists who believed that a sound theory of the earth was a worthy ideal, the realization of which was not quite yet in sight, helped to foster pursuit of a rather broad range of descriptive generalizations, frequently in the form of dispositional regularities. Contemporary discussions of these regularities were sometimes carried on with signs of interest in linking them with possible causal underpinnings; quite often, however, those discussions were remarkably free of overtly causal interpretation, which is to say that they tended to take a form that was not outwardly theoretical. Thus they often appeared in more or less austerely descriptive form. It would be a mistake, nonetheless, to ignore the theoretical motivation behind these generalized descriptions: they were sought out in a spirit of hopeful expectation for orderly resolution of the time's growing wealth of geological knowledge into a meaningful synthesis.

XI

Early Geoscience Mapping, 1700–1830

Introduction

Between the middle of the eighteenth century and the first decades of the nineteenth, scientific understanding of the earth went through dramatic changes, one of the chief results being a distinct geological science with its own goals and procedures. It was not totally coincidental that this era witnessed also a revolution in thematic mapping (Robinson, 1982). The newly emerging science of geology came in the early nineteenth century to express its ideas to a remarkable degree through visual representations, irreversibly taking on what Martin Rudwick (1976a) has aptly called a visual language. Maps were of course a major part of this new language, arguably the most crucial part.

In particular, the Geological Map in the restricted sense, seeking to indicate subsurface structure as well as surface distribution of rocks, was the cartographic centerpiece. The strategic primacy of the geological map rested mainly on its function in giving clues to the position of stratigraphic units, position being the key to recognition of temporal sequence. A method for graphic representation of information vital to constructing a framework of succession was naturally embraced as indispensable to a science coming to define its objectives in essentially historical terms. So thoroughly was the stratigraphic map integrated into geological thinking and procedure that some of the literature of geological mapping's history ignores aspects of geoscience mapping not seen as having contributed to its eventual triumph. Readers of this literature may be misled by a seemingly exclusive concern for the stratigraphic map's development, making it appear that practically every cartographic effort worth notice was focussed on achieving the goal of proper structural representation. I do not think all efforts were in fact so directed. I will try to indicate that some pioneers of mapping had different ideas about the sort of role maps could play in the earth sciences' growth. And I hope to suggest more generally that the development of geoscience mapping is most usefully comprehended as more than a set of changes in technique, since technical alterations are tied to adjustments in ideas about the things it may be profitable to represent cartographically.

Rather than try to trace the growth of early geoscience mapping comprehensively in chronological order, I will find it convenient, in this illustrated promenade, to discuss several topics consecutively. I intend first to rehearse briefly some of the generally well-known steps by which geological maps came into being and assumed a large role in early nineteenth-century geology. Then I plan to double back and comment on three related patterns of development: (1) the possible role of mining plans and sections in stimulating conceptions of geological mapping; (2) the effort to create effective two-dimensional depiction of relief; and (3) the development of certain thematic maps differing from geological maps defined strictly, as significant in early geoscience cartography.

The illustrations are all from original printed materials in the History of Science Collections of the University of Oklahoma Libraries. The Oklahoma History of Science Collections contain no special archive of maps as such, but nonetheless include a representative selection of early geoscience maps by virtue of excellent holdings in scientific books and journals. One of the Collections' areas of particular strength is in the earth sciences to the mid-nineteenth century.

Cartographic Emblems

It is instructive to take first a map over 300 years old, one bearing little resemblance to geoscience maps as we conceive them (figure 1). To our eyes Athanasius Kircher's depiction of the Alpine origins of European river systems, in his *Mundus Subterraneus* (1665), may hardly seem a map at all. It is an emblem, a pictorial expression of a theory of the maintenance of springs, and thus rivers, through an organic circulation of waters over and within the earth – a theory that may incidentally remind us of the then still novel circulation of the blood in the human microcosm.

In Kircher's maps the proportion of emblematic content to empirical data is undisguisedly high. In virtually all the other maps we are about to consider the ratio may be said to be reversed; that is, the maps we are mainly interested in purported in some way to portray a relationship among phenomena by a visual replication of those relations, rather than by an ideograph for the theory.

Although it is not my aim here to examine the reasons why Kircher's map is fundamentally different from those prepared not too much later, we should recognize that Kircher was not engaged in an unsuccessful effort to do what scientists of subsequent generations would accomplish. His intentions were different. Unlike them Kircher had a premodern outlook in which a qualitative grasp of the whole counted most. For him, comprehension of natural processes

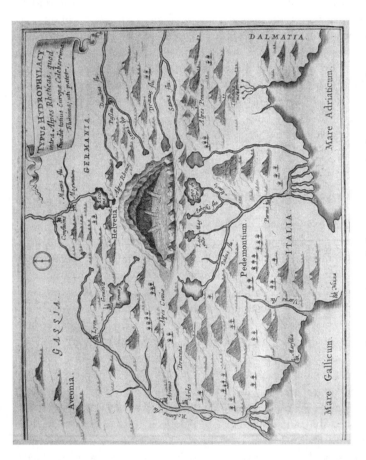

Figure 1. A map revealing Kircher's adherence to a theory of subterranean connections between seas and mountains, to account for continued action of springs (Kircher, 1665, p. 71). This theory tended to give significance to common origins of different rivers within a small district – in this instance the Rhaetian Alps. (Map dimensions: 153 x 196 mm.)

as means of fulfilling discernible ends had as high a conceptual priority as precision in measurement or analytical subdivision of problems into constituent parts would have for his successors (as these were already coming to have for many contemporaries whose intellectual preferences he could not share).

The differences between more recent maps and Kircher's cartographic improvisation may tempt the unwary into the mistaken presumption that modern geoscience maps are atheoretical records of information. Probably few practiced users of such maps are so deceived, although through routine employment a map's status as a theoretical construct may recede from view (Harrison, 1963). For such adepts, the task of explaining to a neophyte how the map works should bring this status back into focus. And it is difficult to consider the historical varieties and changes in conception of geoscience mapping without being aware that different conventions of representation are themselves expressions of theories.

From Mineralogical Maps to Geognosy

Modern notions of geoscience mapping were the natural companions of the new earth science's first stirrings between two and three centuries ago. In a sense they were extensions of the industrial age's nascent mindset. To say this is not to maintain that geology's maturation was a consequence of the Industrial Revolution, nor that the mode of production determines consciousness, only that the interdependence of material and intellectual conditions of existence obtains in earth science as in any human activity and way of thought (Porter, 1973). It happens to be fairly easy to show links between many of the first mineralogical maps and concern about exploitable resources. But, no less significant (if less evident, perhaps because too obvious to our minds), is the network of tacit ideas these early maps reflect – the elevation of dispositional relations among mineral objects above their purposive relations, for example.

The idea of a mineralogical map dates back at least as far as the late seventeenth century; its realization, though, came only in the early 1700's (Eyles, 1972). Naturalists like Luigi Ferdinando Marsigli, in his sumptuously-produced account of the Danube River basin (1726), displayed the locations of certain mineral deposits using traditional chemical and metallurgical symbols. Instances of more elaborate use of mineralogical spot-mapping are seen around mid-century in the work of the French naturalist Jean-Étienne Guettard (figures 2 and 3). Guettard was fairly typical in combining appreciation for the economic merits of knowing where resources lay with the prevailing sensibilities of Enlightenment natural history. The era's mental habits lent themselves to the

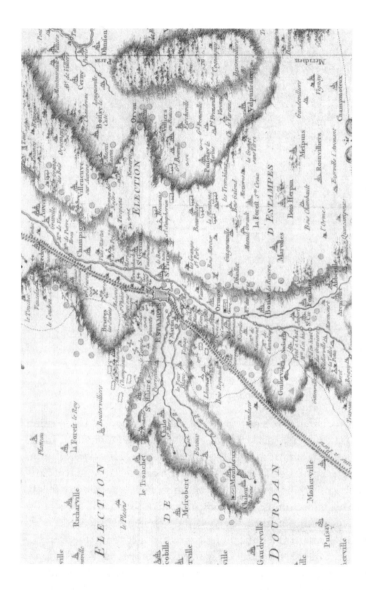

Figure 2. Central portion of a mineralogical map of the country (the *Beauce* region) around Étampes, southwest of Paris, accompanying Guettard's paper on conglomerates (Guettard, 1757, pl. 8, at p. 192: *Carte minéralogique de l'élection d'Estampes*). Sites of mineralogical or lithological interest are indicated by symbols shown in figure 3. In his text (p. 167) Guettard attributed this map to a M. Chardon of Étampes. (Detail shown: about one-third of the map.)

task of describing and classifying the objects furnishing the world, and of finding their mutual relations in the overall order of things.

Figure 3. Legend of the mineralogical map depicted in figure 2.

The best known of Guettard's distributional maps went beyond the locating of scattered points of mineralogical interest, and stipulated continuous "bands" or zones of lithologically similar materials (figures 4 and 5). This extrapolation, indicated on maps prepared by Guettard's geographical colleague in the Paris Academy of Sciences, Philippe Buache, was offered in response to fellow academicians' wishes to see simple graphic expression of Guettard's perception of distributional regularities in the rocks and their contents (Guettard, 1751, p. 363–364). While Guettard's identification of three distinct zones (the "sandy," "marly," and "schistose or metallic," respectively) implied a distributional concept going beyond simple spot-locations, it did not reflect notions of stratigraphy or structure (Rappaport, 1969). Guettard had critics who thought he was rash in extending his broad continuities into regions where reliable

Early Geoscience Mapping, 1700–1830

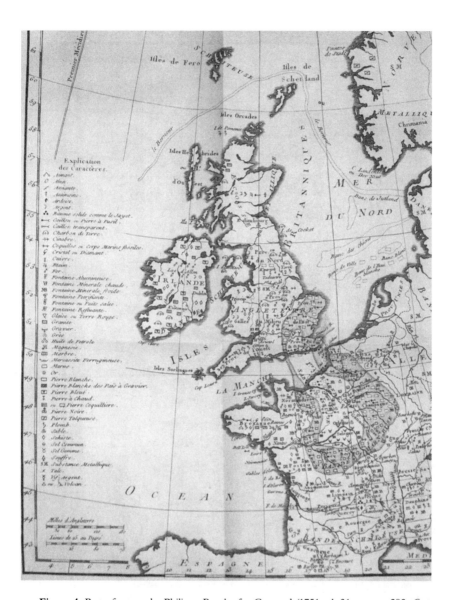

Figure 4. Part of a map by Philippe Buache for Guettard (1751, pl. 31, opp. p. 292: *Carte minéralogique sur la nature du terrein d'une portion de l'Europe*). The darkened area is the *bande marneuse* or marly zone, enclosing the sandy zone, and surrounded by the schisty or metallic zone. (Detail shown: about three-quarters of the map.)

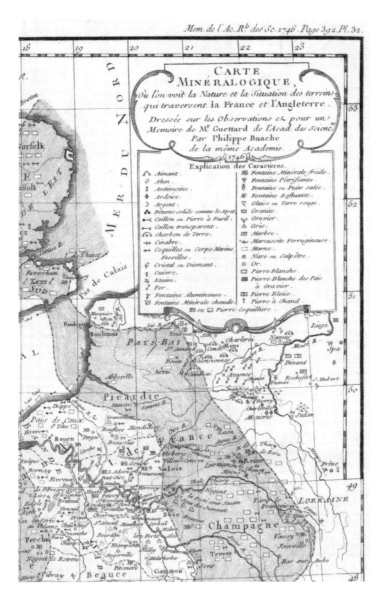

Figure 5. Part of a second map by Buache for the same *mémoire* by Guettard (pl. 32, opp. p. 292: *Carte minéralogique, où l'on voit la nature et la situation des terreins qui traversent la France et l'Angleterre*). Here again the marly zone is shaded. The legend visible here is supplemented by some additional symbols given elsewhere on the map. (Detail shown: about one-third of the map.)

information was scarce, including the Middle East (Guettard, 1755, pl. 8) and North America (Guettard, 1756, pl. 7; Cailleux, 1979).

Meanwhile in Germany, beginning in the 1770's some scientists and engineers linked to the Freiberg mining academy were beginning to produce mineralogical maps that used color to indicate lithological continuity, and that soon showed signs of evoking structural relations. A good example is the petrographic map in Johann Friedrich Wilhelm Charpentier's book on the mineralogical geography of Saxony (1778). This map (figure 6) appeared just a few years after the very first hand-colored geoscience maps, and was one of the first to use boxes for a color key. These are in an arbitrary order, not reflecting a conception of spatial or temporal order. Moreover, the colors are only a general guide to the main rock types; the greater importance of the now conventional spot-symbols is suggested by the fact that four of them are used without corresponding colors, in addition to the eight symbols matched with colors.

The Freiberg school Charpentier represented was soon advocating a more directly spatial sense of rock structure. Abraham Gottlob Werner taught a generation of field naturalists and mining engineers to identify rock types by a combination of lithological character and stratigraphic position. Through Werner geognosy became closely connected with the idea of rock sequences corresponding to temporal succession. While the basic ingredients of such principles had been available to earth science for over a century, it was in Wernerian geognosy that spatial arrangement was first tied to temporal sequence in a coherent system.

Werner advocated mapping as part of his geognostic doctrine, with a view toward exhibiting relations among rocks. His essay on the preparation of maps was appropriated without attribution by his Scottish disciple Robert Jameson (Ospovat, 1971, p. 30). The accompanying schematic map (color plate 1) illustrates many of Werner's cartographic ideas. He wanted colors as nearly as possible to match the corresponding rock units, and he recommended that the bottom junction of each stratum carry an intensified line of color where it is superimposed. Werner also adapted use of arrows for direction of dip, the arrows' lengths varying in accord with the magnitude of the angle.

The Wernerians were those pressing hardest, at the end of the eighteenth century, toward cartographic depiction of structure. But mapping of minerals and soils was developing along roughly similar lines in several countries – not just France, Germany, and Great Britain, but Italy, Sweden, and Russia as well (Eyles, 1972).

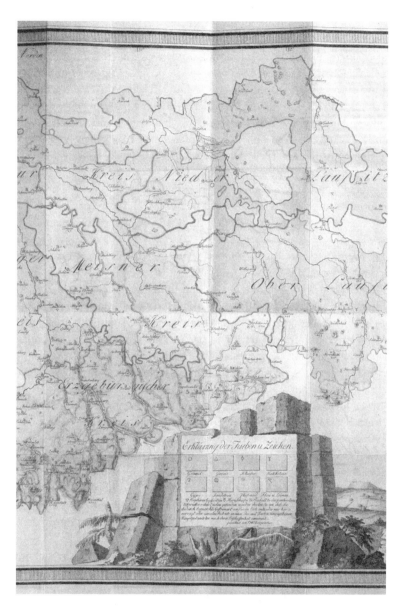

Figure 6. A segment of Charpentier's colored petrographic map (1778, opp. p. xlii: *Petrographische Karte des Churfürstenthums Sachsen und der Incorporirten Lande*). In the legend eight rock types are identified by both color and symbol, four others by symbol only. (Detail shown: about one-third of the map.)

Early Geoscience Mapping, 1700–1830

Stratigraphic Mapping

By steps and degrees, then, rather than in a sudden leap, geological thinking moved toward the conception of mapping strata rather than mineralogy. The epoch-making maps that most clearly accomplished the decisive steps were the work of the French zoologists Georges Cuvier and Alexandre Brongniart, on one hand, and the English civil engineer William Smith on the other.

The Cuvier-Brongniart map (figure 7 and color plate 2) first appeared in 1811 in conjunction with their reconstruction of the fossil succession in the Paris Basin. Ideas about a possible regular relationship between strata and their fossil contents had been floating about for several decades, but with Cuvier and Brongniart the use of fossils as the primary criterion in stratigraphic correlation came into clear focus (Rudwick, 1976b, ch. 3).

Their map is about 61 cm. high and 71 cm. wide, on a scale of 1:200,000. It covers a region from Beauvais and Soissons in the north to beyond Étampes and Fontainebleau in the south, from La Roche-Guyon in the west to Montmirail and Nogent-sur-Seine in the east. The map is plainly designed to assist the text's structurally-informed discussion of the region's rocks. A colored key (figure 7) indicates in sequential order the several formations deposited in the Paris Basin, above the chalk and the "plastic clay": successive limestones, followed by gypsum with freshwater fossils grading up into gypsum with marine fossils, then a series of sandstones and some more freshwater formations near the top. (From the top down, the key follows the formations' temporal order, and would require inversion to suit the modern custom matching the column with spatial order.) Several sections published in the same monograph aid the map's structural interpretation.

In showing the continuity of strata over several dozens of miles, despite horizontal grading of one lithology into another, Cuvier and Brongniart established that fossils could serve as a diagnostic criterion superior to lithology. Their work also showed the inadequacy of Wernerian rock classification, since all these rocks in the Paris Basin lay on top of the chalk, the upper boundary of Werner's secondary. The earth's history must, therefore, be longer and more complicated than was generally thought.

Also, the grading of marine and freshwater sediments into one another bespoke important oscillatory changes in local environment without radical disturbances. Much interested in problems of the origins of each stratum, Cuvier and Brongniart saw their mapping technique as serving historical interpretation of each region where it might be used.

William Smith's map (figure 8 and color plate 3) was of a far larger area, and was more complicated to produce. His *Delineation of the Strata of England and*

XI

Figure 7. An eastern part of the geognostic map of the Paris Basin, published by Cuvier and Brongniart (1811): *Carte géognostique des environs de Paris…* 1810. The map was republished several times; this photograph comes from Cuvier (1812). This portion of the Paris Basin includes the Marne Valley from above Château-Thierry (right) to the eastern outskirts of Paris (left). The boxes on the right are colored keys to the main formations, in inverted spatial order. In this black-and-white photograph, the dark gypsum formations (blue) and shaded silicious limestones (purple, lower left) are the only tones distinguishable. (Detail shown: nearly one-quarter of map.)

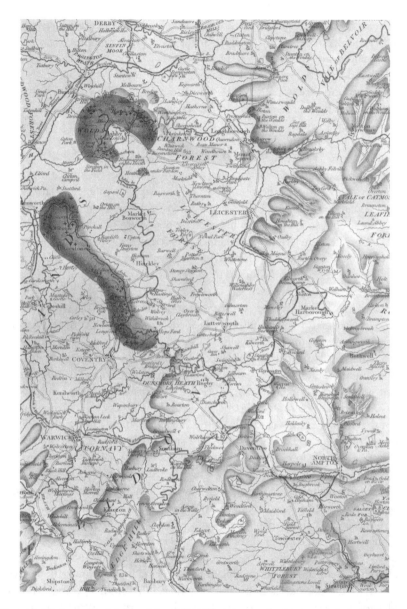

Figure 8. The lower-right portion of sheet VIII of Smith's map, showing part of the Midlands, from Warwick and Northampton north to Leicester and Derby. Formations included are, from the southeast: Forest Marble and Clay (blue), overlying the Oolites (yellow), in turn above the Blue Marl (blue, with a number of prominent lobes at the lower boundary), over the Blue Lias, Red Marl and gypsum, and coal. (Detail shown: about one-fifth of the map.)

Wales finally came out in 1815, years after contemporary geologists in England were already using Smith's methods. It was in fifteen sheets, at five miles to the inch. If mounted together these sheets made an imposing map well over eight feet high and nearly three-quarters as wide.

The hand-coloring was shaded so that the intense tones appear at the bottom of each stratum. Accompanying the sheets was a color key in proper (i.e., spatial) stratigraphic order. Still another key appeared in Smith's explanatory *Memoir* (1815b), and in due course a number of colored sections were produced, further aiding a grasp of structure.

Smith's big project had begun in the 1790's. Recent research indicates that his basic work relied originally on topographic and lithological identification, and that Smith's use of fossil indicators came into his mapping only rather late in the game (Laudan, 1974, 1976). Smith believed the strata were formed pretty much as they are found, so it was left to his successors to work out the consequences of the opposite conviction, that there had been serious dislocations. Smith practically made it a matter of principle to deal with positions of strata without concern for their origins, in contrast to the historical orientation of Cuvier and Brongniart.

Knowledge of Smith's work was passed about within the English geological community well before his map was published. Working independently, Cuvier and Brongniart were in print several years earlier than Smith. They exploited the principles of fossil correlation to notable advantage, and their approach lent itself to interpretation of rock structures in terms of historical conditions of formation, deformation, and removal. This being the case, it is not surprising that some British geologists who began to make effective use of geological mapping were led at least as much by the French example as by Smith's. This was true for Thomas Webster, whose 1814 study of the Isle of Wight and the neighboring coast raised English geology to a new level of visual expression (Englefield, 1816). On the other hand, English leadership was acknowledged in the descriptive techniques of mapping, drawing French and other Continental geologists to England for cartographic tutelage.

Among them, Cuvier, Brongniart, and Smith brought to fruition a set of possibilities that had been gradually assembled over several decades – not least because of Werner and his geognostic school. These geologists provided the conditions for turning stratigraphic mapping into a burgeoning geological industry. In the second decade of the nineteenth century geology moved rapidly toward becoming a thoroughly graphic enterprise. British geologists such as Webster, John Farey, George Greenough, and Henry De la Beche, and Continental ones like Leopold von Buch, J.B.J. d'Omalius d'Halloy, and Ami Boué began to treat mapping as a guiding feature of geological discourse.

Early Geoscience Mapping, 1700–1830

The restoration of peace in 1815 reduced impediments to scientists' travel throughout Europe, and stimulated international comparisons of strata. The preparation of significant geological studies by investigators foreign to the locale is symbolized by the reproduction here of Boué's map of Scotland (figure 9) and De la Beche's rendering of the geology of part of Normandy (figure 10).

Among the most important results of all this activity was the emerging conviction of a definite and universal order to the strata, something that however strongly intimated by local study required general empirical confirmation. Comparisons were sought with remoter parts of the world, such as North America, based in part on information digested in maps of which William Maclure's (1809) was an early example (Schneer, 1981).

Stratigraphy became, then, the hinge upon which much of the new geological science turned in the first decades of the nineteenth century. Sophisticated stratigraphic maps in turn made possible another kind of map – the paleogeographic map. One noteworthy instance (color plate 4) is Charles Lyell's endeavor to "enable the reader to perceive at a glance the great extent of change in the physical geography of Europe which can be proved to have taken place since some of the older tertiary strata were deposited" (1833, p. 312). Basing his map largely on Boué's general European geological map, Lyell observed that his graphic method could readily show which districts had been converted from sea to land at some time during the period at issue, but not vice-versa. Also in the early 1830's, Léonce Élie de Beaumont produced a sketch-map reconstructing some of the land-sea relations in the early tertiary (figure 11).

Mining Projections

I turn now to a particular way the techniques of mining engineers may have helped eighteenth-century naturalists extend their imagination beneath the earth's surface. Since the Renaissance, geometric methods had been commonly used in mining operations. But proto-geological theorists did not for the most part see those local methods as relevant to their interests, until scientists began to act more fully in accord with a rising ideology favoring a convergence of practical and abstract knowledge. Around the middle of the eighteenth century specialists in mining technology and philosopher-naturalists seem to have gotten more closely in touch with one another; the greater profit may well have been to the theoreticians.

Mine plans abounded in the publications of Enlightenment rationalizers of the mining arts. Typical are the plans of mineral veins of the Joachimsthal

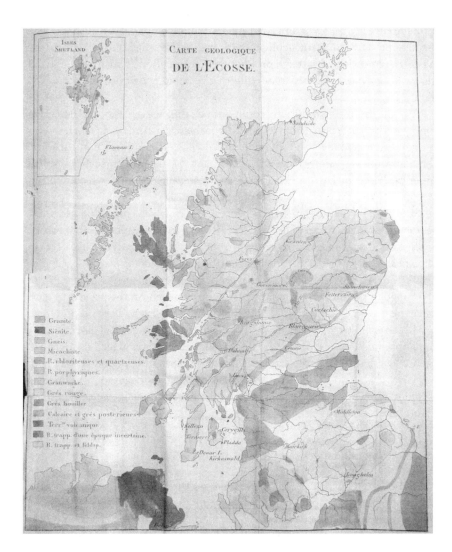

Figure 9. Boué's general geological map of Scotland (1820). The thirteen rock units are hand-colored. An accompanying, uncolored map provided some indication of relief. Boué said (p. 503) that more detailed maps, unlike this one, should show orientation and order of superposition of the rock masses. (Map dimensions: 238 x 195 mm.)

Figure 10. De la Beche's *Geological Map of Portions of the Departments of the Seine Inferieure, the Eure, Calvados and La Manche* (1824, pl. 11). Relief is shown by hachures. Eight colored sections supplement De la Beche's successional interpretation of the fifteen units indicated in spatial order in the colored key. (Map dimensions: 23.5 x 36.5 cm.)

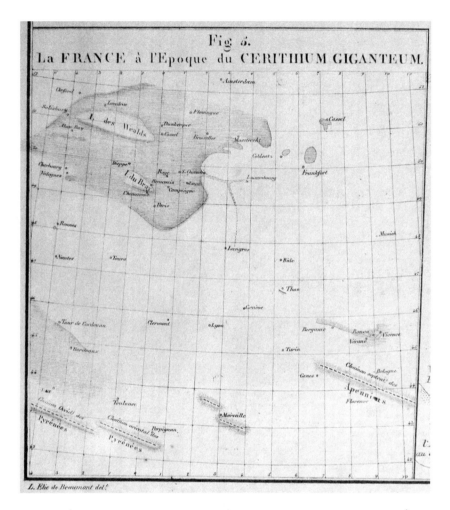

Figure 11. Élie de Beaumont's paleogeographic map of western Europe in the early tertiary (Élie de Beaumont, 1833, pl. VII, fig. 5: *La France à l'époque du Ceritium giganteum*). (Map dimensions: 170 x 117 mm. Part of a fold-out plate engraved by Ambroise Tardieu.) Élie de Beaumont first used such a map in his geology course at the École des Mines (1833, p. 121).

Early Geoscience Mapping, 1700–1830   19

mining district in the works of travelers like Ignaz von Born (1777) and Johann Jacob Ferber (1774b). It is unclear just how much these two-dimensional plans in themselves fostered interest in "subterranean geometry" outside of technological circles. But quite intriguing, from the standpoint of our present subject, is the appearance of plans matched with sections, one a projection of the other, to convey a clearer sense of underground position. How old this practice was I have not found out, but it evidently became fairly common in the second half of the eighteenth century. The French mining engineer Antoine-Gabriel Jars, in his posthumous descriptions of mines he investigated during the 1750's and 1760's, provided such matched profiles and plans, notably those shown in figures 12 and 13. Similarly matched sections and plans can be found in naturalists' writings giving particulars about mines, including some works of Ferber's (1774a, pl. 1).

It is difficult to escape the suspicion that these exercises in three-dimensional representation, becoming habitual parts of literature attended to by scholars more than by craftsmen, in turn had an effect on the geological mentality, which up to about this time had generally kept to its occupation with two-dimensional distribution. In this way technical practices may have helped early geological scientists conceive problems about the earth's crust "in depth."

Representation of Relief

Another interesting aspect of geoscience mapping's development during the eighteenth and early nineteenth centuries, one we have seen something of in the maps reviewed already, was the problem of how to display surface relief. This question had greater importance to geological thought than might be suspected, because before around 1800 no very consistent and clear distinction was made between structure and topography.

Centuries-long custom of pictorial representation was the background out of which eighteenth-century topographers worked. Mountains as mole-hills (figure 1) were standard conventions, and the seeming carelessness of such renderings of mountains lingered on in some maps clear into the nineteenth century.

From throughout the eighteenth century we can find many examples of maps that worked toward a higher standard of exactness for the forms of certain features like rivers and lakes than for hills, mountains, and valleys. Mountains were not yet much comprehended as individuals, but were commonly regarded as a type with little interest. Before mappers would portray mountains as unique objects, with distinct features, a change in attitude would be required (Broc, 1969). Although Johann Jacob Scheuchzer (1723) did exhibit some topographic

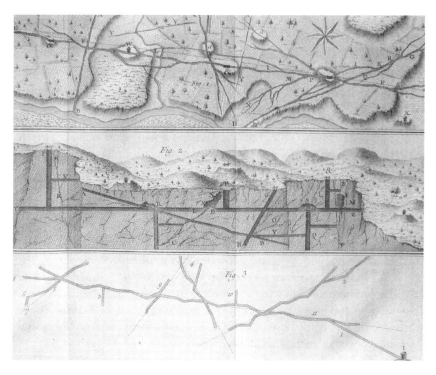

Figure 12. Overhead and side views of veins and a mine in Jars (1780, pl. 1). Figure 1 (top) is a plan of "veins of different orientations and inclinations." Figure 2 (middle) is a "section or profile of a mine." Figure 3 (bottom) is a horizontal plan of the mine works profiled in figure 2 above. The figures are used in Jars' opening section on theoretical and practical subterranean geometry (p. 1–54). (Plate dimensions: 205 x 248 mm.)

Figure 13. Plan and corresponding section of one of the Kongsberg silver mines in Norway, visited by Jars in 1767 (1780, pl. 9). (Plate dimensions: 314 x 213 mm.)

interest in differentiating particular summits, his cartographic rendering of peaks hardly got beyond a style of endless repetition (figure 14).

Semipictorial evocation of relief began in certain cases to verge toward a slightly abstracted technique of shading. This is apparent in Marsigli's map showing the Danube headwaters (figure 15). A transitional state between a bird's-eye view and a shading convention is even more evident in some parts of Marsigli's other maps in the same work. Subtle hachuring was one of two important new techniques introduced at roughly the same time. Hachuring caught on rapidly and reached high levels of refinement by the end of the eighteenth century. The other notable invention was the contour line, but it met with little acceptance until the start of the following century.

The earliest published maps using contour lines were designed to indicate depth in water. Isobathic lines appear on several maps of the 1720's and 1730's. Among the most famous examples from around mid-century are Philippe Buache's English Channel maps (figure 16). But neither for depths nor altitudes did contour lines thrive, for many decades. Hachures, on the other hand, drew the sustained attention of draughtsmen hoping to capture the relief of land.

Early hachuring commonly resulted in an exaggerated flatness of appearance for the greater part of the area portrayed, and an equally exaggerated steepness of declivities. Valleys look as if they were incised into level uplands (figures 2 and 10). An improvement could be made, of course, by multiplying the levels. And in the hands of skilled artists remarkable effects were achieved. A map in Peter Simon Pallas' account of journeys in southern Russia during the 1790's, for example, begins to create an illusion of some continuity from one level to another (figure 17).

A somewhat similar effect is seen in an early nineteenth-century map in André-Jean-Marie Brochant de Villiers' report (1817) on a locality in the St. Gotthard district of the Alps (figure 18). Here the relief comes across by hachured representation of levels, almost as if through combination of contour lines with shading. Conceivably this was even deliberate.

Advances in techniques for portraying relief roughly paralleled the rise of geological sensitivity to the importance of perceiving rock structure in three dimensions. As we have seen, sections were a supplementary device resorted to with increasing regularity.

Geoscience Empiricism and Cartography

It seems to me worth mentioning that the map we just noticed from Pallas' travels (figure 17), although a striking product of the cartographic art, was by no means characteristic of Pallas' work. None of the few maps Pallas published

Figure 14. Scheuchzer's pictorial charting of the Inn's rise in the upper Engadine, and surrounding mountains (1723, fig. 11, opp. p. 450). North is toward the bottom. Lower left: Pontresina, St. Moritz. Upper right: Maloja, Lake of Siglio. (Plate dimensions: 280 x 249 mm.)

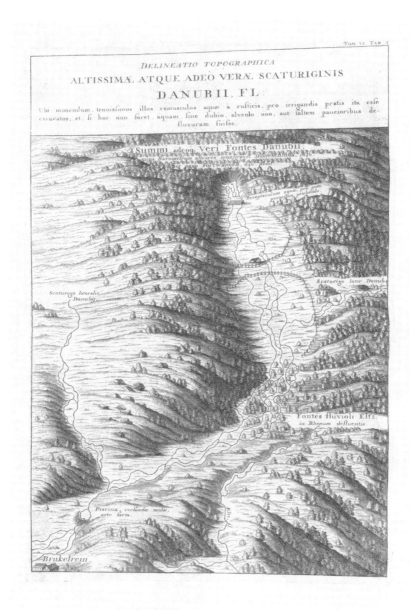

Figure 15. The source of the Danube, in Marsigli (1744, pl. 1, opp. first p. of text). (Plate dimensions: 374 x 256 mm.)

XI

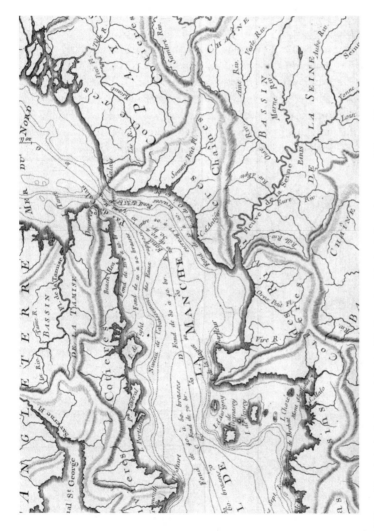

Figure 16. Part of Buache's physical map of the English Channel (1756, pl. 14: *Carte physique et profil du canal de la Manche*). Contour intervals are 10 *brasses* or fathoms. The straight lines mark the location of the accompanying "profile" (not shown). Buache noted that this map was initially presented in manuscript form to the Académie des Sciences in 1737. Note the *chaînes côtières* dividing river drainages, and one chain's submarine crossing of the Channel to England (cf. figure 24). (Detail shown: about two-thirds of the map.)

is tied in any remarkable way to his major points of theoretical interest. A more general remark that I think can be made about maps in eighteenth-century geological writings is that their presence or absence is no reliable guide to the author's reputation as an observer. Déodat Dolomieu and Horace-Bénédict de Saussure were, like Pallas, among the most highly-esteemed geological empiricists of the later eighteenth century, and their work also made slight strategic use of maps. The few maps found in Dolomieu's writings are singularly empty of any valuable geological thought or information. Saussure's map of the Mont Blanc district (figure 19) is not without interest and attraction, but considering the body of information contained in his *Travels in the Alps* Saussure communicated little of scientific moment through maps. (Saussure was also a pioneer alpinist keenly alive to the individuality of mountains, something one might not guess from this map, so little evolved from Scheuchzer's type seen in figure 14.)

Figure 17. A topographic map of a district of the northern Caucasus, from Pallas (1799, pl. 16). The area includes several miles of the valley of the Narzan, a tributary of the Podkuma [Podkumok] River (right). (Map dimensions: 210 x 414 mm. Engraved by Gründling.)

But I would like to take note of two classic instances of geological observation reported through maps in the eighteenth century. One is Marsigli's early-eighteenth-century representation of a "hidden coast" off the French Mediterranean shore (figure 20). Soundings showed what seemed to be a shallow and extensive submarine platform, terminating rather abruptly. This had apparent implications for the true delineation of the continents.

The other case is the survey of the Auvergne volcanic district made under Nicolas Desmarest's direction beginning in the 1760's. His first published map

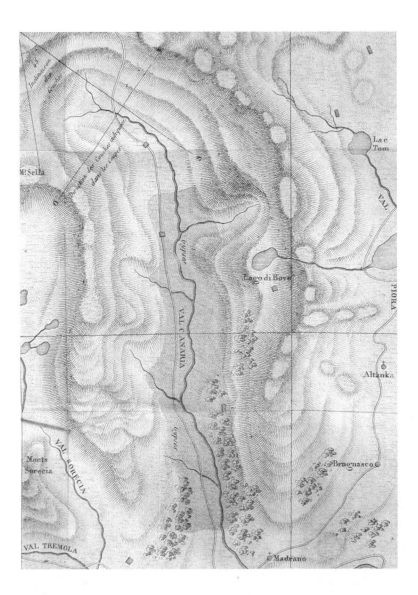

Figure 18. Map of the Val Canaria in the Alps, in Ticino, from Brochant de Villiers (1817, pl. 6, after p. 384). Accompanied by a pair of profiles, not shown. (Map dimensions: 134 x 210 mm.)

Figure 19. A central portion of the "map of the part of the Alps neighboring Mont Blanc," in Saussure (1786, following title page). The glaciers of the Mont Blanc massif are shown draped over the terrain like a bumpy sea. (Detail shown: about one-third of the map.)

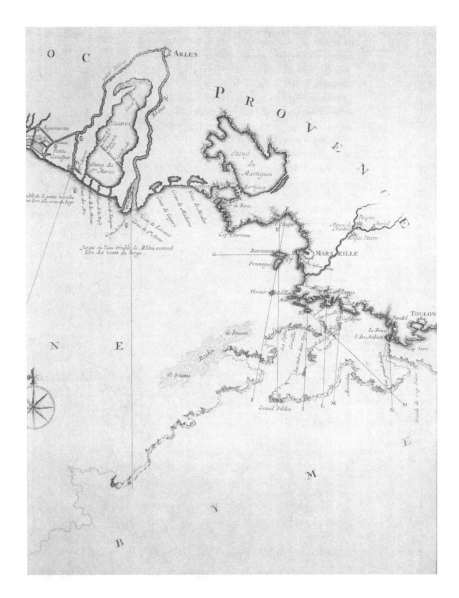

Figure 20. Part of the Golfe du Lion, from Marsigli (1725, pl. 1, between p. 2 and p. 3: *Carte du Golfe de [sic] Lion entre le Cap Sisie en Provence et le Cap de Quiers en Roussillon*). Off the Mediterranean coast near Marseille, the well-defined lines mark the location of a steep edge to the "hidden coast" at sixty to seventy fathoms of depth, established by soundings. The finer line (lower left) represents the shelf margin's more uncertain location, from scant evidence. (Detail shown: nearly all the eastern half of the map.)

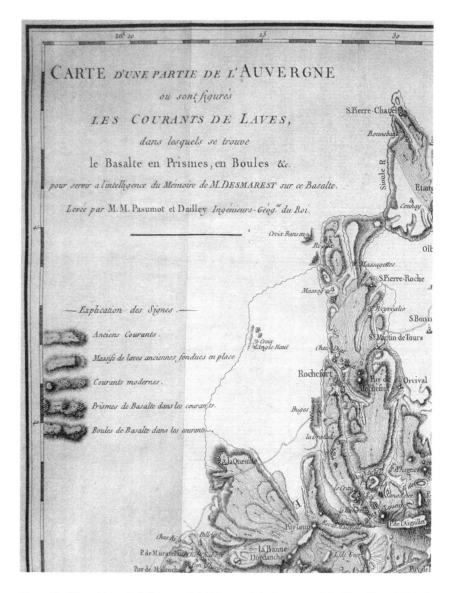

Figure 21. Title and legend of Desmarest's initial map of the lavas around the Mont-Dore (1774, pl. 15). Recent lavas are distinguished from old ones. Also distinguished are old lavas considered to have been erupted in place. This was significant because certain of the older lavas now found separate from the main lava masses were believed to have been cut off by the accumulated effects of erosion. (Detail shown: about one-fifth of the map.)

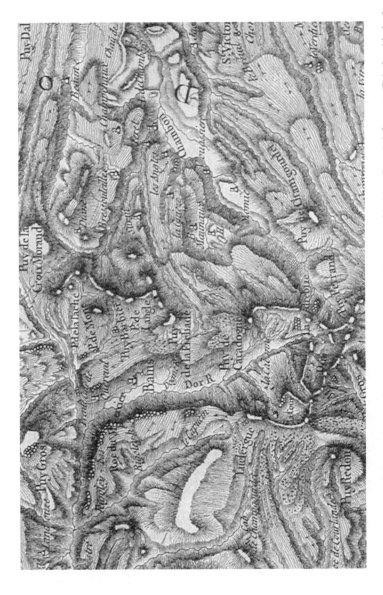

Figure 22. A detail from the southern part of Desmarest's Auvergne map. The town of le Mont-Dore is left center ('Bains'), with the Lac Chambon on the right. Arrows show the directions of lava flows. Modern lava flows cover old ones. Exposed basalt prisms are marked on the margins of some flows. Dissection of old lava flows by stream action is seen in several places. (Detail shown: a little under one-twelfth of the map.)

(figures 21 and 22), presented to the Paris Academy in 1771, shows results of meticulous surveys of the positions of various flows, done in collaboration with two royal geographical engineers. Desmarest and his co-workers succeeded to a degree perhaps unmatched throughout the century in capturing visually the extent of volcanic flows of different ages. Later cartographic results of Desmarest's continuing work were still coming out over sixty years after he began, and were still commanding respect at that time.

A type of field map I have not encountered from before the late eighteenth century is the highly detailed and local chart or plan confined to areas measurable in, say, tens of yards. Specimens I know about include sketches in Pierre-Bernard Palassou's study of the Pyrenees (1784). The map in figure 23 shows rocks at stream level, identified as argillaceous schists with singular variations of stratigraphic orientation. Quite similar large-scale maps done by John Clerk of Eldin in the 1780's were recovered recently with the drawings for James Hutton's *Theory of the Earth* (Craig, 1978).

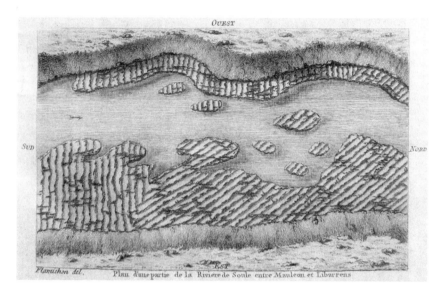

Figure 23. Map of rocks exposed along a stream bed in the western Pyrenees (Palassou, 1784, pl. 2, no. 2, opp. p. 38). The villages identified in the caption are now named Mauléon-Licharre, and Libarrenx; the stream here called the Soule is now known as the Saison or Gave. (Map dimensions: 96 x 153 mm. No scale given.)

Theories of the Earth

The final group of illustrations involves maps governed by theoretical concerns different from the stratigraphic problems that tended to dominate as the geological map came of age.

The French geographer Philippe Buache, whose work we have seen in connection with the early use of isobathic lines (figure 16) and with Guettard's efforts to display the distribution of mineral matter (figures 4 and 5), was perhaps the eighteenth century's most determined advocate of the theory of a global "skeleton" or "scaffolding." Possibly this idea derived from a Renaissance vision of an animated earth. In the eighteenth century, however, it flourished for a time in a mechanism-dominated intellectual climate where the basic framework of geological thinking was the "theory of the earth." These theories reflected the presumption that scientific understanding of the earth should be law-like (preferably expressed in mechanical terms), global in scope, and comprehensive in accounting for whatever changes the earth has undergone. Far from directing attention principally to details of local rock disposition, the tradition of theories of the earth tended to encourage breadth of vision.

For Buache, the existence of a pattern of continuous, interconnecting mountain ranges, manifesting some geometric order, was a basic organizing principle of earth science (figure 24; also figure 16). Much of the credence this idea momentarily received was lost later in the century, as geographical information not amenable to the theory gathered and as simple geometric laws for mountain-chain organization failed to reveal themselves. Nonetheless, the sort of theoretical impulse this exemplified – an ideal of generalized observational law as the proper basis for a theory of the earth – was a tenacious one.

Geometric laws or relations which must simply be described, even though not for the present understood, were also on the mind of the great naturalist Georges-Louis Leclerc, Count Buffon, when he supplemented his 1749 *Theory of the Earth* with maps supposedly showing order in the orientation of the continental land masses (figure 25). Both old and new worlds, the earth's two great continental agglomerations, were determined by Buffon to be disposed along long axes of maximum northern and southern extent. These axes were inclined about thirty degrees to the equator, but in opposite directions. Buffon reckoned that the geographic centers of gravity, as it were, of both these land masses ranged in symmetrical positions across the equator. This information, Buffon believed, could not but be fundamental in ways not yet apparent to comprehension of the earth as a physical object with an ordered, non-accidental history.

XI

Figure 24. Part of Buache's "physical planisphere," centered on the North Pole (Buache, 1756, pl. 13, opp. p. 416). Many of Buache's maps display hi insistence on submarine continuity of mountain-chains, and the function of these as a sort of global skeleton. Cf. figure 16. (Detail shown: about one-half of the map.)

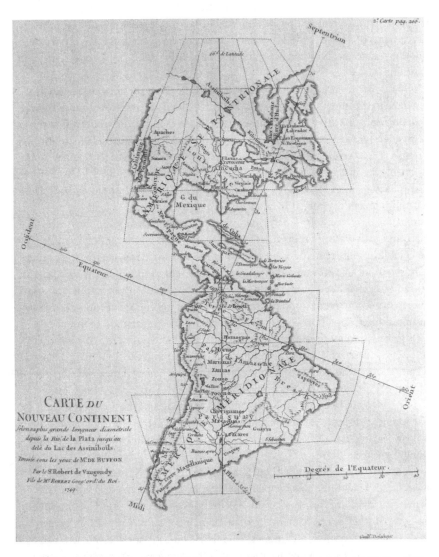

Figure 25. Buffon's map of the new world "according to its greatest diametral length" (Buffon, 1749, opp. p. 206). According to Buffon, nearly equal surface areas exist on either side of the vertical axis inclined about 30 degrees to the meridian. A comparable relation is found for a similar line cutting the old world (Europe, Asia, Africa). The two symmetrical axes mark land extending to the same maximum northern and southern latitudes, but in the old world the land's bulk is centered north of the equator, while the opposite holds in the new. These and other relations, including a notable parallelism of coasts across the Atlantic, were taken as significant of deeper principles not yet discovered. (Map dimensions: 217 x 178 mm. Drawn by Robert de Vaugondy, engraved by Guillaume Delahaye.)

Although we may be tempted to think Buffon's notions amusing or fantastic, it is important to realize that he and Buache were not part of some reactionary movement, but participated in a natural outgrowth, in many ways progressive, of the Cartesian-Newtonian natural philosophy with its mechanism and quantification.

Some of the same patterns of thought involved in theories of the earth are reflected in the ingenious maps of Edmond Halley, half a century before Buache and Buffon. One of the best known is Halley's map of the trade winds thirty degrees either side of the equator (Halley, 1688). With that map Halley identified a regularity in terrestrial phenomena, and was also able to assign it a provisional mechanical cause. He did something much the same with an equally famous map charting lines of equal compass variation. The seemingly so promising isarithmic maps conceived by Halley had hardly any direct emulators, however, until Alexander von Humboldt a century later. Besides inventing the isothermal map, and articulating other natural regularities such as the biological affinities detectable within climatic zones, Humboldt was an enthusiast for geometric patterns in mountain-range constitution and orientation (Cannon, 1978, ch. 3). So far as I have been able to tell, however, Humboldt's "loxodromism" – his belief in geophysically fundamental laws in the geometry of orographic and stratigraphic disposition – was never expressed in the form of maps.

Ideas akin to loxodromism do show up in the maps of some of Humboldt's close predecessors and contemporaries. In mapping the Pyrenees, Palassou (1784) thought he could discern continuous stretches of limestone and schist lying more or less regularly along lines running 73 degrees west (figure 26). Not many years later, in an adaptation of Palassou's work, Louis Ramond de Carbonnières found very similar lines of orientation in the same region, although he now located trends of primary, transition, and fossiliferous secondary rocks along these axes (figure 27).

A persistent tradition of such geometric interests, sometimes but not always manifested in maps, helps make intelligible the maps of Léonce Élie de Beaumont, in some respects an intellectual descendant of the Enlightenment theories of the earth. A member of a French mission, early in his career, to study English mapping techniques, Élie de Beaumont applied himself to geodynamical studies seeking to combine investigations of mountain origins with historical reconstruction of past geological events. A map with some of his initial tectonic work (1829–1830) displays parallelism in contemporary mountain chains as a key to piecing together the earth's crustal history through its episodes of mountain building (figure 28).

It has been argued recently (Greene, 1982) that the course of change in nineteenth-century geological thought followed tectonic-episodic pathways

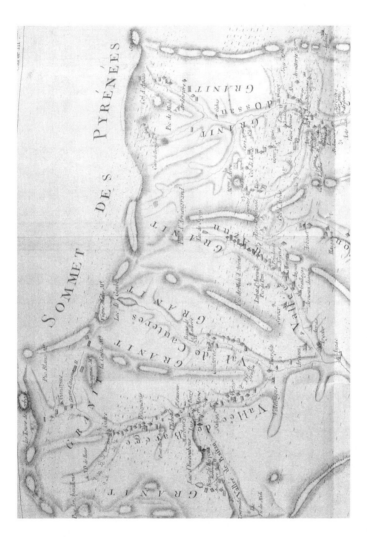

Figure 26. Part of a map of the French side of the Pyrenees, from Palassou (1784, carte 3: *Partie des Montagnes qui dominent les Vallées d'Ossau, d'Asson, d'Azun, de Cauterès de Lavedan et de Barège*). North is to the bottom. The map covers an area south and west of Lourdes. Faintly visible dotted and dashed lines running east-southeast to west-northwest represent the direction of alternating beds of limestone and schist, indicated by rectangular and diamond symbols, respectively. The stretches of beds are interrupted here and there by masses of granite. Rounded stones cover broad areas of the foothills and lower river valleys. (Detail shown: about one-half of the map.)

Figure 27. French side of the high Pyrenees, from Ramond de Carbonnières (1801, pl. 5). North is to the bottom. The valleys of Barèges and Cauterets are near the center. The dotted lines marking the "chaînon collatéral" on either side of the main axis indicate the mean locations where these formations begin, moving outward. But the more far-removed dotted lines show the innermost limits where continuous fossil-bearing beds are found. (Map dimensions: 161 x 229 mm.)

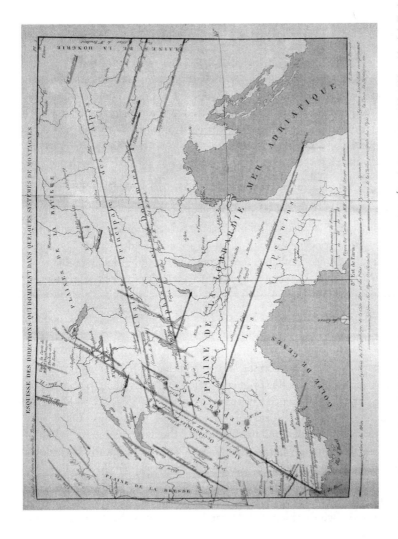

Figure 28. "Sketch of the directions which dominate in some mountain systems," from Élie de Beaumont, 1829-1830, v. 19, pl. 2. Each of the six distinct systems of mountain ranges (marked in different colors) exhibits parallelism. In later development of this work Élie de Beaumont enlarged the number of systems. (Map dimensions: 170 x 246 mm.)

like Élie de Beaumont's rather more than has ordinarily been thought, and stratigraphic-gradualist routes rather less. It may also be that a "loxodromic" element in early geoscience mapping, largely unattended to in the historical literature, was more significant than has been realized.

Conclusion

It is almost impossible to review the changing place of maps in the earth sciences during the century or so before the 1830's without being convinced that they mark a great change in assumed forms of geoscience communication. At the start of the eighteenth century scientific discourse about the earth used visual representation in general, and maps in particular, in little more than ancillary ways. By around 1830, geological communication was reaching a point where it could hardly have proceeded without maps of various kinds. Geoscience had generated a new visual literacy.

While maps became imperative elements in the language of the earth sciences, their development evidently followed more than a single line. This is in keeping with the theoretical pluralism that characterized geology for many years, well beyond its establishment as a recognized science. The most promising approach we can take to further enlarge our understanding of geoscience maps in history remains to examine them "as mirrors of geological theory" (Rappaport, 1964, p. 74).

Acknowledgments

I wish to thank Michelle R. Vosgier and Jun Fudano for their work on the photographs reproduced in the original version of this article, published in the *Proceedings* of the Geoscience Information Society (1985). Kiyoon Kim and Michael N. Keas provided research assistance. The University of Oklahoma Faculty Research Fund supported preparation of the illustrations.

[2007: Digital technology has made it possible to present improved versions of the illustrations in this volume. I thank Melissa Rickman, Mark Hopkins, and Julia Daine, and particularly Kerry Magruder, for their painstaking work on these illustrations. Grants from the University of Oklahoma's College of Arts and Sciences and Vice President for Research make it possible to include four figures in color.]

A Note on Sources

No effort has been made to provide here a comprehensive bibliography of the history of early geoscience mapping. To anyone wanting to explore the subject further I recommend Rudwick (1976a) as an initial reading. Those familiar with Rudwick's brilliant article will recognize my indebtedness to it. References found there, in recent works listed below such as Robinson (1982), and in the bibliographies to entries in the *Dictionary of Scientific Biography* (1970–1980), will lead the way into the literature. See also the thorough book on Irish mappers by Herries Davies (1983). I was unable to consult Dudich (1984).

In addition to my reliance on Rudwick's work, I must mention my equally heavy obligation to Rachel Laudan, who generously lent me a copy of her excellent doctoral dissertation (1974). Her study, on which I depended a good deal, is a fine example of historical scholarship placing geological mapping's evolution within its scientific and cultural context. This broader dimension is lacking in many older and more narrowly-conceived investigations of geoscience mapping's past, such as North (1928) and Ireland (1943), which nonetheless remain worth consulting.

REFERENCES

Born, Ignaz von, 1777, *Travels through the Bannat of Temeswar, Transylvania, and Hungary, in the year 1770*, trans. by Raspe, R.E., London, J. Miller for G. Kearsley, 320 p.

Boué, Ami, 1820, *Essai géologique sur l'Écosse*, Paris, Mme Vve Courcier, 519 p.

Broc, Numa, 1969, *Les montagnes vues par les géographes et les naturalistes de langue française au XVIIIe siècle*, Paris, Bibliothèque Nationale, 298 p.

Brochant de Villiers, A.J.F.M., 1817, Observations sur les terrains de gypse ancien qui se rencontrent dans les Alpes, *Annales des Mines*, v. 2, p. 257–300.

Buache, Philippe, 1756, Essai de géographie physique, où l'on propose des vûes générales sur l'espèce de charpente du globe, *Mémoires de l'Académie Royale des Sciences*, Paris, v. for 1752, p. 399–416.

Buffon, G.L. Leclerc, Comte de, 1749, *Histoire naturelle, générale et particulière*, v. 1, Paris, De l'Imprimerie Royale, [iv], 612 p.

Cailleux, André, 1979, The geological map of North America (1752) of J.-E. Guettard, *in* Schneer, Cecil J. (editor), *Two hundred years of geology in America*, Hanover, New Hampshire, University Press of New England, p. 43–52.

Cannon, Susan Faye, 1978, *Science in culture: the early Victorian period*, New York, Dawson and Science History Publications, 296 p.

Charpentier, Johann F.W., 1778, *Mineralogische Geographie der Chursächsischen Lande*, Leipzig, Siegfried Lebrecht Crusius, 432 p.

Craig, Gordon Y. (editor), 1978, *James Hutton's Theory of the Earth: the lost drawings*, Edinburgh, Scottish Academic Press, 67 p.

Cuvier, Georges, 1812, *Recherches sur les ossemens fossiles de quadrupèdes*, v. 1, Paris, Deterville, vi, 120; 20; viii, 278; 232 p.

Cuvier, Georges, and Brongniart, Alexandre, 1811, Essai sur la géographie minéralogique des environs de Paris, *Mémoires de la Classe des Sciences Mathématiques et Physiques de l'Institut Impérial de France*, Paris, v. for 1810, pt. 1, p. 1–278.

De la Beche, Henry T., 1824, On the geology of the coast of France, and of the inland country adjoining; from Fécamp, Department de la Seine Inférieure, to St. Vaast, Department de la Manche, *Transactions of the Geological Society*, London, ser. 2, v. 1, p. 73–89.

Desmarest, Nicolas, 1774, Mémoire sur l'origine & la nature du basalte à grandes colonnes polygones, déterminées par l'histoire naturelle de cette pierre, observée on Auvergne, *Mémoires de l'Académie Royale des Sciences*, Paris, v. for 1771, p. 705–775.

Dudich, E. (editor), 1984, *Contributions to the history of geologic mapping: proceedings of the Xth INHIGEO symposium*, Budapest, Akademiai Kiado, 442 p.

Élie de Beaumont, Léonce, 1829–1830, Recherches sur quelques-unes des révolutions de la surface du globe, *Annales des Sciences Naturelles*, v. 18, p. 5–25, 284–416; v. 19, p. 5–99, 177–240.

Élie de Beaumont, Léonce, 1833, Observations sur l'étendue du système tertiaire inférieur dans le nord de la France, et sur les dépots de lignite qui s'y trouvent, *Mémoires de la Société Géologique de France*, v. 1, pt. 1, p. 107–121.

Englefield, Sir Henry C., 1816, *A description of the principal beauties, antiquities, and geological phoenomena, of the Isle of Wight, with additional observations of the strata of the island, and their continuation in the adjacent parts of Dorsetshire, by Thomas Webster*, London, William Bulmer for Payne and Foss, 238 p.

Eyles, Victor A., 1972, Mineralogical maps as forerunners of modern geological maps, *The Cartographic Journal*, v. 9, p. 133–135.

Eyles, Victor A., and Eyles, Joan M., 1958, On the different issues of the first geological map of England and Wales, *Annals of Science*, v. 3, p. 190–212.

Ferber, Johann Jacob, 1774a, *Beschreibung des Quecksilber-Bergwerks zur Idria in Mittel-Cräyn*, Berlin, Christian Friedrich Himburg, 76 p.

Ferber, Johann Jacob, 1774b, *Beyträge zu der Mineral-Geschichte von Böhmen*, Berlin, Christian Friedrich Himburg, 162 p.

Greene, Mott T., 1982, *Geology in the nineteenth century: changing views of a changing world*, Ithaca and London, Cornell University Press, 324 p.

Guettard, J.-É., 1751, Mémoire et carte minéralogique sur la nature & la situation des terreins qui traversent la France & l'Angleterre, *Mémoires de l'Académie Royale des Sciences*, Paris, v. for 1746, p. 363–392.

Guettard, J.-É., 1755, Mémoire sur les granits de France comparés à ceux d'Egypte, *Mémoires de l'Académie Royale des Sciences*, Paris, v. for 1751, p. 164–210.

Guettard, J.-É., 1756, Mémoire dans lequel on compare le Canada à la Suisse, par rapport à ses minéraux, *Mémoires de l'Académie Royale des Sciences*, Paris, v. for 1752, p. 189–220.

Guettard, J.-É., 1757, Mémoire sur les poudingues, seconde partie, *Mémoires de l'Académie Royale des Sciences*, Paris, v. for 1753, p. 139–192.

Halley, Edmond, 1688, An historical account of the trade winds, and monsoons, observable in the seas between and near the tropicks, with an attempt to assign the physical cause of the said winds, *Philosophical Transactions*, no. 183, v. for 1686, p. 153–168.

Harrison, J.M., 1963, Nature and significance of geological maps, in Albritton, Claude C., Jr. (editor), *The fabric of geology*, Reading, Massachusetts, Palo Alto, London, Addison-Wesley, p. 225–232.

Herries Davies, Gordon L., 1983, *Sheets of many colours: the mapping of Ireland's rocks, 1750–1890*, Dublin, Royal Dublin Society, 242 p.

Ireland, H. Andrew, 1943, History of the development of geologic maps, *Bulletin of the Geological Society of America*, v. 54, p. 1227–1280.

Jameson, Robert, 1811, On colouring geognostical maps, *Memoirs of the Wernerian Natural History Society*, v. 1 (for 1808-1810), p. 149–161.

Jars, Gabriel, 1780, *Voyages métallurgiques*, v. 2, Paris, L. Cellot, C.A. Jombert, & L.A. Jombert, 612 p.

Kircher, Athanasius, 1665, *Mundus subterraneus*, v. 1, Amsterdam, Joannem Janssonium et Elizeum Weyerstraten, 346, [+ vi] p.

Laudan, Rachel [Rachel Bush], 1974, *The development of geological mapping in Britain, 1795–1825*, PhD dissertation, University of London, 279 p.

Laudan, Rachel, 1976, William Smith: stratigraphy without palaeontology, *Centaurus*, v. 20, p. 210–226.

Lyell, Charles, 1833, *Principles of geology*, 2nd ed., v. 2, London, John Murray, 338 p.

Marsigli, Luigi Ferdinando, 1725, *Histoire physique de la mer*, Amsterdam, Aux Dépens de la Compagnie, 173 p.

Marsigli, Luigi Ferdinando, 1726, *Danubius pannonico-mysicus*, 6 vols., The Hague: P. Gosse, R.C. Alberts, P. de Hondt; Amsterdam: H. Uytwerf & F. Changuion, 96, 149, 137, 92, 154, & 128 p.

Marsigli, Luigi Ferdinando, 1744, *Description du Danube*, v. 6, La Haye, Jean Swart, 128 p.

North, F.J., 1928, *Geological maps: their history and development with special reference to Wales*, Cardiff, National Museum of Wales and University of Wales, 133 p.

Ospovat, Alexander M., 1971, Introduction, *in* Werner, Abraham Gottlob, *Short classification and description of the various rocks*, translated with an introduction and notes by A.M. Ospovat, New York, Hafner, p. 1–35.

Palassou, P.B., 1784, *Essai sur la minéralogie des Monts-Pyrénées*, Paris, Didot jeune, 331 p.

Pallas, Peter Simon, 1799, *Kupfer zu P. S. Pallas Neuen Reisen in die südlichen Statthalterschaften des Russischen Reichs, in den Jahren 1793 und 1794*, 2 vols. in 1, Leipzig, G. Martini, 52 pl., 3 maps.

Porter, Roy, 1973, The Industrial Revolution and the rise of the science of geology, in Teich, Mikuláš, and Young, Robert (editors), *Changing perspectives in the history of science*, Dordrecht and Boston, D. Reidel, p. 320–343.

Ramond de Carbonnières, Louis F.E., 1801, *Voyages au Mont-Perdu et dans la partie adjacente des Hautes-Pyrénées*, Paris, Belin, 392 p.

Rappaport, Rhoda, 1964, Problems and sources in the history of geology, 1749–1810, *History of Science*, v. 3, p. 60–77.

Rappaport, Rhoda, 1969, The geological atlas of Guettard, Lavoisier, and Monnet: conflicting views on the nature of geology, in Schneer, Cecil J. (editor), *Toward a history of geology*, Cambridge, Massachusetts, and London, M.I.T. Press, p. 272–287.

Robinson, Arthur H., 1982, *Early thematic mapping in the history of cartography*, Chicago and London, University of Chicago Press, 266 p.

Rudwick, Martin J.S., 1976a, The emergence of a visual language for geological science 1760–1840, *History of Science*, v. 14, p. 149–195.

Rudwick, Martin J.S., 1976b, *The meaning of fossils: episodes in the history of palaeontology*, 2nd ed., New York, Science History Publications, 287 p.

Saussure, Horace-Bénédict de, 1786, *Voyages dans les Alpes*, v. 2, Geneva, Barde, Manget & Compagnie, 641 p.

Scheuchzer, Johann Jacob, 1723, *OYPE IOITH Helveticus, sive itinera per Helvetiae alpinas regiones*, v. 3, Leyden, Vander Aa, p. 353–520.

Schneer, Cecil J., 1981, William Maclure's geological map of the United States, *Journal of Geological Education*, v. 29, p. 241–245.

Smith, William 1815a, *A delineation of the strata of England and Wales, with part of Scotland*, London, J. Cary, 16 maps.

Smith, William, 1815b, *A memoir to the map and delineation of the strata of England and Wales with part of Scotland*, London, J. Cary, 51 p.

Plate 1. Jameson's schematic geognostic map (Jameson, 1811, opp. p. 160). The paper in which this appeared was read to the Wernerian Society in April 1808. Presumably this is an imaginary locality – if not, it is unidentified. A dozen colored rock units are shown. In addition to giving some indication of structure by darkened lines of contact in the color of the overlying rock, the map prescribes arrows showing direction of dip. Several specified ranges of steepness in dip correspond to certain arrow lengths – the shorter, the steeper. Crossed arrows signify vertical or nearly vertical formations. In addition, the small cross marks a high point of relief. (Map dimensions: 187 x 250 mm.)

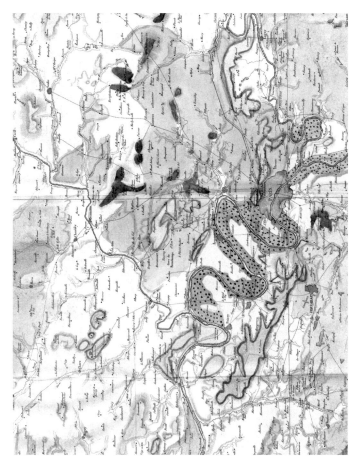

Plate 2. A west-central portion of the Cuvier-Brongniart map, an eastern part of which is shown in figure 7 (Cuvier, 1812). Paris is in the lower center. The stippled markings along the Seine and lower Marne represent pebbles rounded in transport. Areas of sandstone (millstone grit) and sands without shells are shown in light orange, appearing as enclosed by the gypsum (dark blue) above which they lie. Other formations shown include the *calcaire grossier* or coarse limestone (light yellow), siliceous limestone (lavender), marine marls (light blue), and freshwater deposits (light green). (Detail shown: about one-quarter of the map.)

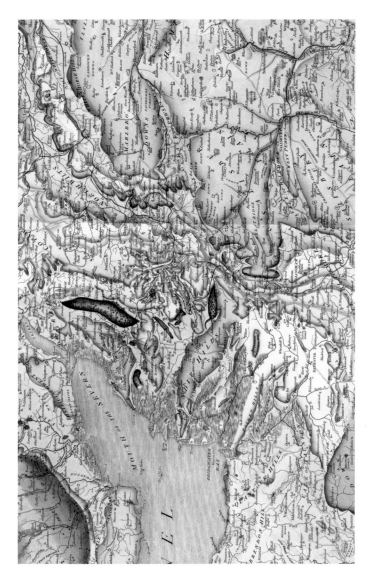

Plate 3. The central part of sheet XI of William Smith's *Delineation of the Strata of England and Wales* (1815a). Parts of southwest and southwest-central England, from the Vale of Taunton and the Salisbury Plain in the south to the Severn estuary and southern Cotswold Hills in the north. Clay and chalk in the east and southeast are shown overlying a sequence of some ten other formations. The conspicuous dark areas in the northwest belong to the coal measures. The University of Oklahoma copy of Smith's map shown here (bound together with Smith's *Memoir*, 1815b) is thought to be a slightly later issue, perhaps late 1816 or early 1817 (see Eyles and Eyles, 1938). (Detail shown: about one-half of the map.)

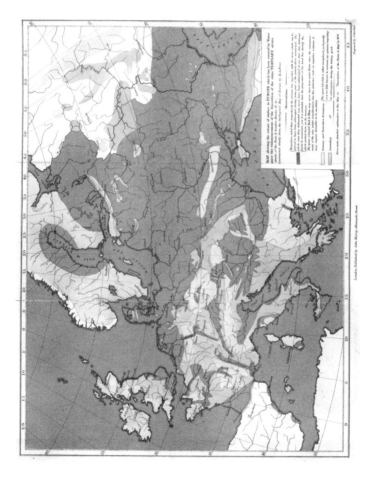

Plate 4. Lyell's *Map shewing the extent of surface in Europe which has been covered by water since the commencement of the deposition of the older tertiary strata* (1833, opp. p. 1). Areas left blank in Iberia, Norway, Asia Minor, and Russia were deemed inadequately known for inclusion. Dark areas other than the present seas are known "to have been submerged during some part of the period" since the early tertiary. Pink areas represent primary and transition formations; blue areas represent secondary formations.

XII

Geology in 1776:
Some Notes on the Character
of an Incipient Science

What was the condition of geology 200 years ago? Perhaps the first point to establish is that if we take geology to be a certain form of inquiry and knowledge developing and changing in the hands of its most original practitioners, in 1776 this was still an Old World preserve. The point may need to be made as an apology for not looking, in this paper, at American geology, since our conference is devoted mainly to a bicentennial celebration and retrospective of geology in America. I think it will not be disputed, however, that in the late eighteenth century, whatever interest may attach to the activities of scientists elsewhere, Europe remained the locus of innovation and change in geology, as in the sciences generally.

One way of recounting what was going on in the world of geology in 1776 might consist of a record of the activities and statuses of important figures. For example, in 1776 Abraham Gottlob Werner had recently published his first book, *On the External Characters of Minerals*, and was completing his first year on the faculty of the Bergakademie Freiberg. James Hutton, aged 50, having long since given up scientific farming for the amenities of Edinburgh, was occupied with diverse commercial and intellectual interests, but the first publication

XII

of the theory of the earth he had been developing privately still lay nine years in the future. Horace Bénédict de Saussure was finishing up his two-year service as Rector of the Geneva Academy, and was already well embarked on the series of Alpine investigations that would be the substance of his *Voyages dans les Alpes*. That summer 200 years ago, in fact, Saussure took the diplomat and volcanic enthusiast William Hamilton with him on one of his two trips to Chamonix.

The deficiencies of this sort of heroic chronicle are obvious. Knowledge of such things is interesting, even essential, in our historical consideration, but it is far from sufficient. And the insufficiency does not stem merely from the incompleteness of the account. It will not suffice to continue down the roster of geologists, no matter how far we go past all the famous figures and in among the geologically obscure. To understand satisfactorily what was going on in geology in 1776, even the most exhaustive inventory of persons and their doings will not answer.

What is needed is to generalize appropriately about the problems, methods, assumptions, aspirations, accomplishments, and failures of the age's geologists. Like all disciplines, historical inquiry is a search for an order binding together a diversity of events, conferring on them a meaning they would not otherwise display. An article of the historian's faith is that a vision of such an order can be formulated which will give focus and significance to otherwise designless recitations of particulars.

There have been numerous scholarly attempts to provide parts of the historical vision I speak of.[1] They have given us the capability of making general historical formulations about the state of geology around 1776, some of which might be stated in this way:

(a) By 1776 a sense of geological time was being worked out. A bold few had directly challenged the Mosaic chronology, but perhaps at least as importantly a larger number of scientists were proceeding cautiously with the opinion that the evidence was soundly interpretable only through admission of a terrestrial age exceeding the Biblical allotment, perhaps by a large but as yet uncertain margin.[2]

(b) The organic origin of certain fossils having long since been widely acknowledged, a stratigraphic geology linking fossils to specific formations was being conceived.[3]

(c) The majority of writers on geological dynamics continued, as they had done for some time, to look on the aqueous processes of crystallization, sedimentation, and erosion as the chief agents by which the earth's features were fashioned. Nevertheless there had been and continued to be a smaller number who championed the dynamic role of a supposed internal heat.[4]

(d) In a movement on the whole distinct from the doctrinal concerns of the central-fire advocates, there was developing in 1776 an awareness of effects of volcanic action far broader than had before been suspected. This fostered an inquiry into the causes and effects of volcanoes, but did not at this time lead most participants of the movement to abandon a general preference for a neptunistic version of geological dynamics. The seeds of the basalt controversy had thus been sown, but they were only now beginning to germinate, and would not produce fruit for over a decade to come.[5]

(e) After a long period of relative neglect, geomorphology was again receiving the attention of imaginative thinkers. One consequence in the offing was Hutton's brilliant resolution of the denudation dilemma.[6]

(f) A long tradition of attempting to improve the classification of minnerals was leading by 1776 to a keener sense of chemistry's relevance to a theory of the earth, and to greater use of a historical criterion to understand the place of minerals in the earth's economy.[7]

(g) The advent of new cartographic techniques for portrayal of geological phenomena was beginning to impinge on the geologist's sense of how he should approach his task, and was helping to give to geology a practical expressibility that enhanced its prospects for subsidy.[8]

These sketchy summaries of only a few among many generalizations already made available by scholars, although they represent a significant legacy, for which we are in debt, are not all we have. For example, many valuable investigations have been made, in historical context, of individual figures or works from eighteenth-century geology.[9] All the same, I think that none of those to whom the debt is held would claim that the story of which they have illuminated some part is ready to be fully told.

Yet it is not just historical information we need; what is also missing is a better perspective on how historical information should be sought and used. Probably more than in other areas of the history of science, terminalistic or Whiggish perspectives have continued to prevail. There has been a less thorough realization than elsewhere in the history of science of the disadvantages in presuming that the task is to show where modern scientific convictions come from, to identify and trace the sources of the present, to use the way things turned out as the criterion for judging what went on before. I think there has been in the history of geology an inadequate reformulation of historical tasks in terms of gaining an understanding of the past on its own ground. To be sure, in recent years some excellent scholarship has shown how the history of geology can avoid the terminal fallacy. But even in such cases there is a tendency to fall easily into the habit of using labels and categories of questionable validity. Recall the terms used or implied in the above generalizations: *geological time*, *stratigraphic geology*, *geological dynamics*, *volcanic geology*, *geomorphology*. Terms like these are often used with little historical discrimination, as if there were no problem in their intrusion into any historical period. Perhaps there is a problem. In what follows I want to discuss one way we may be able to throw a clearer light on the eighteenth-century developments that concern us: by considering with greater care the terms in which we approach them.

Let us return to the question with which we began: What was the condition of geology in 1776? A good first step, I suggest, is to recognize that, in a significant way, the question is wrongly put. Speaking strictly, there was *no* geology in 1776. The term itself had not yet been coined, at least not with the scientific meaning it was shortly to acquire. Not until 1778 did Jean André DeLuc perform that service, and only in the early nineteenth century did the word gain something approaching universal recognition and acceptance.[10]

The absence of the term geology from the lexicon in 1776 was not accidental. We should not allow ourselves to assume casually that geology was actually there, but had not yet been named. Words of all kinds, not excluding scientific ones, have a way of making their appearances in response to perceived needs. The need that DeLuc saw in 1778 had quite arguably been recognized, in different ways and degrees, by some few others before him. Nonetheless, the need had not been responded to in just this way, and the response itself became a historical act with its own consequences.

It would be excessive, of course, to argue that the concept of geology appeared only with the word. Ideas emerge slowly, by stages. This is likely to be especially true of the idea of a discipline, which encompasses purposes, methods, procedural rules. There may occur, in the evolution of a disciplinary concept, substitutions of symbolic terms in parallel with shifts in the idea. So we should be surprised if close relatives to the concept of geology were not to have existed before it. Indeed they did, frequently in association with organized knowledge about the objects to be found in the earth.

Consider mineralogy. The idea of dealing with mineral substances as a realm of natural history dates back to ancient times. The English *mineralogy*, however, may have come into use only in the seventeenth century; Boyle employed the term in 1690.[11] French use of *minéralogie* appeared no later than 1732;[12] Louis Bourguet referred to the studies of *minéralogues* in 1742;[13] and in 1753 Baron d'Holbach (Paul

Henri Thiry) suggested calling one learned in this field a *minéralogiste*.[14] As names for recognized avenues of knowledge, mineral natural history and mineralogy came in the eighteenth century to have widely acknowledged meaning—primarily as descriptive sciences but in some cases bearing a broader historical sense of the mineral realm as a totality.

Several somewhat more exotic suggestions were made during the eighteenth century for terms to designate knowledge of fossils (in the traditionally wide sense of things dug out of the earth), stones, and rocks. A work of Johann Jacob Scheuchzer, *Specimen Lithographiae Helveticae Curiosae* (1702), spurred the Welshman Edward Lhuyd to use the Anglicized *lithography* in 1708 for a description of stones and rocks.[15] *Lithology* was used in English in 1716,[16] and *lithologie* by the French as early as 1742, in a work by Antoine Joseph Dezallier d'Argenville.[17] *Oryctologie* (the natural history of fossils) was in the title of another of his books in 1755.[18] For a time both *lithologie* and *oryctologie* made a bid for fairly extensive use in France, together with their variants *lithographie* (1752) and *oryctographie*, whose suffixes emphasized the descriptive function of these disciplines.[19] The *Encyclopédie* included brief definitions and remarks on each of these words in 1765, calling *oryctologie* and *oryctographie* synonyms for *minéralogie*.[20] *Lithologue*, as a name for a practitioner of lithology, was proposed in 1752 and accepted by the Académie Française, together with *lithologie*, a decade later.[21] Jacques Christophe Valmont de Bomare referred to a *lithologiste* in 1762 as one "who possesses the science of stones, and teaches it,"[22] and Emanuel Mendes da Costa spoke of lithologists in 1746.[23] The Supplement to Chambers' *Cyclopaedia* (1753) defined oryctography as "that part of natural history wherein fossils are described" and oryctology as "that part of physics which treats of fossils."[24] An even more advanced neologism was devised by Johann Heinrich Pott, who offered *lithogeognosie* in the title of a 1746 book as a label for knowledge of the earth's parts.[25] In a 1776 book George Edwards put forward *fossilogy* as his preferred name for the science of nonextraneous mineral bodies, or minerals not of organic origin.[26] For Edwards a specialist in this field was a fossilist, a word used previously by Mendes da Costa and Thomas Pennant in reference to a student of extraneous or organic fossils.[27]

The descriptive nature of these earth sciences, especially as understood before the latest part of the eighteenth century, is striking. Lithology, oryctology, mineral geography, orography (knowledge of the physical arrangement of mountains), and many other terms were usually agreed to fall within the province of descriptive natural history. Mineralogy and physical geography were often treated in a similar manner, although in some cases they tended more easily to acquire developmental overtones. Also within the tradition of natural history lay the practice of recording the phenomena of springs, mineral ore deposits, earthquakes, volcanoes, and the like, all studied with the general assumption that their foremost characteristics are to be considered as fixed aspects of an essentially static world. Insofar as they conformed to this natural history outlook, the men who contributed their efforts to these endeavors were inclined to believe that they were dealing with "the immanent properties and processes of the physical earth and its constituents."[28]

The ancestry of geology is not exclusively in descriptive natural history. Relevant activity was taking place also in already existing analytical sciences, like chemistry. For example, many of the most prominent French students of a theory of the earth living in 1776 had received their first instruction during the 1750's and 1760's from the influential chemist Guillaume François Rouelle.[29] It is well known that chemists over a long period of time have commonly taken an interest in minerals and earths, and it is not rare to find chemists of the eighteenth century who took a generalized interest in an understanding of both the earth's constitution and its history. Since we all have an idea of what chemistry and geology are, it is tempting to suppose that such chemists were indulging their additional interest in mineralogy or geology. It is a temptation to resist, however, in the name of better history, and I would like to expand on the point with further comment on Rouelle.

The question is whether Rouelle, in imparting to his students rather definite and detailed ideas about the earth's constitution and past, was getting into something beyond the scope of his chemistry. I think not. In my opinion Rouelle did not think, as he dealt with the earth, its parts, and its historical development, that he was doing something apart from chemistry. On the contrary, he regarded this terrestrial inquiry as a significant dimension within the broad scope properly entailed by chemistry. To Rouelle, as to many chemists before him, chemistry was the understanding of the formation of material, not just in principle, but also in the specific instances the world affords. And what grander or more important instance was there than the composition and differentiation of the earth itself? Chemistry was knowledge of the material world's specific products as well as of chemical operations; it was knowledge of history as well as of process.

These assertions require historical demonstration, something I can do no more than indicate in rough outline here. Our knowledge of

Rouelle's ideas about the theory of the earth is derived almost entirely from what some of his students had to say about it, and these students created a record of two different kinds.[30] First, there exist numerous copies of detailed notes taken at Rouelle's chemistry course, a part of which he evidently devoted regularly to a discussion of how the earth's crustal parts should be classified and subjected to historical interpretation. Secondly, several of Rouelle's distinguished pupils who later became geologists or engaged in further geological investigations (notably Nicolas Desmarest, Antoine Laurent Lavoisier, Antoine Grimoald Monnet, and Jean Darcet) left commentaries on their teacher's views. This second type of evidence about Rouelle's opinions has been given close attention in trying to reconstruct Rouelle's ideas. The pupils who later made good in geology had more reason than others to give consideration to the value of what Rouelle had taught in an area that came to have special importance to them. But without impeaching the significance and usefulness of these sources, should we not also consider that they may reflect an evaluation of Rouelle's teachings in light of the fact that geology had subsequently become an identifiable field of investigation? If we wish to know how Rouelle categorized his discussion of the theory of the earth, it may well be better to rely on the first type of source, contemporary notes which cannot reflect the bias of a subsequent reorganization of disciplines.

As far as I can see, the notes of Rouelle's *Cours de chimie* do not justify any argument that his theory of the earth lies outside his chemistry. On the contrary, the organization of the course in parts corresponding to the three realms (vegetable, animal, mineral) reflects a broad, all-encompassing definition of the science. The course notes display a conception of chemistry that one observer (Jean Mayer) calls "très ambitieuse" (meaning both ambitious and pretentious). Mayer adds that "if [Rouelle] could have, he would have required of [chemistry] the explanation of the genesis of the world and the answer to the riddle of life."[31] While Mayer sounds condescending, historically he is close to the mark.

One more illustration of the breadth of historical linkages in science may assist our understanding of how eighteenth-century thinkers could have had notions we find strange on how to proceed in studying the earth. Recent scholarship in the history of Renaissance chemistry (or chemical philosophy) has revealed with greater clarity than ever before that the conception of chemistry held in the sixteenth and seventeenth centuries, especially by those influenced by traditions of alchemy, iatrochemistry, and Paracelsianism, was that of a global inquiry into the whole of nature.[32] Chemical philosophy tended to hold all of nature to be "a vast chemical laboratory," and chemistry, if properly cultivated, was thought capable of illuminating the order, structure, and processes of the terrestrial sphere, among other things. Now, the chemistry of the eighteenth century was obviously more than just the direct outgrowth of this Renaissance version. It involved extensive new developments, and overturned portions of its Renaissance antecedent. But for all that, Rouelle's conception of the scope and role of chemistry may have owed something to alchemical and Paracelsian influences. Furthermore, the chemical articles of the *Encyclopédie*, many of them written by Rouelle's pupil Gabriel François Venel, reveal an alchemical ideal of gaining knowledge of the whole of nature through chemistry.[33] Although innovative developments were in the process of transforming chemistry, an old tradition of the science's universality did not die easily.

A conclusion we can draw from this discussion of Rouelle's views on the place within science of a theory of the earth is that not much over 200 years ago some important ideas that we view as geological were nurtured as an integral part of an already existing science. Natural as it is for us to refer to these conceptions as having been developed by a chemist reaching beyond his field, that is not how Rouelle saw it. And if Rouelle's chemistry embraced a theory of the earth, other sciences of the time could have similarly included considerations that from our vantage point appear to have been geological.

Another distinct and long-standing branch of learning that encompassed serious investigation of the earth during the eighteenth century was geography. The importance of the activities and ideas of geographers in eighteenth-century developments concerning a science of the earth has been underestimated by most historians of geology. Opportunities for rectifying this situation are now enhanced by the appearance of two extensive studies of French geography.[34]

It seems clear that at least among French geographers of the eighteenth century a growing emphasis on physical geography was marked by some imaginative attempts to reduce empirical knowledge of the earth's surface features to general physical principles. Philippe Buache, for example, formulated a theory of an interlocking global network of mountain ranges that had important consequences not only for French traditions in physiography, but also for the ideas of such a scientist as Desmarets on the theory of earthquakes and the general organization of land masses. From 1730 until his death in 1773 Buache was *adjoint géographe* in the Royal Academy of Sciences of Paris, the only member who held a position explicitly identifiable with a science of the earth. Not until the 1785 reorganization, with the creation of a class for natural history and mineralogy, would it become possible for any Paris academician other than the *adjoint géographe* to have so official an identity with the scientific study of the earth.

There may have been other sciences in addition to chemistry and geography that possessed components later observers would feel inclined to separate out as geological. A possible example is botany. I suspect that for some of the eighteenth century's botanists who concerned themselves with earth science, this was a pursuit less thoroughly divorced from botany than has been supposed. The study of minerals and soils was relevant to knowledge of plant life, and the understanding of fossil plant forms in historical terms was accepted by some botanists as part of their work.

These comments on how some eighteenth-century scientists looked upon their own concern for scientific understanding of the earth can be summarized as a challenge to what I think is a rather widespread misconception about the organization and structure of eighteenth-century earth science. That misconception can be put like this: Throughout the eighteenth century, scientists who carried out investigations of a sort we would call geological generally assumed the existence of a scientific enterprise corresponding in nature and scope, though not in name, to the geology that came to be recognized around the end of the century. These scientists would have had little difficulty in grasping geology as a concept if it could have been presented to them. With or without an appropriate label, they shared a sense of the enterprise that was not significantly altered by the advent of the word geology.

Our brief inquiry into some facets of the organization of geological endeavor some 200 and more years ago suggests that the science of geology as it came to be conceived in the train of the term's invention was not a direct, lineal descendant of a more primitive eighteenth-century version of the same thing. Instead, the late eighteenth-century developments that set the stage for the emergence of geology as a new discipline included a reordering or reorganization of the sciences, making room in a new way for a distinct science of the earth.

I suggest that the devising and use of the new term geology in the late eighteenth century did mark a change, identifying a growth toward a common theoretical understanding of a scientific goal engaging thinkers of diverse types. We have seen that the interests of these diverse types had been focused by men who considered themselves to be students of chemistry, geography, mineralogy, and other sciences. If there was a genuine change involved in the production of a new name, it was mainly bound up in an increasing recognition that various groups of scientists practicing relatively distinct disciplines had come to be united by a set of ideas held in common. To some extent, geology came into existence as a reorientation of existing disciplines, or parts of disciplines, under a new umbrella. It was the fashioning of a new relationship among activities that had already been going on for some time.

It remains to identify the essential ingredients of the new, integrated relationship I am claiming was brought about late in the eighteenth century. I offer my opinion briefly as a conjecture in need of further study. It is that there was a growing acceptance of *development* or *history* as the framework within which the descriptively studied objects and events of the earth were to be explained in terms of *process*. Fossils, minerals, strata, ore deposits, and mountains, as well as earthquakes, volcanoes, floods, and shifting shorelines—these phenomena were increasingly looked upon as data to be understood through the processes that caused them, processes that were the mainsprings of terrestrial change. Geology came to denote a collection of activities or disciplines subjected to the unifying themes of law-like process and historical development. There was still ample room within this framework—there had to be—for major differences of opinion regarding the nature of the main historical events and their temporal extent, but in the main controversies of the day the disputants could generally operate as geologists because they agreed at least that the information they dealt with had to be rendered intelligible within a matrix of orderly events occurring in time. They expected their science to involve "the determination of configurational sequences, their explanation, and the testing of such sequences and explanations."[35]

It has been my central concern to argue that key terms used by scientists in specific historical conditions are reflections of distinctive scientific ideas, and that the historian of geology can get useful historical information from study of such terms. I have also argued that the pattern of development of words for earth-related disciplines supports the thesis that geology hardly existed as a distinct science until late in the eighteenth century. This is not to say that scientists before then made no valuable contributions to knowledge of the earth, or that they did not possess understanding that was in some way meaningful. Rather it means that they made their contribution or gained their understanding within a conceptual framework foreign to that of later geologists. They would not have recognized their work as it would later be fitted into an order understood by, say, Lyell. Lyell is, in fact, a figure with whom it is hard to juxtapose most of our eighteenth-century geologists. Several of the contributions to the Lyell symposium in 1975 made clearer than ever the degree to which conventional historical interpretations of geology's past have been governed by Lyell's use of history to advance his own geological ideas.[36] We are still in the process of releasing ourselves historically from a Lyellian bondage.

It should go without saying that if our purpose is to understand eighteenth-century science on its own terms, we shall have to make an effort to overcome the habit of imposing our categories (or Lyell's) on it after the fact. But even if our purpose is—less historically—to trace more fully and accurately the paths by which we have arrived at our present condition, we still need to make the same effort. A faithful account of how the past led to the present will be acquired best by acting as though the past had value only for itself. By excluding consideration of later developments as fully as possible, we can come closest to reconstructing a past sustained by the evidence, rather than by our expectations of it.

Acknowledgment

I am grateful to Rhoda Rappaport and Alexander M. Ospovat for useful comments on this paper, and to John H. Eddy, Jr., for research assistance.

Notes

1. For general discussion of eighteenth-century geology see Rhoda Rappaport, "Problems and Sources in the History of Geology, 1749–1810," *History of Science* (1964), 3:60–77; V. A. Eyles, "The History of Geology: Suggestions for Further Research," ibid. (1966), 5:77–86; and V. A. Eyles, "The Extent of Geological Knowledge in the Eighteenth Century, and the Methods by Which It Was Diffused," in *Toward a History of Geology*, ed. Cecil J. Schneer (Cambridge, Mass., and London, M.I.T. Press, 1969), 159–183. Also important are Jacques Roger's introduction and notes to his critical edition of Buffon's *Les Époques de la nature*, Mémoires du Muséum National d'Histoire Naturelle, Ser. C, "Sciences de la terre," Tome X, Paris, Éditions du Muséum (1962), and John C. Greene, *The Death of Adam: Evolution and Its Impact on Western Thought* (Ames, Iowa State University Press, 1959). Still

86 Geology in 1776

useful despite its historiographical limitations is Sir Archibald Geikie, *The Founders of Geology*, 2nd ed. (London and New York, Macmillan, 1905).

Mention must also be made of the following items published since this paper was written: Roy Porter and Kate Poulton, "Research in British Geology 1660–1800: A Survey and Thematic Bibliography," *Annals of Science* (1977), 34:33–42; W. R. Albury and D. R. Oldroyd, "From Renaissance Mineral Studies to Historical Geology, in the Light of Michel Foucault's *The Order of Things*," *British Journal for the History of Science* (1977), 10:187–215; and especially Roy Porter, *The Making of Geology: Earth Science in Britain 1660–1815* (Cambridge, Cambridge University Press, 1977).

Each of these sources has relevance to several of the points raised below in items a through g, and some of the works cited below in notes 2–8 treat general issues in addition to the ones which they annotate and should be consulted accordingly.

2. Francis C. Haber, *The Age of the World: Moses to Darwin* (Baltimore, Johns Hopkins Press, 1959). Also Stephen Toulmin and June Goodfield, *The Discovery of Time* (New York, Harper and Row, 1965). On two of the bolder advocates of a broad time scale, see Albert V. Carozzi, "De Maillet's Telliamed (1748): An Ultra-Neptunian Theory of the Earth," in Schneer, ed., *Toward a History of Geology*, 80–99, and Carozzi's introduction to his translation of *Telliamed, or Conversations between an Indian Philosopher and a French Missionary on the Diminution of the Sea* (Urbana, University of Illinois Press, 1968); and Jacques Roger, "Un Manuscrit inédit perdu et retrouvé: *Les Ancedotes de la nature*, de Nicolas-Antoine Boulanger," *Revue des sciences humaines* (1953), 71: 231–254.

3. Although it pays relatively little attention to the eighteenth century, the best point of departure is Martin J. S. Rudwick, *The Meaning of Fossils: Episodes in the History of Paleontology* (London, Macdonald, and New York, American Elsevier, 1972).

4. Rappaport, "Problems and Sources," 66–69; John G. Burke, "Romé de l'Isle and the Central Fire," XIIe Congrès International d'Histoire des Sciences, *Actes* (Paris, Albert Blanchard, 1971), Tome VII, "Histoire des sciences de la terre et de l'océanographie," 15–17.

5. Robert Michel, "Les Premiers recherches sur les volcans du Massif Central (18e–19e siècles) et leur influence sur l'essor de la géologie," *Symposium Jean Jung: Géologie, géomorphologie et struc-*

ture profonde du Massif Central français (Clermont-Ferrand, Plein Air Service Ed., 1971), 331–344; Kenneth L. Taylor, "Nicolas Desmarest and Geology in the Eighteenth Century," in *Toward a History of Geology*, 339–356; Otfried Wagenbreth, "Abraham Gottlob Werner und der Höhepunkt des Neptunistenstreites um 1790," in *Bergbau und Bergleute: Neue Beiträge zur Geschichte des Bergbaus und der Geologie*, Freiberger Forschungshefte, D 11 (Berlin, Akademie-Verlag, 1955), 183–241; Alexander M. Ospovat, "Abraham Gottlob Werner and His Influence on Mineralogy and Geology," Ph.D. dissertation, University of Oklahoma, Norman, 1960; Albert V. Carozzi, "Rudolf Erich Raspe and the Basalt Controversy," *Studies in Romanticism* (1969), 8:235–250.

6. Gordon L. Davies, *The Earth in Decay: A History of British Geomorphology, 1578–1878* (New York, American Elsevier, 1969).
7. Introduction and notes by Alexander M. Ospovat accompanying his translation of Abraham Gottlob Werner, *Short Classification and Description of the Various Rocks* (New York, Hafner, 1971); Charles Spencer St. Clair, "The Classification of Minerals: Some Representative Mineral Systems from Agricola to Werner," Ph.D. dissertation, University of Oklahoma, Norman, 1965; D. R. Oldroyd, "Mechanical Mineralogy," *Ambix* (1974), 21:157–178; D. R. Oldroyd, "Some Phlogistic Mineralogical Schemes, Illustrative of the Evolution of the Concept 'Earth' in the 17th and 18th Centuries," *Annals of Science* (1974), 31:269–305; Seymour H. Mauskopf, *Crystals and Compounds. Molecular Structure and Composition in Nineteenth-Century French Science*, Transactions of the American Philosophical Society, New Series, Vol. 66, Pt. 3 (Philadelphia, American Philosophical Society, 1976), esp. 7–16.
8. Rhoda Rappaport, "The Geological Atlas of Guettard, Lavoisier, and Monnet: Conflicting Views of the Nature of Geology," in *Toward a History of Geology*, 272–287; and Rappaport's Ph.D. dissertation, "Guettard, Lavoisier, and Monnet: Geologists in the Service of the French Monarchy," Cornell University, Ithaca, 1964. A stimulating study of the question of geology's links with practical motives for science in Britain is Roy Porter, "The Industrial Revolution and the Rise of the Science of Geology," in *Changing Perspectives in the History of Science: Essays in Honour of Joseph Needham*, ed. Mikuláš Teich and Robert Young (Dordrecht and Boston, D. Reidel, 1973), 320–343.
9. For example, Rhoda Rappaport on Lavoisier: "Lavoisier's Geologic Activities, 1763–1792," *Isis* (1967), 58:375–384; "The Early Disputes between Lavoisier and Monnet, 1777–1781," *British*

Journal for the History of Science (1969), 4:233–244; and "Lavoisier's Theory of the Earth," ibid. (1973), 6:247–260. Also Alexander M. Ospovat on Werner: in addition to works already cited, "Reflections on A. G. Werner's 'Kurze Klassifikation'," in *Toward a History of Geology*, 242–256, and "The Work and Influence of Abraham Gottlob Werner: A Reevaluation," *Proceedings of the XIIIth International Congress of the History of Science* (Moscow, Editions Naouka, 1974), Sec. VIII, "The History of Earth Sciences," 123–130; as well as others. Also Jacques Roger on Buffon's *Les Époques de la nature* (above, note 1).

10. Jean André DeLuc, *Lettres physiques et morales, sur les montagnes et sur l'histoire de la terre et de l'homme* (En Suisse: Chez les Libraires Associés, 1778), [vii]-viii n. See also Frank Dawson Adams, "Earliest Use of the Term Geology," *Bulletin of the Geological Society of America* (1932), 43:121–123; "Further Notes on the Earliest Use of the Word 'Geology'," ibid. (1933), 44:821–826; *The Birth and Development of the Geological Sciences* (Baltimore, Williams and Wilkins, 1938), 165–166; and Arthur Birembaut, "La Minéralogie et la géologie," in *Histoire de la science*, ed. Maurice Daumas (Paris, Gallimard, 1957), 1104–1105, 1113. Birembaut calls attention to Diderot's use of "géologie" in 1751 as one of four subheadings of "cosmologie" (the others being "uranologie," "aerologie," and "hydrologie") in the "Système figuré des connoissances humaines" for the *Encyclopédie* (Vol. I, facing p. xlvii). The organizing principle here appears to be of cosmic regions rather than types of investigation, and this is substantiated by the appearance of "minéralogie" elsewhere in the scheme.

11. *Oxford English Dictionary*, VI, 467.

12. Ferdinand Brunot, *Histoire de la langue française des origines à nos jours*, 13 vols. (Paris, Armand Colin, 1966–72), Vol. VI: *Le XVIIIe Siècle*, 625; Adolphe Hatzfeld and Arsène Darmesteter, *Dictionnaire général de la langue française du commencement du XVIIe siècle jusqu'à nos jours, précédé d'un traité de la langue*, 2 vols., 6th ed. (Paris, Delagrave, 1920), II, 1524.

13. Louis Bourguet, *Traité des petrifications*, 2 pts. in 1 vol. (Paris, Briasson, 1742), I, 1.

14. Brunot, *Le XVIIIe Siècle*, 627.

15. *Philosophical Transactions*, 1708–1709 (1710), 26:161.

16. *Oxford English Dictionary*, VI, 347.

17. *L'Histoire naturelle éclaircie dans deux de ses parties principales, la lithologie et la conchyliologie, dont l'une traite des pierres et l'autre des coquillages* (Paris, De Bure l'Aîné, 1742).

18. *L'Histoire naturelle éclaircie dans une de ses parties principales, l'oryctologie, qui traite des terres, des pierres, des métaux, des minéraux, et autres fossiles* (Paris, De Bure l'Aîné, 1755).
19. Paul Robert, *Dictionnaire alphabétique et analogique de la langue française. Les Mots et les associations d'idées*, 6 vols. (Paris, Société du Nouveau Littré, 1954–1964), IV, 282; Hatzfeld and Darmesteter, *Dictionnaire général de la langue française*, II, 1413. Scheuchzer used "oryctographia" in his *Helvetiae historia naturalis*, 3 vols. (Zürich, In der Bodmerischen Truckerey, 1716–1718), III: *Meteorologia et oryctographia Helvetica*.
20. *Encyclopédie*, IX (1765), 587; XI (1765), 677.
21. Hatzfeld and Darmesteter, *Dictionnaire général de la langue française*, II, 1414.
22. Valmont de Bomare, *Minéralogie, ou nouvelle exposition du règne minéral*, 2 vols. (Paris, Vincent, 1762), II, 348.
23. Mendes da Costa, "A Dissertation On Those Fossil Figured Stones Called Belemnites," *Philosophical Transactions*, 1747 (1748), 44 (pt. 2):398, 406. The communication was dated 1746.
24. *A Supplement to Mr. Chambers's Cyclopaedia: Or, Universal Dictionary of Arts and Sciences*, 2 vols. (London, printed for W. Innys [et al.], 1753).
25. . . . *Chymische Untersuchungen welche fürnehmlich von der Lithogeognosia oder Erkäntniss und Bearbeitung der gemeinen einfacheren Stein und Erden ingleichen von Feuer und Licht handeln* (Potsdam, C. F. Voss, 1746).
26. Edwards, *Elements of Fossilogy: Or, An Arrangement of Fossils, into Classes, Orders, Genera, and Species; With Their Characters* (London, printed by B. White, 1776).
27. Mendes da Costa, "Dissertation," 406; Pennant, *British Zoology*, 4 vols. (London, printed for Benjamin White, 1768–1770), I, 41.
28. The words are those of George Gaylord Simpson, "Historical Science," in *The Fabric of Geology*, ed. Claude C. Albritton, Jr. (Reading, Mass., Addison-Wesley, 1963), 25.
29. Rhoda Rappaport, "G.-F. Rouelle: An Eighteenth-Century Chemist and Teacher," *Chymia* (1960), 6:68–101; "Rouelle and Stahl— The Phlogistic Revolution in France," ibid. (1961), 7:73–102; and Jean Mayer, "Portrait d'un chimiste: Guillaume-François Rouelle (1703–1770)," *Revue d'histoire des sciences* (1970), 23:305–332.
30. Extant original sources on Rouelle's courses are indicated in the articles by Rappaport and Mayer, and in Rappaport's article on Rouelle in *Dictionary of Scientific Biography* (1975), XI, 562–564.
31. Mayer, "Portrait d'un chimiste," 332.

32. See, for example, publications of Allen G. Debus, including *The English Paracelsians* (London, Oldbourne, 1965); "Renaissance Chemistry and the Work of Robert Fludd," *Ambix* (1967), 14: 42–59; "Edward Jorden and the Fermentation of the Metals: An Iatrochemical Study of Terrestrial Phenomena," in *Toward a History of Geology*, 100–121; "The Chemical Debates of the Seventeenth Century: The Reaction to Robert Fludd and Jean Baptiste van Helmont," in *Reason, Experiment, and Mysticism in the Scientific Revolution*, ed. M. L. Righini Bonelli and William R. Shea (New York, Science History Publications, 1975), 19–47; and *The Chemical Philosophy: Paracelsian Science and Medicine in the Sixteenth and Seventeenth Centuries*, 2 vols. (New York, Science History Publications, 1977).
33. Jean-Claude Guédon, "Le Lieu de la chimie dans l'*Encyclopédie* de Diderot," *Proceedings of the XIIIth International Congress of the History of Science* (1974), VII ("The History of Chemistry"), 80–86; also Guédon's doctoral dissertation, "The Still Life of a Transition: Chemistry in the *Encyclopédie*," University of Wisconsin, Madison, 1974.
34. Numa Broc, *Les Montagnes vues par les géographes et les naturalistes de langue française au XVIIIe siècle: Contribution a l'histoire de la géographie* (Paris, Bibliothèque Nationale, 1969), and *La Géographie des philosophes: Géographes et voyageurs français au XVIIIe siècle* (Paris, Éditions Ophrys, 1974?).
35. Again, the words of George Gaylord Simpson (above, note 28). It is understood, of course, that the conceptions of process and development were not created by scientists of the late eighteenth century. What is proposed here is the advent at that time of a more or less collective recognition of these concepts' parallel applicability to diverse terrestrial phenomena.
36. *British Journal for the History of Science*, Lyell Centenary Issue (1976), 9: Part 2. See especially Roy Porter, "Charles Lyell and the Principles of the History of Geology," 91–103, and Alexander M. Ospovat, "The Distortion of Werner in Lyell's *Principles of Geology*," 190–198.

XIII

The beginnings of a French geological identity

During the latter part of the eighteenth century geology's emergence as a distinct scientific discipline was marked not only by important conceptual innovations but by institutional acknowledgement of geologists' special roles and by the specific self-identification of some scientists with this new field. In the particular case of the French-speaking world (therefore including notably the Genevese naturalists) the emerging geological identity was strongly associated with empiricism, secularism, and practical utility. Major national organizations of French science experienced adjustments (in particular the Académie des Sciences and the Jardin du Roi) or were created (especially the Ecole des Mines) to accommodate geology's new importance. Four individual scientists in whom the geological identity was evident were de Saussure, Dolomieu, Faujas de Saint-Fond, and Lamétherie.

***

XIII

The historical emergence of geology as a distinct science occurred, most historians of science agree, during the late eighteenth and early nineteenth centuries. Since the appearance of a new science can be seen as amounting to the genesis of a new kind of knowledge, a different way of understanding, recent historical study of geology's origins and early development has quite appropriately included much emphasis on the conceptual or theoretical changes this innovation constituted, and on the changing cultural circumstances attending it (1). In this paper I would like to call attention to some historical changes in scientists' self-consciousness that accompanied and reflected the epistemic adjustments through which geology began to become an acknowledged scientific field. For just as geology's emergence brought into being a science that in some real sense had not existed before, so also a body of recognized practitioners of this novel enterprise began to form and identify themselves with it, eventually forming, in the nineteenth century, a geological community. I hope here to sketch out a few features of the early development of a geological identity in one important segment of this multinational movement, specifically among scientists using the French language.

Whatever may be said about the very impressive accomplishments of eighteenth century investigators and thinkers in the earth sciences, it was only rather late in the century that a specific field of endeavor aimed at scientific understanding of the earth came to be known widely as geology, and that practitioners of this specialty were called geologists or *géologues* (Dean, 1979). But the use of these specific terms (or others related to them, including geognosy) is not the only significant indicator of a changed awareness of scientific status. A variety of tangible signs indicate the beginnings, by the late eighteenth century, of a group sense among French-speaking earth scientists whereby a number of them acknowledged their participation in a specific enterprise whose objectives were not satisfactorily subsumed within traditional scientific categories. This emerging consciousness of commitment to some separate, but as yet incompletely worked out, ordering of scientific objectives, methods, and theories about the earth's organization, processes, and development was ripe enough by the turn of the century, I believe, to warrant our calling it a geological identity (2).

---

(1) Albury & Oldroyd, 1977; Ellenberger, 1975-1977; Guntau, 1978; Oldroyd, 1979; Porter, 1973 & 1977; Roger, 1974; Rudwick, 1971. Among these authors, Porter has given the most attention to the question of a disciplinary self-awareness on the part of geologists.
(2) As originally delivered in somewhat different form at the 26th International Geological Congress, this paper was entitled « The Beginnings of a French Geological Community. » I have changed the title both to give a more accurate idea of its content and also to avoid seeming to imply a necessary connection between the informally developing community I describe here and the rather more specialized meaning some sociologists and historians of science have attached to the idea of a scientific community.

I

If we try to put into historical context the various manifestations of the earth sciences and techniques before, say, the 1770's, we see a host of significant endeavors and accomplishments that at the time had far less obvious coherence, fewer perceived interconnections, than would later become evident to those possessing a geological perspective. Habituated though the historian of geology is to select early instances of scientific study of the earth as if they can be imagined as embryonic stages of an eventually mature geological science, might it not also be historically useful to try to bear in mind a discrimination among these proto-geological endeavors, one conforming more nearly to contemporary ideas of the organization of science ? (3). A rough and no doubt incomplete review of earth-science activities and developments in the early and middle parts of the century might include the following :

— continuations of the established seventeenth century tradition of theories of the earth, characterized by attempts to account for the earth's features within a framework dominated by the operations of fixed natural laws (the best-known examples include the theories of Bourguet, de Maillet, Boulanger, and Buffon) (4);

— enthusiastic expansion of natural history studies of fossil and mineral bodies, often motivated by confidence in the ultimate humanitarian as well as commercial profitability of such knowledge, manifested not only in a virtual mania for collections, inventories, natural history courses, and systematic texts, but also in sober and scrupulously empirical reports by naturalists of high competence (such as Réaumur, Guettard, de Sauvages) (5);

— notably increased popularity of travel narratives paying significant attention to the distribution and detailed local circumstances of minerals, mountain structure, and other terrestrial features (cf. the publications of Bouguer and La Condamine, translations of naturalist-voyagers' accounts such as J. Anderson's and J.G. Gmelin's,

---

(3) Cf. Taylor, 1979.
(4) See Jacques Roger's introduction to Buffon, 1962 (ix-clii), and Roger, 1973, where he argues that theories of the earth constituted the « intellectual frame of the sciences of the earth from the end of the xvııth to the beginning of the xıxth century, [and were] not a preliminary sketch to modern geology » (23). In this view the theories of the earth represented a mode of thought different from that assumed in due course by geology, and so nineteenth century geology was not so much an extension of these theories as a repudiation of them.
(5) Birembaut, 1964 ; Laissus, 1964b ; Lamy, 1930 ; Mornet, 1911. In some respects geology's formation as a distinct science corresponded to a transition away from the conception of studying mineral objects as one constituent area within the global programme of natural history, toward a more independently theoretical-analytical-historical treatment of mineral phenomena in which classical natural history became only one constituent part. Seen this way, geology's elevation to scientific status occurred not as a rise *within* natural history, but rather as a move *beyond* it. Cf. Monnet, 1779, 1-58, « Précis historique sur les progrès de la minéralogie en France. »

and the compendious *Histoire générale des voyages* of the abbé Prévost d'Exiles) (6);

— enlarged pursuit of more formal geographical enterprises, including some attempts to reduce physical geography to a more systematic order (seen especially in Buache), and important pioneering of novel cartographic methods to represent topography and mineral distribution (7);

— fascinated attention, sometimes speculative and in some cases quite descriptive and analytical, to certain specific types of terrestrial phenomena or processes, such as springs and earthquakes (8);

— and (not least) growing regard for information derived from the practical investigations of specialists in mining, canal and road construction, mineral water analysis, and the like, sometimes set forth in forms of detailed inquiry going beyond any immediate practical need or application, and often displaying the Enlightenment's typical belief in the benefits of putting the craftsman's empirical knowledge into the rational order cherished in science (9).

Few if any of these activities stood in isolation. They fit in with one another in many ways, and engaged the attention of people of diverse backgrounds, purposes, and inclinations. The professional habits and traditional viewpoints peculiar to natural history, natural philosophy, mining and engineering, geography and hydrography, medicine, assaying and chemical analysis, antiquarian scholarship, and other influences as well, were here mixed and blended, sometimes in ingenious ways. But they were seldom perceived as united, still less in the way the modern historian of geology is almost irresistibly tempted to gather them together. It is hindsight alone, the knowledge of the historical outcome of events, that allows and seemingly urges us to fuse together this multiplicity of endeavors and make them into proto-geology, an ancestral stage of the science. To the extent that mature geological science was distinguished by a certain state of mind, constituted by a group of conceptual components holding specific relations among themselves, that state did not exist through most of the century, even if many of the components themselves may be clearly discernible.

When Antoine de Jussieu referred in 1718 to his studies of fossil plant impressions in the Lyonnais as « herborizing », he revealed something about the way he mentally ordered his scientific researches (Jussieu, 1721). Prefigured in part by the example of Foucault (1966), historians of eighteenth century science have been showing increased sensitivity to an evident shift in the prevailing conceptual matrix toward the end of the century, and to the importance of the

---

(6) Broc, 1969 & 1975; De Beer, 1949.
(7) Broc, 1971; Rappaport, 1968, 1969a, 1969b, 1973; Rudwick, 1976.
(8) On springs see Desmarest, 1757; for earthquakes, Taylor, 1975.
(9) Birembaut, 1964; Ellenberger, 1975-1977; Gille, 1947; Rouff, 1922.

distinctive character of the organization of ideas prior to that shift for an understanding of the period's science (10). In keeping with this approach, one should be wary of ascribing too modern a historical view of nature to Fontenelle, for example, in his memorable passage calling on the *physiciens* to be the *historiens* of the earth's revolutions (Fontenelle, 1723a, 3-4). Elsewhere in the same volume Fontenelle drew a direct analogy between botany and a science dealing with fossils (earths, stones, minerals, and metals), the point of the analogy being that this fossil science, like botany, must focus first on the « arrangement » and « distribution » of its objects (Fontenelle, 1723b, 12). These were the formal concerns that so preoccupied Enlightenment science. It may even be that the high priority of « arrangement and distribution » in eighteenth century science is one of the main reasons for the notable theoretical caution exhibited by a sequence of early French geological figures, from Réaumur and Guettard through Rouelle and the young Desmarest, a prudent tendency which Ellenberger (1980, 39) has called « a sort of geological agnosticism ». At any rate these naturalists were not only theoretically circumspect, they could not be identified by their contemporaries or by themselves (before about 1760 or so) as geologists, or for that matter even as devoted principally to earth science.

II

During the second half of the eighteenth century, and accelerating as it drew toward a close, a significant pattern of change can be seen in the manner French earth sciences and techniques were recognized and encouraged. Two features of this development were (1) a marked increase in career opportunities and acknowledged roles for engagement in various aspects of earth science, and (2) a growth of personal recognition, among the enhanced numbers of identifiable earth scientists, of mutuality of interest, intellectual interdependence, and common commitment to shared specialized goals.

The first feature just mentioned fits into the larger historical movement toward a greater degree of scientific specialization, a movement focussed with special intensity in the late eighteenth and early nineteenth centuries. Disciplinary boundaries were taken a good deal more seriously at the end of the eighteenth century than at its beginning, a result that went hand in hand with the heightened prestige of science and its deeper permeation of the fabric of culture. Dramatic increases were seen in the numbers of scientists, standards of research, sense of professionalism, and scope of application of scientific methods. Along with these general conditions there were other factors that particularly favored the study of the earth, inclu-

---

(10) E.g. Albury & Oldroyd, 1977; Oldroyd, 1979; Moravia, 1980.

ding not only a demand for better-organized exploitation of mineral resources and a growing intellectual orientation toward understanding things « historically » by tracing them to their origins, but also a kind of environmentalism that demanded comprehension of the earth, man's abode, as a means toward better understanding of mankind (Moravia, 1967).

The institutions of French science adapted at first informally, and eventually in formal ways, to pressures for a larger role for earth science. Around the middle of the eighteenth century the individuals devoting part of their energies to earth sciences and techniques were sustained by a scattered variety of positions. Rouelle and Daubenton held appointments at the Jardin du Roi, in chemistry and natural history respectively. Both were also making their way up the ladder in the Royal Academy of Sciences (Rouelle in the class of chemistry while Daubenton began in botany, switching into the anatomy category in 1759). Also at the Academy, Buache had been the first holder of the special post of *adjoint géographe* since 1730, and Guettard had begun investigations in mineral geography within his botanical denomination. Buffon, Intendant of the Jardin as well as an academician (first in *mécanique*, changing to botany in 1739), had begun to work out his views respecting earth science in the opening volume of his *Histoire naturelle* (11).

Aside from the abidingly important roles in French scientific life played by the Academy and the Royal Garden, other main avenues for individual employment using earth-science expertise lay in private emolument through the giving of courses or through private patronage, and in royal appointments for mining inspection, metallurgical assaying, and the like. Following the inauguration in 1745 of Daubenton's mineralogical course at the Jardin du Roi, Valmont de Bomare commenced giving courses in his *cabinet* in 1756. Sage's course dates from 1760. Rouelle's chemistry courses, his private ones beginning perhaps as early as 1737 and the public courses at the Jardin after his appointment there in 1742, included treatment of mineralogy and the theory of the earth, and are known to have had a considerable influence on a number of French geologists of the later part of the century (12).

Demand for publications on minerals, fossils, lithological observations, classification schemes, and so forth kept busy a corps of authors, cataloguers, and translators. Notable in this last category was the baron d'Holbach, who led the way in making available in French the important systematic and observational and technical writings of important German and Swedish authors. Thanks in part to d'Holbach and his successors (such as Dietrich), the French geo-

---

(11) Institut de France, 1968; Hahn, 1971; Laissus, 1964a.
(12) Birembaut, 1964; Guerlac, 1975; Rappaport, 1960. Inclusion of earth science in chemistry courses and texts remained common through most of the century (e.g. Bucquet, Chaptal).

logical scene of the second half of the eighteenth century was, in the context of the time, relatively well exposed to ideas and information from abroad.

By the middle part of the century the government had taken initiatives to develop and employ a small cadre of mining specialists, the most successful being individuals like Hellot and Genssane who blended theoretical and practical talents. The arrival on the scene of A.-G. Jars and J.-P.-F. Guillot-Duhamel in the later 1750's continued this pattern. For some time the aim of training mining engineers was attempted as an adjunct function of existing institutions such as the Ecole des Ponts et Chaussées. The government's interest in fostering accurate cartographic representation of the land and its mineral resources similarly brought together men of different technical and scholarly backgrounds and inclinations. It is noteworthy that a positive attitude toward the collaborative interaction of philosophical or theoretical understanding and the experience of the practical arts had deep roots in eighteenth century French thought. This attitude yielded, in the geology that emerged in France at the century's end, a coherence of theory and practice that stood in contrast to the more distant relations seen in Britain between geological science and technical knowledge (13).

The enlargement of acknowledged roles for earth scientists, beginning simply with flexible use of existing disciplinary categories but leading late in the century toward eventual formalization of newly designated fields, can be seen in the Paris Academy of Sciences. The Academy admitted Jars in 1768 and Desmarest in 1771, in chemistry and *mécanique*, respectively. (Aside from his researches in Auvergne, Desmarest was known mainly as a student of industrial processes.) Haüy was admitted in 1783, in the botany section. The Academy also named a small host of recognized earth-science specialists as correspondents, including Genssane (1757), De Luc (1768), Dietrich (1775), Guillot-Duhamel (1775), Dolomieu (1778), Picot de Lapeyrouse (1780), Palassou (1781), and Lamanon (1783). The 1785 reorganization at last made possible the designation of a section for natural history and mineralogy. Desmarest, Sage, and Gua de Malves were the first *pensionnaires*, Darcet and Haüy the *associés*, the former moving up the next year upon Gua's death and being replaced by Guillot-Duhamel. With the 1795 reconstitution of the Academy as the First Class of the Institut National, the six resident members in natural history and mineralogy were Darcet and Haüy (both named by the Directory) and Desmarest, Dolomieu, Guillot-Duhamel, and Lelièvre (elected) (14).

---

(13) Birembaut, 1964 ; Gille, 1947 ; Rappaport, 1969b ; Rouff, 1922.
(14) Institut de France, 1968. Geologically-oriented non-resident associates elected before the turn of the century included Valmont de Bomare, Patrin, and Sage (all in 1796) and Gillet de Laumont (1799). It should perhaps be mentioned that two *honoraires* of the Academy in the old régime who showed special inte-

Meanwhile, at the Jardin du Roi, Faujas de Saint-Fond was added as an assistant naturalist in 1778, and in the 1793 reorganization he was made the Muséum's professor of geology. The Muséum's personnel, including Lamarck, Cuvier, and after 1802 Haüy, made it a great center in earth science as well as the life sciences at the start of the nineteenth century.

New positions exploited in the interest of earth science came into being elsewhere. At the Collège Royal Darcet and Daubenton assumed professorships in the 1770's and taught mineral science. Sage contrived the establishment of a chair of mineralogy and docimastic metallurgy that he occupied in 1778, with Guillot-Duhamel as his assistant. This was a critical step toward the creation of the Ecole des Mines de la Monnaie in 1783, where professors such as Sage, Guillot-Duhamel and Hassenfratz were occupied during the 1780's. A generation of students began to pass through the Ecole des Mines who were destined to make up an important segment of the emerging geological community in the period right around the turn of the century. These included Guillot-Duhamel's son Jean Baptiste Duhamel, Lefèbvre d'Hellencourt, Besson, Lelièvre, Muthuon, Baillet du Belloy, and Brochant de Villiers. The reorganized Ecole des Mines de la rue de l'Université, operating from 1794 to 1802, took on an especially significant role in helping to define the geological identity. Among its scientific personnel were Guillot-Duhamel, Baillet du Belloy, Hassenfratz, Vauquelin, Haüy, Dolomieu, and Alexandre Brongniart. One sign of the continuing impact of the Ecole des Mines on the institutionalization of geology in nineteenth century France is the fact that of ten elected officers in the first year (1830) of the Société Géologique de France, five had studied at the Ecole des Mines (Brongniart, Cordier, Brochant de Villiers, Dufrénoy, and Elie de Beaumont). Three of these — Brochant de Villiers, Dufrénoy, and Elie de Beaumont — continued at the Ecole des Mines as teachers at that time (15).

With the expansion in numbers of active and articulate specialists in earth science and mining engineering, publications grew accordingly. Existing journals accommodated a part of this new literature. A singularly receptive journal, one in which is preserved a record of the increasingly conscious groping for a distinctive geological identity, is *Observations sur la physique, sur l'histoire naturelle et sur des arts*, or *Journal de physique*. Especially during the long editorship of Lamétherie starting in 1785, articles on mineralogical and geological subjects, both theoretical and practical, were welcomed. Also, the founding of the *Journal des Mines* in 1794 (succeeded by the *Annales des Mines* in 1816) was a landmark, providing

---

rest in the earth sciences were Malesherbes and the duc de La Rochefoucauld d'Enville.
   (15) Torlais, 1964; Aguillon, 1889; also Bulletin de la Soc. Géol. de France, no. 1, 1830, 8.

a specialized journal that deliberately brought together information on the technique and supervision of mining with geological science.

III

In addition to the rise of new opportunities for careers and scientific recognition in the earth sciences, changes were also occurring in the ideas and attitudes of the scientists themselves. The second abovementioned feature in the general pattern of change in geological recognition and consciousness was a growing sense of mutual intellectual interdependence, amounting to a nascent community spirit. This spirit entailed a conscious striving, late in the century, for an integrated earth science, an anticipation of some so far unclear synthesis. The presence of such a climate of opinion can be seen in the statements and behavior of a number of the more highly visible representatives of the period's earth science. Four of them I select to comment upon here are Horace-Bénédict de Saussure, Déodat de Dolomieu, Barthélemy Faujas de Saint-Fond, and Jean-Claude de Lamétherie. These four men were well acquainted with one another, corresponded among themselves, and displayed a commitment to a geological science they thought belonged especially to men like themselves.

DE SAUSSURE

In the literature of late-eighteenth century geology few scientists were treated with the kind of respect that was accorded to this Genevese patrician. Greatly appreciated as an energetic and judicious observer, this inveterate Alpine traveller was also steeped in the publications of his time in natural history and the theory of the earth. Seeking to build inductively a true geological system, he realized to his dismay that in his own work his success at amassing the empirical stuff for an inductive process was not proportionately matched by his contribution to bringing that process to a culmination. One senses that de Saussure was frustrated by his inability to find his way through the myriad sorts of information necessary to his purpose, toward a coherent and comprehensive theory. It was widely understood that facts were his great strength, but then he was no mere narrative describer either, but possessed a strong competitive instinct for creative scientific synthesis (16).

---

(16) An anecdote in Freshfield (1920, 162) illustrates de Saussure's competitive involvement in the attempt to generalize out of observation and collection. At the St. Gotthard Hospice he met Charles Greville, on his way to Naples to visit his uncle William Hamilton. De Saussure relates that he was at first suspicious that Greville might want « to get the benefit of my observations, but I found with a pleasure which was perhaps ignoble that he was not a serious student and did not attempt to generalize. He was on the look-out for curious specimens

XIII

One of the earliest users of the terms *géologie* and *géologue*, de Saussure also referred often to the *théorie de la terre* that must be the eventual product of sound researches, yet made plain his opinion that no comprehensive theory truly worthy of that title had yet arisen. With de Saussure, as with some his contemporaries, there are grounds for suspicion that the meaning he attached to the word *théorie* was somewhat different, more narrow, than that which it had carried in traditional theories of the earth. The way the terms *naturaliste* and *philosophe* are handled in his *Voyages dans les Alpes* suggests that he was sensitive to the special connotations they tended to bear — of curiosity-collector and systematizer ; it is as if he conceived of the geologist as not precisely either one of these, but rather a new type of investigator drawing his methods selectively from both of these traditions. De Saussure's wide correspondence, and his assiduous attention to his « Agenda », a sort of guide for geological observers, are among the indications of his strong belief in a collective geological enterprise, one to which he believed any prepared voyager could make at least a small contribution (17).

DOLOMIEU

Of all French scientists of the eighteenth century, perhaps none had the geological identity as clearly as did Dolomieu. A high-born Knight of Malta who was known for his many journeys, especially in the Mediterranean volcanic districts and the European mountain ranges, he explicitly related practically all his scientific interests and activities, including those in mineralogy and chemistry, to his sense of an emerging, coherent science providing a theoretical and developmental view of the earth. In the final years of his abbreviated career he deliberately promoted the term *géologie*, and pointedly differentiated it from both analytical-descriptive natural history and comprehensive earth systems (18).

De Saussure is said to have looked upon Dolomieu as his successor. If this meant that Saussure saw in Dolomieu a ten-year-younger

---

for his collection, without any consideration for grouping them. I recognized that he was in no sense a formidable rival ». Carozzi (1976) has argued that de Saussure's failure to produce original interpretations to match his excellent observations should be ascribed mainly to his increasing preoccupation, from 1774 to the end of his life, with educational and political reform in Geneva. These involvements were undoubtedly of capital importance to de Saussure, but there remains room for doubt that he would have succeeded in reaching the theoretical synthesis to which he aspired even if these concerns had not distracted him. See also Carozzi, 1975.

(17) De Saussure, 1779-1796, 1796. De Saussure was by no means the only Genevese scientist with a passion for geology and a commitment to its collective pursuit. Others included Jean-André De Luc (although he was not in Geneva after 1773) and Marc-Auguste Pictet, whose Geneva-based *Bibliothèque britannique* had a strong geological bias.

(18) Dolomieu, 1798, 402-403. See also Lacroix, 1921, and Taylor, 1971.

form of himself, this was not unreasonable, since Dolomieu's posture respecting observation, system, and specific theories shared much with de Saussure's. For his part Dolomieu wrote of de Saussure as geology's greatest champion. Like de Saussure, Dolomieu believed that geological understanding would be served best by special concentration on the study of mountain regions. Although Dolomieu outlived de Saussure by less than three years, his impact on French geological consciousness carried far, and is perhaps sometimes underestimated. Dolomieu tended to be regarded in his time as more successful than de Saussure in reducing his findings to a general and comprehensive form, and in applying them to outstanding geological problems (19).

As impressive in a way as his observations and ideas are Dolomieu's connections with so many geologists of his own and the next generation. Among his field companions were Picot de Lapeyrouse, Pictet, Fleuriau de Bellevue, Brongniart, and Brochant de Villiers, to name just a few. He was outspokenly proud of his role in helping to prepare a new generation of geologists and mineralogists. His students at the Ecole des Mines and student-collaborators in the field include a number of the leaders of early nineteenth century geology, such as Cordier, Brochant de Villiers, Gillet-Laumont and Lelièvre (20).

Faujas de Saint-Fond

The greatest contemporary geological reputations, late in the century, seem to have been enjoyed mainly by men known for their travels, field observations, and caution against overeager generalization or systematizing. Saussure and Dolomieu were among these, along with Pallas and Ferber. Faujas de Saint-Fond was sometimes placed within this group, although generally somewhat nearer its margin. Two things that affected Faujas' reputation were the fact that he had been Buffon's protégé, not an unmixed blessing in the eyes of admirers of fieldwork, and the impression he had formed early in his career as a rather ruthlessly ungenerous contender for scientific priority. Dolomieu, for whom Faujas was the man who had introduced him to the study of volcanoes, considered Faujas a good friend whose high potential was not altogether fulfilled because of his susceptibility to bureaucratic, political, and social distractions. But by the turn of the century, Faujas had, figuratively and literally, a quite respectable place in French geology. This lawyer-turned-

---

(19) Dolomieu, 1794; Lacroix, 1921, *1*, 59.
(20) In addition to documentation in Lacroix, 1921, further information on Dolomieu's traveling companions is in unpublished notebooks held by the Archives de l'Académie des Sciences. References to his students are contained in notes from his courses at the Ecole des Mines, now in possession of the Bibliothèque de l'Ecole des Mines (esp. MS 175/B3).

naturalist had compiled a record of serious field research, especially concerning volcanic rocks, and he held the most visible French teaching position for geology. Having served for a period as Royal Commissioner for mines and quarries, he was attuned to practical as well as theoretical facets of geology. He shared with de Saussure and Dolomieu an expressed commitment to the caution and restraint necessary for proper geological theorizing, and like them he accepted the ideas that mountains hold especially important data for geologists and that the earth's evidence indicates a long history marked by periodic revolutions or relatively abrupt upheavals (21).

In his opening lecture for his geology course at the Muséum in 1802, Faujas expressly set out to define geology's current state, rationalize the science's separate existence, and urge on what he saw as growing French support and encouragement for it. He emphasized the grandness of conception appropriate to geology, likening it to a philosophy of nature. The picture he gave of geology incorporated information and perspectives from zoology and botany rather more than is the case in de Saussure's and Dolomieu's work, yet while insisting on geology's close alliances with other disciplines he supported its autonomy (22).

Lamétherie

Of the characters one can connect with an emerging French geological identity, Lamétherie is no doubt in some ways a less suitable representative than de Saussure, Dolomieu, or Faujas. Unlike these others Lamétherie had little field experience, and he was more at odds than any of them with the existing scientific establishment. A contentious and visionary thinker with a medical background, he had some of the tendencies ones sees in his contemporary Lamarck — toward nature-philosophizing when the main current was moving in the direction of narrower specialization, and toward construction of a grand system of nature on the basis of supposed fundamental properties of matter that were not considered wholly established (one of Lamétherie's enthusiasms was for crystallization as the basic process of all organized bodies).

---

(21) Challinor, 1971 ; Offner, 1956. For Dolomieu's opinion of Faujas, Lacroix, 1921, I, 48-49. A scathing denunciation of Faujas's grasping for priority is in Bachaumont, 1780-1789, *24*, 97-98. Other documentation regarding the controversies he precipitated can be seen in Guettard, 1768-1786, *5*, 637-668.
(22) Faujas de Saint-Fond, 1809, *1*, 1-44, « De l'état actuel de la géologie. Introduction. Discours prononcé au Muséum d'histoire naturelle, pour l'ouverture du cours de géologie, le premier mai, 1802 ». In his earlier book on the volcanoes of Vivarais and Velay (Faujas de Saint-Fond, 1778, iii) he had expressed reservations about the direct and unrestricted applicability of an existing science like chemistry to geological problems : « Il existe une Chymie de la nature bien supérieure à celle de l'art... » Emphasis on differences between the scientist's laboratory and the laboratory of nature were common in the writings of late-eighteenth century geologists.

XIII

But if it may be clear from our vantage point that Lamétherie's genius was not of the same order as Lamarck's, it is also apparent that Lamétherie had a great talent for stirring up effective scientific dialogue and for interest in others' work. He became in fact a kind of promoter and publicist of geological and mineralogical research (but not these fields exclusively), and although he held strongly to his opinions he worked to air geological ideas of various doctrinal stripes. A decidedly pro-Wernerian thinker, Lamétherie was nonetheless a leader among Continentals in calling attention to Hutton's ideas. As editor of the *Journal de physique* from 1785 to 1817 he tried to foster a geological synthesis by bringing together earth-science reports of all kinds — theoretical and practical, sweepingly broad and narrowly-focussed. Similarly, in his several specifically geological books, and as *professeur-adjoint* of mineralogy and geology at the Collège de France from 1801, he gathered diverse geological information and views. Less reserved than most of his contemporaries about comprehensive systems, his tendency to organize material around the « theories of the earth » of various scientists is somewhat deceptive, since in many cases the theories in question were not actually of such cosmic scope. Lamétherie was strongly tied to some of the more curious undercurrents in French science at the end of the century, but he was taken seriously by a number of geologists (including Dolomieu, who said Lamétherie was his best friend). Part of the reason for the standing Lamétherie had was his advocacy of a geological identity (23).

IV

I have been concerned here to emphasize the overt sensibility, among a significant group of French-speaking scientists late in the eighteenth century, of being geologists and doing geology. The historical existence of that sensibility does not, I think, hang critically on whether or not all individuals involved in this movement favored use of the same words to designate it. The aging Desmarest, for example, was unenthusiastic about the term *géologie*, saying he did not know what it meant, and consistently preferred *géographie physique* to name the most comprehensive earth science he thought attainable (24). Faujas apparently felt equally strongly that *géologie* must be pushed and defended, and he was to experience much annoyance over what he considered Cuvier's perverse disregard of the

---

(23) Taylor, 1973. Starting in 1791, Lamétherie's annual « Discours préliminaire » in the *Journal de physique* typically devoted extensive coverage to *géologie*, as well as *minéralogie*.
(24) Desmarest, 1794-1828, *1*, 842. But Desmarest did make occasional use of *géologie* and *géologue* nonetheless, e.g. *4*, 22, 163, 653-654.

word (25). That the status of *géologie* remained less than totally settled well into the nineteenth century is indicated by the *Dictionnaire des sciences naturelles*, in which it was claimed in 1820 that *géognosie* was replacing *géologie* in France as an accepted scientific term, the latter bearing the connotation of unscientific speculation and hypothesis (26). Whatever the disagreements on terminology, though, by 1800 a consensus was beginning to take form that a synthetic science of the earth should be recognized, embracing but not identical to the ongoing natural history activities aimed at inventorying the mineral realm, aided by but not limited to the analytical methods that had been transforming mineralogy and giving birth to crystallography, and involving a dose of history, a focus on natural process acting in time (27).

If a geological identity existed, there is a strong presumption that it had some conceptual underpinnings. To just what extent the French geological world at the turn of the century operated on a coherent set of widely-accepted tenets is a question beyond the scope of this paper, but an important one. To venture an opinion briefly, I think it could be shown that French geologists of the early nineteenth century had come to share some fundamental convictions to a degree sufficient to give their work a consistency not altogether incomparable to that discerned by Rudwick (1971) among pre-Lyellian British geologists. And if that should be so, the point made by Rudwick — that geological thought was already well organized early in the nineteenth century, and not in need of being established out of disorder by Lyell — would be strengthened.

Yet one must also take care not to make nineteenth century scientists out of the eighteenth century figures I have described here as fashioning a geological identity. For Enlightenment characters they were for the most part. There is even something encyclopedic about them and their enterprise; the very product they seemed to envision was a collective work, built not in the bold vision of one heroic thinker but constituted by the patient and accurate studies of many

---

(25) Cf. Faujas letters to Ménard de la Groye, 28 Feb. 1808, 30 Jan. 1812, in Archives de l'Académie des Sciences, dossier Faujas. Faujas had unfriendly relations with Cuvier, who called him Faujas Sans Fond.
(26) Dictionnaire des sciences naturelles (1816-1830), *18*, 437-438.
(27) It seems possible that a greater concern somehow to cope with the earth's development, a subtle shift of emphasis from understanding of processes toward understanding of sequences of events, may be connected with the shift Ellenberger (1980, 66-67) has noticed from actualistic to « catastrophist » geology among the French in about the last fifteen years of the eighteenth century. The basically ahistorical character of uniformitarian ways of thinking is well known. Although post-Lyellian historians of geology tend naturally to appreciate the applicability of uniformitarianism to the historical tasks of ordering geological events in time, this use of a rigorously focussed attention on processes was far less clearly understood in the eighteenth century. So it may be that what looks to us rather like late-eighteenth century catastrophism was not so much a deliberate change of geodynamical policy among the period's geologists as it was a by-product of an increased attention to the task of ordering events in time.

dedicated observers. Not that all contributions were to be valued equally, but no commanding figure dominated the scene, nor does it appear that one was expected to. Whatever synthesis was anticipated was expected to arrive through fairly ordinary investigation. Suspicious of most inherited systems, these geologists tended to feel that the main problem was to get good information and organize it in the right way.

The spirit of the emergingly self-aware network of francophone geologists at the turn of the century was highly empirical, sure of its broad utility, open to contributions of knowledge from the practical arts, wary of the too-comfortable rational unity of traditional theories of the earth, yet confident of an imminent resolution of outstanding theoretical difficulties through some kind of normalizing synthesis. It also tended to be quite secular. If the relevance of scripture to geology was an issue, this was so only at the periphery; providentialism in geology had come to be regarded as a somewhat puzzling, rather British aberration (28). The earth had come to be seen as a material system subject to natural laws, susceptible of comprehension by the secular tools of understanding bequeathed by the Enlightenment. Prominent among these tools, of course, were the analytical and organizing procedures of natural philosophy and natural history, but also making its presence felt in a perhaps less obvious but potentially revolutionary way was the cultivation of historical modes of explanation.

In closing, it may be observed that two great figures of French science stand astride the subject of this paper, without entering deeply into it : Buffon and Cuvier. So great was their influence that in the conventional periodization of French science it could be said that the concerns treated here begin in the era of Buffon and culminate in the start of the era of Cuvier. It is interesting that both these men were evidently drawn into their respective concerns for geological problems through their greater preoccupation with living things (29). Yet it is not at all so clear that the issues that were central to the leaders in creating a French geological identity, before Cuvier came to have a serious impact on geology, focussed with special intensity on life forms.

REFERENCES

AGUILLON L., L'Ecole des Mines de Paris, notice historique, in *Annales des mines*, *15*, ser. 8, 433-686, 1889.

---

(28) In 1802 Faujas called the English scriptural orientation in geology « une sorte d'obstination » (Faujas de Saint-Fond, 1809, *1*, 20). On the relatively high degree of French naturalists' liberation from orthodox restraints in the eighteenth century, see Rappaport, 1978.
(29) Roger, introduction to Buffon, 1962 ; Coleman, 1964.

XIII

ALBURY W.R., & OLDROYD D.R., From Renaissance mineral studies to historical geology, in the light of Michel Foucault's The Order of Things, *Brit. J. for the Hist. of Sci.*, *10*, 187-215, 1977.

BACHAUMONT L.P. de (ed.), *Mémoires secrets pour servir à l'histoire de la république des lettres en France, depuis MDCCLXII jusqu'à nos jours ; Ou, journal d'un observateur*, 36 vol., Londres, 1780-1789.

BIREMBAUT A., L'enseignement de la minéralogie et des techniques minières, in *Enseignement et diffusion des sciences en France au XVIIIe siècle*, 365-418, Ed. R. Taton, Paris, 1964.

BROC N., *Les montagnes vues par les géographes et les naturalistes de langue française au XVIIIe siècle*, Paris, 1969.

— Un géographe dans son siècle : Philippe Buache (1700-1773), *XVIIIe siècle*, 3, 223-235, Paris, 1971.

— *La géographie des philosophes : Géographes et voyageurs français au XVIIIe siècle*, Paris, 1975.

BUFFON G.L.L., *Les époques de la nature. Edition critique par J. Roger*, Paris, 1962.

CAROZZI A.V., Horace Bénédict de Saussure, in *Dictionary of sci. biography*, 12, 119-123, New York, 1975.

— Horace Bénédict de Saussure : Geologist or educational reformer ?, in *J. of Geol. Education*, 24, 46-49, 1976.

CHALLINOR J., Barthélemy Faujas de Saint-Fond, in *Dictionary of sci. biography*, 4, 548-549, New York, 1971.

COLEMAN W., *Georges Cuvier, Zoologist*, Cambridge, Massachusetts, 1964.

DEAN D.R., The word « geology », in *Annals of sci.*, 36, 35-43, 1979.

DE BEER G.R., *Travellers in Switzerland*, London, New York, Toronto, 1949.

DESMAREST N., Fontaine, in *Encyclopédie, ou dictionnaire raisonné des sciences, des arts et des métiers, par une société de gens de lettres*, 7, 81-101, Ed. D. Diderot et J.L. d'Alembert, Paris, 1757.

— *Encyclopédie méthodique, Géographie-physique*, 5 vol., Paris, 1794-1828.

DICTIONNAIRE DES SCIENCES NATURELLES, 60 vol., Strasbourg et Paris, 1816-1830.

DOLOMIEU D., Discours sur l'étude de la géologie, in *J. de physique*, 2 [45], 256-272, 1794.

— Rapport fait à l'Institut National, par Dolomieu, sur ses voyages de l'an cinquième et sixième, *J. de physique*, 3 [46], 401-427, 1798. [Also in *J. des mines*, 7, n° 41, an 6, 385-432, 1798.]

ELLENBERGER F., *A l'aube de la géologie moderne : Henri Gautier (1660-1737)*, Paris, 1975-1977. [Extrait de *Histoire et nature*, n°^{s} 7 and 9-10.]

— De l'influence de l'environnement sur les concepts : L'exemple des théories géodynamiques au XVIIIe siècle en France, *Rev. d'hist. des sci.*, 33, 33-68, 1980.

FAUJAS DE SAINT-FOND B., *Recherches sur les volcans éteints du Vivarais et du Velay*, Grenoble et Paris, 1778.

— *Essai de géologie, ou mémoires pour servir à l'histoire naturelle du globe*, 2 vol., Paris, 1809.

FONTENELLE B. de, Sur les pétrifications trouvées en France, *Hist. de l'Acad. Royale des Sci.*, *1721*, 1-4, 1723a.

— Sur la formation des cailloux, *Hist. de l'Acad. Royale des Sci.*, *1721*, 12-16, 1723b.

FOUCAULT M., *Les mots et les choses : Une archéologie des sciences humaines*, Paris, 1966.

FRESHFIELD D.W., *The life of Horace Benedict de Saussure*, With the collaboration of H.F. Montagnier, London, 1920.

GILLE B., *Les origines de la grande industrie métallurgique en France*, Paris, 1947.
GUERLAC H., Balthazar-Georges Sage, *Dictionary of sci. biography*, 12, 63-69, New York, 1975.
GUETTARD J.-E., *Mémoires sur différentes parties des sciences et arts*, 6 vol., Paris, 1768-1796.
GUNTAU M., The emergence of geology as a scientific discipline, *Hist. of sci.*, 16, 280-290, 1978.
HAHN R., *The anatomy of a scientific institution : The Paris Academy of Sciences, 1666-1803*, Berkeley, Los Angeles, London, 1971.
INSTITUT DE FRANCE, *Index biographique des membres et correspondants de l'Académie des Sciences du 22 décembre 1666 au 15 décembre 1967*, Paris, 1968.
JUSSIEU A. de, Examen des causes des impressions des plantes marquées sur certaines pierres des environs de Saint-Chaumont dans le Lyonnois, *Mém. de l'Acad. Royale des Sci.*, 1718, 287-297, Paris, 1721.
LACROIX A., *Déodat Dolomieu, membre de l'Institut National (1750-1801)*, 2 vol., Paris, 1921.
LAISSUS Y., Le Jardin du Roi, in *Enseignement et diffusion des sciences en France au XVIIIe siècle*, 287-341, éd. R. Taton, Paris, 1964a.
— Les cabinets d'histoire naturelle, in *Enseignement et diffusion des sciences en France au XVIIIe siècle*, 659-712, éd. R. Taton, Paris, 1964b.
LAMY E., *Les cabinets d'histoire naturelle en France au XVIIIe siècle et le cabinet du Roi (1635-1793)*, Paris, 1930.
MONNET A.-G., *Nouveau système de minéralogie, ou essai d'une nouvelle exposition du règne minéral, auquel on a joint un supplément au traité de la dissolution des métaux, avec des observations relatives au Dictionnaire de Chymie*, Bouillon, 1779.
MORAVIA S., Philosophie et géographie à la fin du XVIIIe siècle, in *Studies on Voltaire and the Eighteenth Century*, 57, 937-1011, 1967.
— The Enlightenment and the sciences of man, *Hist. of sci.*, 18, 247-268, 1980.
MORNET D., *Les sciences de la nature en France au XVIIIe siècle*, Paris, 1911.
OFFNER J., Faujas de Saint-Fond, naturaliste dauphinois (1741-1819), *Procès-verbaux mensuels de la Société Dauphinoise d'Ethnologie et d'Archéologie*, 32, n°s 251-253, 30-41, 1956.
OLDROYD D.R., Historicism and the rise of historical geology, *Hist. of sci.*, 17, 191-213, 227-257, 1979.
PORTER R., The industrial revolution and the rise of the science of geology, in *Changing perspectives in the history of science : Essays in honour of Joseph Needham*, 320-343, Ed. M. Teich & R. Young, Dordrecht and Boston, 1973.
— *The making of geology : Earth science in Britain, 1660-1815*, Cambridge, 1977.
RAPPAPORT R., G.-F. Rouelle : An eighteenth century chemist and teacher, *Chymia*, 6, 68-101, 1960.
— Lavoisier's geologic activities, 1763-1792, *Isis*, 58, 375-384, 1968.
— The early disputes between Lavoisier and Monnet, 1777-1781, *Brit. j. for the hist. of sci.*, 4, 233-244, 1969a.
— The geological atlas of Guettard, Lavoisier, and Monnet : Conflicting views of the nature of geology, in *Toward a history of geology*, 272-287, ed. C.J. Schneer, Cambridge, Massachusetts, and London, 1969b.

XIII

— Lavoisier's theory of the earth, *Brit. j. for the hist. of sci.*, 6, 247-260, 1973.
— Geology and orthodoxy : The case of Noah's flood in eighteenth century thought, *Brit. j. for the hist. of sci.*, 11, 1-18, 1978.

ROGER J., La théorie de la terre au XVIIe siècle, *Rev. d'hist. des sci.*, 26, 23-48, 1973.
— Le feu et l'histoire : James Hutton et la naissance de la géologie, in *Approches des lumières : Mélanges offerts à Jean Fabre*, 415-429, éd. B. Guyon, Paris, 1974.

ROUFF M., *Les mines de charbon en France au XVIIIe siècle, 1741-1791. Etude d'histoire économique et sociale*, Paris, 1922.

RUDWICK M.J.S., Uniformity and progression : Reflections on the structure of geological theory in the age of Lyell, in *Perspectives in the history of science and technology*, 209-227, ed. D.H.D. Roller, Norman, Oklahoma, 1971.
— The emergence of a visual language for geological science 1760-1840, *Hist. of sci.*, 14, 149-195, 1976.

SAUSSURE H.B. de, *Voyages dans les Alpes, précédés d'un essai sur l'histoire naturelle des environs de Genève*, 4 vol., Neuchâtel, 1779-1796.
— Agenda, ou tableau général des observations et des recherches dont les résultats doivent servir de base à la théorie de la terre, *J. des mines*, 4, n° 20, Floréal an IV, 1-70, 1796. [Also in *Voyages dans les Alpes*, 4, 467-529.]

SCHNEER C.J. (ed.), *Toward a history of geology*, Cambridge, Massachusetts, and London, 1969.
— *Two hundred years of geology in America*, Hanover, New Hampshire, 1979.

TATON R. (ed.), *Enseignement et diffusion des sciences en France au XVIIIe siècle*, Paris, 1964.

TAYLOR J.G., Eighteenth century earthquake theories : A case-history investigation into the character of the study of the earth in the Enlightenment, *Univ. of Oklahoma Ph.D. dissertation*, Norman, 1975.

TAYLOR K.L., Déodat de Dolomieu, *Dictionary of sci. biography*, 4, 149-153, New York, 1971.
— Jean-Claude de Lamétherie, *Dictionary of sci. biography*, 7, 602-604, New York, 1973.
— Geology in 1776 : Some notes on the character of an incipient science, in *Two hundred years of geology in America*, 75-90, ed. C.J. Schneer, Hanover, New Hampshire, 1979.

TORLAIS J., Le Collège Royal, in *Enseignement et diffusion des sciences en France au XVIIIe siècle*, 261-286, ed. R. Taton, Paris, 1964.

XIV

THE *ÉPOQUES DE LA NATURE* AND GEOLOGY DURING BUFFON'S LATER YEARS

INTRODUCTION

Buffon's treatise *Des Époques de la Nature* is not only one of the best-known scientific books of the second half of the eighteenth-century, it is also among the most accessible. Besides the existence of many editions of Buffon's *Histoire Naturelle*, for which the *Époques* may be regarded as providing a culminating general framework, several separate editions of the *Époques* have appeared within the last century. A recent one intended for a large public, and still available, was ably edited by our colleague Gabriel Gohau.[1] The definitive scholarly edition with extensive commentary and analysis was published in 1962 by Jacques Roger. It is good news that this volume, long out of print, has now been reissued.[2]

In fact, a great deal that anyone might wish to know about the *Époques* and its place in the science of its day has already been said by Professor Roger. Together with my keen appreciation of the honor of being invited to speak to this gathering about Buffon's *Époques*, comes the knowledge that this famous text is inescapably linked in our minds with our colleague who has so superbly presented and interpreted it. Inevitably, a considerable part of what I have to say must be a partial digest of what we have already learned through Jacques Roger, supplemented by valuable insights from others such as Dr. Gohau.[3]

For the historian considering Buffon's *Époques de la Nature* in relation to geology's contemporary development, a number of seemingly contradictory thoughts present themselves. The *Époques* represents perhaps the ultimate example of the *genre* of the Theory of the Earth, yet at the same time it was a work on some counts out of touch with the latest pertinent scientific trends and developments. The *Époques* is hailed, with justification, as a landmark in the history of geological discourse, yet some of its fundamental theses were repudiated by most of the leading practitioners of the nascent geological science. Certain elements of the *Époques* certainly exercised a large influence on geological thinking in the years and decades subsequent to its publication, yet one senses that the ablest of those involved in the kinds of problems Buffon addressed tended to regard the *Époques*, taken as a whole,

---

1. Buffon [9].
2. Buffon [8].
3. See especially Buffon [8], but also Roger [36], [37], and [38]; and Gohau [21]. For the sake of economy, references are omitted in this paper for a number of fundamental points concerning the *Époques de la Nature*, which the reader can readily find established in Jacques Roger's modern critical edition [8].

as an artistic masterpiece worthy of admiration and preservation, but not of emulation. One is even drawn to the idea that the author of the *Époques de la Nature* was, during the century's closing decades, for geology at least, at once both a central and a marginal figure. Buffon was credited with a grand achievement in evoking a coherent account of the Earth, and was complimented with manifest sincerity through widespread adoption of certain critical parts of his conception and terminology; yet at the same time he was widely viewed, among peers best situated to judge, as clinging to some indefensible theories and as following an outmoded style of scientific inquiry.

Just how much real contradiction there is in this sort of assessment of Buffon's *Époques* can be debated. I believe one finds, in taking a close look at the historical situation, that the apparently antithetical claims I have just mentioned are generally confirmed, but also reconciled; or if not altogether reconciled, at least the seemingly contradictory nature of these observations is greatly diminished. This resolution comes chiefly through our recognition that Buffon and his geological contemporaries spoke to rather different constituencies, and did not share quite the same views on some fundamental questions, such as the ways geological knowledge should be sought and framed, or the character of original and meritorious geological research. It will be part of my objective today to identify some of the main points that differentiated Buffon from many of the geologists of his time, as well as to recognize the ground they held in common.

THE *ÉPOQUES DE LA NATURE* IN BUFFON'S GEOLOGICAL ŒUVRE

The *Époques de la Nature* was undoubtedly Buffon's most thoroughly-considered geological work, the result of his mature reflection. With a 1778 imprint and placed on sale in 1779, the book came three decades after the *Théorie de la Terre,* and embodied some momentous shifts of perspective and emphasis from that earlier book, which was itself a major landmark in geological literature. While the *Théorie* and the *Époques* are Buffon's two geological works that are usually remembered, it is worth bearing in mind that about one quarter of the 36 quarto volumes of *Histoire naturelle* have a geological character. Two volumes of the *Supplément* (I, 1774; II, 1775) were presented as expansions on the *Théorie de la Terre,* and can be seen even more aptly as preludes to the *Époques*. These included treatments of the action of heat and water on mineral substances, and Buffon's studies of the cooling of molten metals, with application of the results to the cooling of the planets, including Earth. Then, several years after the appearance of the *Époques*, the *Histoire naturelle des minéraux* was published in five volumes (1783-1788), articulating detailed views on minerals and rocks, as well as the processes of their formation and alteration. So the *Époques*, which constitutes only part of one volume, is not, strictly speaking, Buffon's last word geologically.[4]

I think it is also worth noting that the subject matter of the *Époques de la*

---

4. Buffon [7]. Besides *Des Époques de la Nature* (pp. 1-254), *Notes justificatives des faits rapportés dans les Époques de la Nature* (pp. 495-599), and *Explication de la carte géographique* (pp. 601-615), this volume includes *Additions et corrections aux articles qui contiennent les Preuves de la théorie de la Terre, depuis la page 127 jusqu'à la fin de ce volume* (pp. 255-494).

While in a certain sense virtually all of Buffon's *Histoire Naturelle* was a work to which his subordinates contributed anonymously, it is thought that the *Histoire naturelle des minéraux* in particular was the result of collaborative effort.

The Époques de la Nature and Geology during Buffon's later Years

Nature is cosmogonical, botanical, zoological, and anthropological, as well as geological. Buffon's intentions, I dare say, transcended these categories. His main aim was to reset natural history in a global framework enabling comprehension of the mineral, plant, and animal furniture of the world, including humankind. It is appropriate to see the *Époques* as both more and less than what one might expect of most late eighteenth century geological works. It is more, in the sense of being a complete sequential account of the whole of nature, from the planets to man. It is less, by virtue of the way numerous geological topics and issues, and especially details of investigatory procedures, are subordinated in the presentation of his sequential tableau.

I have found a somewhat surprisingly low level of discussion of Buffon's *Époques* in the geological literature from the score of years following its publication. And apart from those few scientists like Romé de l'Isle, who took Buffon to task on a single issue (the Earth's central heat),[5] the public reactions of critics in the geological arena were generally lacking in passion. Some of Buffon's most avid critics (notably Royou, Feller, and Barruel) were principally concerned about the religious implications of the *Époques*, and resorted to the scientific arena of debate for essentially strategic reasons. Geological writers who mentioned Buffon directly were often careful to mix their criticism with praise. Such is the tenor of even the head-on critiques by Philippe Bertrand and by Marivetz and Goussier in 1780.[6] Somewhat later, when Desmarest devoted a large space in his *Géographie physique* to Buffon, he expressed practically no judgments upon the lengthy extracts from the *Époques* printed there;[7] one has to turn to other articles of the same compendium to see Desmarest's hostility to, for instance, Buffon's theory that submarine currents were important geomorphological agents.[8] One senses in Desmarest, and I think in a number of his contemporaries, a genuine but somewhat diffuse and qualified respect for Buffon, coupled not only with opposition to particular parts of the Buffonian theory but also with reservations about the speculative totality of the *Époques*.[9] Because of this tendency toward a certain reserve regarding Buffon, long

---

5. This is not to say that Romé de l'Isle's private opinion was equivocal. In a 1784 letter to de Saussure he referred to «*l'ignorante éloquence*» of Buffon (Romé de l'Isle [2]).

6. In his assessment of the *Époques*, Deluc offered his judgment that Buffon's *Histoire naturelle*, «*en tant que GÉNÉRALE est défectueuse; mais elle n'en est pas moins, comme PARTICULIÈRE, un trésor de faits & de beautés*» (Deluc [11], T. V : p. 611).

7. Desmarest [13], T. I : pp. 72-121. No doubt Desmarest implied a judgment, however, in referring to the *Époques* as «*cette espèce de cosmogonie*» (p. 72). The major parts of the excerpts of Buffon's works, in Desmarest's collection, are from *Théorie de la Terre* (pp. 72-86) and *Époques de la Nature* (pp. 94-121). The rest (pp. 86-94) comes from Buffon's discussions of the geographical distribution of quadrupeds, in Vol. IX of the *Histoire naturelle* [1761], *Animaux communs aux deux continens*. Desmarest's selections from Buffon's work were republished in Naigeon [29], T. III : pp. 801-836.

8. Desmarest [13], for example T. I : p. 325 (*Maillet*), and T. II : pp. 292-295 (*Allier*).

9. Indeed, one can find denigrations of Buffon's place in science, recorded privately by reputable contemporaries. Freshfield records this passage of a letter by H.B. de Saussure (unfortunately not dated in Freshfield's biography) : «*I have often had occasion to speak of M. de Buffon with members of the Academy. They do justice to the beauty of his style, but they think nothing of him as a man of science : they look on him neither as a physicist, nor a geometrician, nor a naturalist. His observations they account very inexact and his systems visionary. Perhaps jealousy enters into their judgment. M. de Buffon has, no doubt, excited it by his brilliancy : but it is certain his character also arouses hostility;...*» (Freshfield [17] : p. 93). Another highly adverse view of Buffon's standing in the intellectual world of his time was expressed by Marmontel [28], T. II : pp. 240-241. Such opinions, especially of a Marmontel, need certainly to be judged carefully in light of prevailing intellectual frictions and factionalism. While this kind of testimony to Buffon's inferior status among intellectual peers must not be dismissed, it seems to me that it tends to underestimate Buffon's role in forming the period's climate of scientific discourse.

established as an icon in French science, the explicitly-stated opinions of geologists on the *Époques* are probably an insufficient basis for gauging the place of the *Époques* in the geology of the time. The fact, for example, that a number of geological writers quickly adopted the notion of identifying particular terrestrial phenomena with distinct epochs counts for something.[10] But so also does the refusal of most in the geological world, for another generation, to follow Buffon in seeing the primary rocks as igneous in origin. In the end, what other geologists *did* once *Des Époques de la Nature* was loose in the world matters as much as, if not more than, what they said about Buffon's treatise.

The main outlines of the *Époques* are probably familiar to most in this audience. The elegantly phrased *Premier Discours* acknowledges the difficulties in acquiring reliable knowledge of Nature's past, and compares the procedures of natural history with those of civil history. Buffon holds that, in keeping with sound philosophy, we must base our reasoning on what we know to be true from immediate experience and reliable evidence. And it is only in our own days, he says, that a sufficient knowledge has been gained of the Earth's materials and organization to reveal clearly nature's past stages, and to permit «*the night of time*» to be penetrated.[11] He offers a group of selected facts about the globe's present condition, and about monuments or relics of the past, in a manner suggesting a sort of inductive inference of historical consequences from empirical knowledge. The chief conclusion is that the Earth was originally an incandescent ball of hot material, and is still in a long process of progressive cooling; the generation and differentiation of the globe's visible parts have resulted naturally from that irreversible cooling process. The *Premier Discours* ends with remarks on Genesis, where Buffon seeks to disarm religious opponents, insisting on the irrelevance to scientific inquiry of the Bible's literal language, and offering assurances that his system is consistent with properly interpreted Scripture.

The seven epochs which follow the *Premier Discours* tell the Earth's story from its generation out of matter torn from the sun by a comet, down to the present age, which witnesses mankind's rise to a position of major influence in nature. In the second epoch, the Earth's first physiographic inequalities appeared in the vitrescible

10. Philippe Bertrand and Étienne-Claude Marivetz, for instance. See also Haidinger [23], Launay [25], Soulavie [41] and [40], esp. T. I and IV.

Roger [8] (pp. XL-XLI) has noted the employment, during the 1750s, of the term *époques* in a geological sense by Boulanger, Deslandes, Elie Bertrand, and Buffon himself; and he has called attention to Desmarest's apparent effort to claim credit for the expression by his use of it in a 1775 presentation (Desmarest [12]). Roger's point, that naturalists were coming during this period to regard it as normal to speak of events in the Earth's history as marking *époques*, is illustrated in a letter from Desmarest to the Duc de La Rochefoucauld in 1769 : discussing the means of determining whether the Baltic Sea level has really been in decline, Desmarest prescribed a systematic combination of relevant data from both the Mediterranean and the Baltic Seas :

«*Or ces preuves Etant combinées avec celles qu'on pourroit recueillir sur les bords de la baltique seroient bien plus convainquantes que la simple dispute de Suede. Mais pour recueillir ces preuves, il faut savoir distinguer leur caractere propre : Savoir demeler cet ordre de faits d'avec les autres, Savoir les chercher ou ils sont et cela tient à la distinction des Epoques qu'on n'a pas Encore pense a introduire dans l'étude de l'histoire naturelle*» (Desmarest [1])

Although Desmarest's remark suggests that the assemblage of geological data according to chronological order was still a novelty, evidently he and La Rochefoucauld already shared a vocabulary for it.

One pretender to invention of the idea of geological *époques* who seems to have been overlooked is the Belgian physician Robert de Limbourg. In a complaint published in 1780 (Limbourg [27] : p. 335n), Limbourg indicated his belief that Desmarest and Buffon had stolen the notion of geological epochs from a paper he had presented in 1774 (Limbourg [26]).

11. *Époques de la Nature, Premier Discours* [1778], in Buffon [7] : p. 5.

or glassy materials congealing on the surface. In the third epoch, as cooling progressed, condensing vapors created a universal ocean in which animal life grew, and great horizontal beds of calcareous rock were formed from organic debris. Systematic currents further shaped the submarine terrain. These currents, to a large extent governed by tidal action, continued to form the submarine terrain and the forms of shorelines as, in the fourth epoch, the universal ocean gradually retreated, and volcanoes came into action. In the fifth epoch, large animals suited to tropical conditions lived in northern lands which have since become far too cold to sustain them, a result of the globe's continuing refrigeration. The separation of the continents took place during the sixth epoch, forming two major land masses, the old and new continents, divided by two great oceans oriented north-to-south. The general east-to-west movement of the seas further affected the configuration of the land. Finally, in the seventh epoch, man, who had been witness to convulsive changes in the Earth brought on by inundations, earthquakes, and volcanoes, succeeded in establishing a high and learned culture. All this was lost in an age of barbarism, but with the restoration of the sciences over the last thirty centuries has at last come human knowledge and power sufficient to master nature, perhaps to the extent of soon controlling the climate.

The ways in which this remarkably ambitious and coherent work differs from the earlier *Théorie de la Terre* have tended to overshadow the very real continuities in Buffon's geological thought. And just as the *Époques* reveals notable consistencies as well as some important reversals in Buffon's own geological thinking, there is a mixture of the conventional and the exceptional in the *Époques* when compared with what was going on in geological science at the time. In what remains of this brief talk, I will try to identify a few of the more interesting aspects of the geological outlook displayed in the *Époques*, to comment on how Buffon's approach and positions related to the ideas discernible among other geologists of the time, and to remark where appropriate on the consistency or change of course in Buffon's own views. I hope this will contribute a little to our understanding of how Buffon was situated respecting the significant changes that were occurring in geology during the later part of his life.

HISTORY AND THE *ÉPOQUES DE LA NATURE*

The key to the conceptual and rhetorical unity of *Des Époques de la Nature* is its historical character. Although in the *Premier Discours* there is at least a *pro forma* line of reasoning from effects to causes, this book is essentially a narration of events, a continuous story of connected epochs from beginning to present. Perhaps it is true that, logically speaking, the causes of the events are as important as their sequential narration. But it is the sequence that dominates and distinguishes the presentation. Past events, their causes, and evidence of the surviving effects of change are impressively integrated. A telling measure of the supremacy of the historical narrative is that geological evidence or processes are not raised for their own sake; they serve to illuminate a set of events.

Buffon's cosmogonical hypothesis explaining the main line of events in the *Époques* had appeared long before, in the first article of the *Preuves* to the *Théorie de la Terre*. But that idea –of the planet's origins in stellar material torn away from the sun by a comet– remained isolated in the *Théorie*; it found no natural connection there with a geological vision focused on the relations between observed regular effects and the natural processes that produced them, a vision centered not on events but on an almost atemporal network of ordered causes and effects. Now, in the *Époques*, the comet hypothesis is fully exploited. Also brought forward to play a vital role is its implicit corollary, that the Earth has progressively cooled from its initial molten state. In the *Théorie*, the transformation of Earth from an incandescent ball to a watery, fertile globe had been a neglected interlude, as if the Earth's fiery origins were divorced from that book's real business. In the *Époques* Buffon retains much that he had said in the *Théorie* about the geological potency of water. So the *Époques* chronicles a world in which the progressive diminution of an intrinsic terrestrial heat shares the stage, from the third epoch on, with water. Both heat and water function geologically in important ways.

One naturally asks what brought Buffon to shift his perspective so radically between these two famous treatises, from a position akin to that of a cyclic steady-state to an emphasis on directional change, and from an evocation of geological processes almost completely oriented around water to a story of terrestrial change anchored in dissipation of the Earth's supposed central heat. Jacques Roger has argued persuasively that a central reason lies in Buffon's study of living things, where he increasingly found difficulty in recognizing an underlying reality of natural kinds seated in the synchronic relations among individuals; so Buffon turned instead to continuity in time as the unifying basis of species. Our attention has been directed to Dortous de Mairan's revisions of his study on terrestrial heat as the probable source of Buffon's embrace of the secular cooling of the globe.[12] Buffon's geological reorientation, on this interpretation, was guided mainly by non-geological considerations. This is entirely plausible. Indeed, the hypothesis of the Earth's inner heat, although an age-old option for theorists of the Earth, was widely seen in the later eighteenth century as having important geological and mineralogical disadvantages, and it is easy to believe that Buffon needed reasons of another kind to adopt it. However, there were things afoot in geology, and perhaps elsewhere besides, that would lend support to his commitment to directional change in the Earth.

The traditions of the Theory of the Earth had, since the seventeenth century, embraced a chronicling of the Earth's origins and subsequent changes, to account for its present state. Buffon's focus, in the *Théorie de la Terre*, on cyclic change and maintenance of equilibrium in terrestrial processes, had constituted in some ways a departure from the center of those traditions. As has been mentioned, the *Théorie* had included something almost like a lapse of logic in positing the Earth's igneous origins but in neglecting to link that initial condition with the water-dominated sequel. These ideas could hardly be made whole without an adjustment toward a directionalist position. Furthermore, the activities and ideas of other geological thinkers during the third quarter of the century included prominent historical components. Some of these figures had a discernible influence in Buffon's formulation of the *Époques*. Nicolas-Antoine Boulanger is a notable instance; an outstanding feature of Boulanger's unpublished *Anecdotes de la Nature,* from which Buffon appropriated substantial chunks, is its explanation of particular physiographic changes in terms of specific catastrophic inundations.[13] Buffon's exploita-

---

12. Roger, *in* Buffon [8] : pp. XXII ff.; also Haber [22] : pp. 115-117.
13. See Roger [35]; also Bledstein [6].

tion of Boulanger's work, incidentally, was one case among many where Buffon utilized others' ideas while ignoring important ways in which they contradicted his own theory.[14]

Buffon was also clearly impressed by the temporal schematization of distinct rock masses advocated by Johann Gottlob Lehmann and others.[15] While Lehmann is cited in the *Époques*, one notices that Buffon did not refer to his chemist compatriot Guillaume-François Rouelle, whose development of a somewhat similar scheme dividing rocks in terms of character and presumed age was important for some of Buffon's younger contemporaries.[16] To the extent that one can say Lehmann's and Rouelle's views represented historical systems, these implied the fundamental structuring of the landforms during a primitive period, with subsequent additions and changes of secondary significance. The *Époques* of Buffon took up a similar view. While I cannot tell whether Buffon paid attention to Rouelle, one can say that he was not apt to be ignorant of what was well known among others interested in similar problems. (In fact, a review of all the geologically-oriented volumes of the *Histoire Naturelle* makes it hard to avoid the conclusion that Buffon was extraordinarily well informed.) In any case, Buffon was responsive to the work of at least some geological writers who tried to put things in chronological order and sought to relate known effects to past events.[17]

To this it can be added that Buffon's adoption of a temporal perspective may also have been supported by an interest in antiquarian and anthropological ideas. The framing in chronological terms of accounts of human beliefs and institutions evidently held some appeal for Buffon; this is exemplified again in his unacknowledged borrowing from Boulanger, and, eventually, and quite openly, from Jean-Sylvain Bailly.

Before leaving the theme of history, let me add a few observations on the sort of temporal understanding one finds in the *Époques*. One of the more startling features for which the book has ever since been noted, is its pretension to an absolute chronology. Each of the seven epochs was assigned a more or less precise duration. The Earth's age was set at 75 000 years, and terrestrial life could be expected to freeze out after 93 000 more. Study of the surviving manuscript of the *Époques* has shown that Buffon privately entertained a chronological vision far bolder than the one he published. But even the conservative figures printed in the *Époques* were unusual. Among the many geological writers of the late eighteenth century who considered that the Earth's history demanded a relatively lengthy time scale, the

---

14. Boulanger made frequent use of the term *époque*, sometimes referring to a period of time, but on other occasions using the word's older meaning of a fixed moment or chronological reference point. Boulanger, unlike Buffon, did not believe the Earth would submit to any attempt to date its past; we must be content with the less bold but more nearly attainable goal of fixing the relative order of some comparatively recent events. The empirical methods at our disposal for establishing successive events, or revolutions, Boulanger thought, become weaker as we press further back in time. In addition, Boulanger believed the Earth passed through enormous cycles of generation and decay, so the recovery of real origins was to him impossible.

15. Buffon's references to Lehmann are in the *Notes justificatives* —2nd epoch, n. 16; 3rd epoch, n. 23 (Cf. *Notes justificatives des faits rapportés dans les Époques de la Nature* [1778], *in* Buffon [7] : pp. 495-599).

16. On Rouelle, see Rappaport [31], and Ellenberger [14].

17. While the *Époques* obviously displayed sensitivity to temporal ordering of terrestrial phenomena, Buffon here showed significantly less emphasis upon the *distribution, arrangement or disposition* of the Earth's surface components and features than one sees in the earlier *Théorie de la Terre*. It seems to me, however, that among many late-eighteenth-century geologists, distribution and disposition of materials and structures were the object of a great deal of attention.

normal thing was to speak in euphemisms, such as *«a long series of ages»*.[18] The replacement of such euphemisms with specific, large numbers, in the *Époques* was not soon copied by many others. But it is not unlikely that Buffon was the object of some silent gratitude, within the geological brotherhood, for demonstrating that the conventional time boundaries could be contradicted openly in a publication of such high visibility and repute. Such gratitude need not have required agreement that Buffon or anyone else really possessed yet any precise means of chronometric calculation.

The *Époques*, with its status as an official publication written by a distinguished public figure, thus served the important role of legitimizing candidly naturalistic treatment of subjects that were still sensitive. This applied not just to the time scale, but more generally to an exclusively physical, nonprovidentialist discussion of the Earth's formation and subsequent alteration.[19]

The history found in Buffon's *Époques* is irreversible, continuous, and complete. This might all seem unremarkable, but in contemporary geology options of a different kind were receiving increasing attention. The main phases of change in Buffon's *Époques* do not repeat themselves. The great universal ocean of the third and fourth epochs starts to form, advances, and then retreats; these things happen once, and are done with. But some of the more narrowly-focussed geological investigations of the period were turning up evidence of more complex sets of changes than would readily fit into Buffon's system of universal phases of change. Some evidence made it seem necessary to envision successive retreats and advances of the sea. The suggestion was being made, in other words, that changes in the relations of land and sea were historically reversible.[20] It is true that such a conclusion was not a challenge to a temporal *logic* of unrepeated, distinct events. But it was seen at the time as undercutting, or at least as complicating, systems like Buffon's that included the progressive diminution of a universal Ocean.[21]

While geological investigations of the 1770s and 1780s were beginning to indicate a succession of marine incursions and retreats, alternatives to a strictly continuous geological history were also being developed. The best known case was Desmarest's analysis of the Auvergne lavas, in three successive epochs.[22] Desmarest's paper, presented in the 1770s, gave evidence of temporal distinctions among geological events without connecting them historically. That is, the geologically identifiable epochs were separated by intervals unaddressed by this geological

18. This point was made by Rappaport [32] : p. 65.
19. Our esteemed colleague Goulven Laurent quite correctly pointed out, at the Colloque Buffon, that no assessment of Buffon's *Époques* would be complete without recognizing also its remarkable integration of the history of living beings with the history of the Earth (including its emphasis on the correlation of faunal and climatic change, and the implication of organic extinctions). One may find this facet of Buffon's framework of thinking, along with some others, receiving greater appreciation a generation later, in the early nineteenth century, than immediately following publication of the *Époques*.
20. Besides the case of Desmarest [12] mentioned below, see Fortis [16], Bertrand [5], Soulavie [39], and Lamanon [24]. Another interesting instance is that of Lavoisier; see Rappaport [33] : pp. 254-257.
21. A reviewer of Fortis' report on the sequence of volcanic and marine rocks in the Roncà Valley called attention to the way this observation contradicted claims made by Leibnitz *« & d'autres Savans, après lui»*, to the effect that volcanoes acted only after the Earth's universal inundation. With evidence that the theoretically expected sequence was reversed, the reviewer took a jab at geological systems : *«Après ces observations curieuses & plusieurs autres que l'Abbé Fortis a faites dans cette vallée, on ne peut nier que les flots de la mer n'ayent recouvert ces débris des volcans. Systèmes, que devenez-vous?»* (Anonymous [3]).
22. Desmarest [12].

research.

This novel procedure –of reconstructing past geological events in their proper chronological order, without fusing them into a historical continuity– is one we can see retrospectively as holding the promise of advancing geology to a historical level not found in Buffon's *Époques*. Although it is not entirely clear how conscious anyone was of it at the time, the determination of past events on the basis of field evidence represented a significant step beyond Buffon's sort of history. In Buffon's narrative, events are brought about not so much by the combination of circumstances that precede them, as by the laws governing the processes underlying them. This outlook has been called *genetic*, or even *embryological*, as distinct from *historicist*.[23]

History which is the outcome of rules effecting changes that are, as it were, programmed, is presumably best learned about by consideration of those very rules and their logical consequences. So it is that in spite of some important gestures in the *Époques* towards a recognizably historical method of research, the mode of geological investigation revealed there is far from fully historical. Notwithstanding Buffon's engaging comparison of terrestrial phenomena with archives, and his discussion of natural monuments as historical data, his method was not basically an archival method. As Gabriel Gohau has argued, Buffon makes use of geological vestiges as archives in order to supplement an incomplete knowledge of the laws governing the order of events.[24] Otherwise, these vestiges are not archives, but illustrations of the rules. In the *Époques*, the monuments (that is, the geological evidence), are usually not so much *analyzed* to determine specific events to which they testify, as they are *assigned* to the epochs and operations that theory says produced them.

The *Époques* is far from atypical in these respects. Geological research of an archival type which resulted in historical propositions involving discontinuity and complex repetitions of events was relatively new. Buffon's *Époques* was in certain ways a distillation of ideas and attitudes just beginning to be displaced. But another aspect of Buffon's history, namely its completeness, had been on its way out of style among geological specialists for a bit longer. To the majority of the serious geological investigators of the 1770s and 1780s, an assured outline of the Earth's entire history must have appeared premature, if not presumptuous. To those trying to extract some orderly conclusions from the rocks of the Massif Central, the Pyrenées, the Italian Pre-Alps, or the Montmartre quarries, the fullness of the *Époques'* account of the Earth's history was not a very useful model.[25] The determinedly narrative character of the *Époques* almost certainly did encourage a historical conception in the overall scientific climate where these field studies were done, but to most it was evident that only a more modestly limited historical reconstruction was feasible, for the present.

I wish to mention one last aspect of Buffon's manner of talking about historical change, a feature of his approach that was situated within the mainstream of eighteenth century geology. I am thinking here of the distinction between *regular* and *accidental* processes, which corresponds pretty closely to the difference between

---

23. Oldroyd [30] : esp. pp. 193 and 227-28; Gohau [18], [20] and [21].

24. Gohau [21] : pp. 103-105.

25. Investigators whose work bears consideration from this viewpoint include Desmarest, Faujas de Saint-Fond, Pasumot, Soulavie, Palassou, Picot de Lapeyrouse, Ramond, Dietrich, de Saussure, Dolomieu, Pralon and Lamanon.

constructive and destructive geological change.

Something from the heart of this idea survives even to this day, I think, in the peculiar mentality of a certain tourist who, on seeing the Grand Canyon for the first time, was heard to exclaim : «*My God, look at all the damage that river has done!*» A comical notion to us, an attitude very much like this apparently was current in the eighteenth century. Buffon gave expression to it in speaking of the horizontal strata or the corresponding angles of valleys as regular features, while treating certain others pointedly as disruptions or even disfigurements. The causes of the one were ordered, uniform, and normal, those of the other were disorderly, irregular, and accidental. Volcanoes were commonly thought of as producing confused, distorted effects. A favorite geographical theme of the period, to the effect that natural barriers formerly existing at places like Gibraltar and the Bosporus were breached at some time with important topographic consequences, frequently was discussed in language suggesting a deterioration in an original continuity, a break in the natural order. In treating volcanoes or physiographic rifts this way Buffon was certainly not excluding these things from nature, but he was demarcating accidental processes from regular ones, putting them in a separate category. Whether these processes of disfigurement were regarded by Buffon as of a different level of interest to science than the processes of order, or were seen only as especially problematic because less susceptible to scientific understanding in terms of a fixed order, is unclear to me.

Buffon's profession of this deep-seated distinction between regular and accidental processes is evidently tied to his commitment to an essentially a-tectonic geology : the Earth's surface was figured in its basic form by regular causes, and whatever later alterations have occurred are either superficial or destructive. A truly tectonic system required that deformations be comprehended as constructive, something Buffon and many others of his time were unready to accept. Here it is pertinent to note also that the concept of an *époque* as a sustained period bounded at each end by dislocating events could serve as a convenient way of minimizing the discomfort many scientists felt in the presence of natural change, inevitably associated with disorder. Naturalists committed to the highly problematic aim of discovering the Earth's history might understandably find it easier to deal with relatively stable intervals of time than to come to grips with intervening episodes of change. In any case, the interconnections of this idea of geological disfigurement with other components of the period's scientific mentality have not yet been fully explored.[26]

VOLCANOLOGY IN THE *ÉPOQUES DE LA NATURE*

It may be that by a closer look at one particular aspect of the *Époques* some useful points of comparison with contemporary geology can be made. Let us consider briefly, then, what Buffon says about volcanoes –their causes, their effects, and their place in the structure and history of the Earth.

We learn about the place of volcanoes in nature's economy and history in the fourth epoch. Buffon held volcanoes to be a phenomenon of significant but somewhat restricted scope in the Earth's history. They are caused by inflammation of pyrites and bituminous materials through fermentation requiring the presence of water.

---

26. On the idea of accidents in eighteenth century geology, see Rappaport [34]. On the association between a focus on accidents and a historical or archival geology, see Gohau [19], and [21] : p. 104.

No volcano could exist until after the generation of these fermentable and inflammable materials, in large subterranean cavities. Their generation necessitated, first, the prolonged existence of the universal ocean, furnishing acidic components to form salts and iron for pyrites, and secondly the slow, progressive lowering of the universal ocean, revealing lands where plant life came to flourish. In course of time, vast quantities of decaying vegetable matter accumulated, destined for volcanic fuel.

Buffon accorded also a minor, ephemeral role in volcanic action to mineral substances sublimated from the Earth's interior by its intrinsic heat. He suggested that the first volcanic action came about by fermentation of these sublimated minerals in combination with detritus from the ocean. And Buffon thought that continuous emanations of electricity, the base of which he identified with the Earth's inner heat, had a role as a *«general cause»* of subterranean explosions associated with earthquakes and volcanoes, complementary to the *«particular causes»* seated in the effervescence of pyritic and combustible matter.

Notwithstanding his remarks about this combination of general and particular causes, however, the chapter makes plain that for Buffon the main run of volcanic activity owed nothing to the Earth's internal heat; the Earth's original incandescence and its later volcanic episodes are two entirely distinct things.

Quite important to Buffon's volcanology were his presuppositions regarding subterranean structures in mountain regions. Volcanoes *occur in* mountains, and with negligible exceptions do not *produce* them. Mountain ranges owe their existence primarily to the blistering and wrinkling of the globe's surface upon its initial congelation from a molten state. The elevations so formed overlie groups of huge internal cavities, associated with networks of vertical cracks *(fentes perpendiculaires)* and horizontal passages. These cracks and passageways serve several ends, such as the accumulation of fermentable minerals, the percolation of water to sustain the minerals' fermentation, and the downward transmission of volcanic force, from the relatively shallow vacuities where volcanic fires burn, into deeper chambers. The great force of volcanic explosions, Buffon said, comes not directly from a volcanic fire but from the juncture of that fire with large volumes of subterranean water.

As the ocean retreated progressively, revealing ever larger masses of land, volcanoes went extinct through removal of their water supply, while new ones were kindled in turn. In this way volcanic activity has migrated with the shoreline; in principle there has been a historical correlation between a region's emergence from the ocean and its period of volcanism. Buffon classed volcanic rocks as derivative, the result of alterations to one degree or another of other rock types acted upon by the heat and force of the volcano. Worldwide, the total mass of volcanic ejecta is small by comparison with calcareous rocks, and minute when compared with the overwhelming bulk of vitrescible rock. Locally, however, volcanic products can cover considerable areas, in places even filling entire valleys.

Volcanoes represented a growth area in geology since the 1750s, especially as a result of new recognition of the effects of extinct volcanoes.[27] What correspondence is there between the volcanologies of the *Époques* and of Buffon's contemporaries concerned about these phenomena ? On looking at the *components* of the foremost theories in late eighteenth century volcanologies, the correlation is high. Prevalent opinion was pretty much in agreement with Buffon on the ephemeral causes (that is, exhaustible fuel), the shallow focus, and the limited extent of volcanic action.

---

27. See especially Ellenberger [15].

The same is true regarding the derivative character and limited mass of volcanic ejecta, the probable subterranean connections among volcanoes, and the necessary involvement somehow in the volcanic process of water; hence the general arrangement of volcanoes near the sea. Nor was there anything remarkably uncommon in Buffon's tendency toward a somewhat functionalist interpretation of volcanic structures and operations.[28]

So although Buffon's volcanic theory no doubt has some peculiarities of detail, it was rather conventional when analyzed into its components. Real differences between Buffon and his serious volcanological contemporaries lay mainly in the big picture, not in minutiae, and over the tenability of an integrated volcanological system that depended on the truth of each of a number of essential components. Buffon took the main elements of common, but often rather provisional ideas about volcanoes, and built them into a broad dynamic and historical scheme that accounted for the distribution of observed volcanoes and volcanic products. In doing this Buffon could not easily treat individual components of the theory as uncertain; yet that was precisely what it seems to me other volcanological observers of the late eighteenth century often did do.[29]

What we encounter in this volcanological case is, I think, representative of the broader situation in comparing the *Époques* with the work of geologists active in the years around 1779. Volcanological research was being done, frequently, in a spirit of uncertainty about volcanic dynamics and the situation of volcanoes in the larger rock structure. It was not uncommon for investigators to pose their questions narrowly, thus limiting the dependence of field-based conclusions on speculations not susceptible to test. Reports on the disposition and composition of locally-delimited lavas, for example, were frequently formulated deliberately to minimize their reliance on any particular causal hypothesis. These research reports thus tended, evidently, to be only pieces of a larger puzzle still in need of solution. Buffon's book, on the other hand, presented volcanological knowledge as a puzzle that had been solved.

CONCLUSION

Buffon's *Époques* presents geological knowledge in a form reflecting a certain conception of what readers needed or wanted to know. It is unsurprising that this form, which was in a sense encyclopedic, emphasized answers more than unresolved questions. The *Époques* was an enormously stimulating, integrated synthesis. By its nature it was not, however, an invitation to research on fundamentally uncertain difficulties. The drift among the emerging group of volcanological investigators, and I think within francophone geology generally, was in the opposite direction, toward formulation of relatively narrow problems which could be studied empirically. Without rejecting Buffon's commitment to an interpretive framework of historical development subject to the operation of fixed natural laws, many geological thinkers of Buffon's later years did not think that this approach had yet yielded up

---

28. Volcanological writers of the immediate period whose works stand as a basis of comparison include Della Torre, Fougeroux de Bondaroy, Desmarest, Brydone, Hamilton, Strange, Ferber, Fortis, Faujas de Saint-Fond.

29. Examples include Desmarest, Dietrich, Gioeni, Montlosier, Soulavie, Spallanzani, Breislak, and especially Dolomieu.

very many reliable answers. In many cases, the body of evidence pertaining to the narrower problems being posed came to be addressed through procedures of geological fieldwork. Fieldwork of course was practiced by people who placed a high value on new facts. The *Époques de la Nature*, however, had manifestly been written by an author who looked on the discovery of new facts as an inferior basis for claim to scientific merit, when compared with the originality involved in putting together an imaginative and comprehensive synthesis.[30]

In certain ways the differences between Buffon and many of his geological contemporaries, when the *Époques* appeared, reflect the differences in concept, tone, and audience between the ambitious grandeur of a general, philosophical Natural History on one hand, and a group of inquisitive, geographically focussed studies on the other. There was a perceptible distance between the *Époques'* sense of science as a coherent story, and the investigative arena of the emerging geology. The geological world of 1779 was perhaps better prepared to see as its model the Buffon who had earlier written the empirically cautious, highly phenomenological *«De la Manière d'étudier et de traiter l'histoire naturelle»*, than the Buffon of the confidently rational *Des Époques de la Nature*.

ACKNOWLEDGEMENTS

Research support for this paper was provided by the University of Oklahoma Office of Research Administration and by the National Science Foundation, research grant number SES- 8719713.

BIBLIOGRAPHY *

UNPUBLISHED SOURCES

(1)†   DESMAREST (N.), Letter to Duc Louis-Alexandre de La Rochefoucauld, 15 May 1769, Bibliothèque Municipale de Besançon, Ms 1441, fol. 297-298.

(2)†   ROMÉ DE L'ISLE (J.B.L.), Letter to Horace Bénédict de Saussure, 6 December 1784, Bibliothèque Publique et Universitaire de Genève, Archives de Saussure, n° 9, fol. 175-176.

PUBLISHED SOURCES

(3)†   ANONYMOUS, [Notice on FORTIS, Della Valle vulcanico-marina di Roncà], *Observations sur la physique*, 14 (1779) : p. 507.

(4)†   ACADÉMIE IMPÉRIALE DES SCIENCES DE SAINT-PÉTERSBOURG, *Mémoires présentés à l'Académie Impériale des Sciences pour répondre à la question minéralogique proposée pour le Prix de MDCCLXXXV*, St Pétersbourg, De l'Imprimerie de l'Académie Impériale des Sciences, 1786, [II]-101-161p. (2

---

30. Cf. Roger's remarks on Buffon's idea of *«genius»*, and on Buffon's intellectual isolation (Roger, in Buffon [8] : resp. p. XCIII and p. CXIII).

* Sources imprimées et études. Les sources sont distinguées par le signe †.

prefatory pages, unpaginated, followed by Launay [25] and Soulavie [41]).
(5)† BERTRAND (P. M), «Mémoire sur les volcans de Tourves en Provence», *Observations sur la physique*, 15 (1780) : pp. 36-38.
(6) BLEDSTEIN (M. A.), *Nicolas-Antoine Boulanger : An Eighteenth Century Naturalist and Historian*, Ph.D. Dissertation, New York University, 1977 (University Microfilms International, n° 77-21268). Two parts : 226p.; III-612-IIp.
(7)† BUFFON (G.L. Leclerc de), *Histoire naturelle, générale et particulière... Supplément*, T. V Paris, De l'Imprimerie Royale, 1778, VIII-615-XXVIII p.
(8)† BUFFON (G. L. Leclerc de), *Les Époques de la Nature*, Édition critique avec le manuscrit, une introduction et des notes par J. Roger. In : *Mémoires du Muséum National d'Histoire Naturelle, Série C, Sciences de la Terre*, Tome X, Paris, Éditions du Muséum, 1962, CLII-343p. Réimpression : Paris, Éditions du Muséum, 1988.
(9)† BUFFON (G. L. Leclerc de), *Des Époques de la Nature*, Introduction et notes par Gabriel Gohau. Paris, Éditions Rationalistes, 1972, XXVI-227p.
(10) CIANCIO (L.), «Alberto Fortis and the Study of Extinct Volcanoes of Veneto (1765-1778)», paper delivered at *XIIIth Symposium of the International Commission on the History of Geological Sciences*, Padua, 1 October 1987.
(11)† DELUC (J.A.), *Lettres physiques et morales sur l'histoire de la terre et de l'homme, adressées à la Reine de la Grande Bretagne*, La Haye, De Tune, and Paris, V. Duchesne, 1779, 5 vol. in 6°.
(12)† DESMAREST (N.), «Extrait d'un mémoire sur la détermination de quelques époques de la nature par les produits des volcans, & sur l'usage de ces époques dans l'étude des volcans», *Observations sur la physique*, 13 (1779) : pp. 115-126. [Read in 1775.]
(13)† DESMAREST (N.), *Encyclopédie méthodique. Géographie-physique*, Paris, Chez H. Agasse, an III [1794]-1828, 5 vol.
(14) ELLENBERGER (F.), «L'Enseignement géologique de Guillaume-François Rouelle (1703-1770)», *in Comunicaciónes Cientificas, V Reunión Cientifica, International Commission on the History of Geological Sciences*, Madrid, Ibérica, 1974 : pp. 209-221.
(15) ELLENBERGER (F.), «Précisions nouvelles sur la découverte des volcans de France : Guettard, ses prédécesseurs, ses émules clermontois», *Histoire et nature*, n° 12-13 (1978) : pp. 3-42.
(16)† FORTIS (G. B. [Alberto]), *Della Valle vulcanico-marina di Roncà nel territorio Veronese. Memoria orittografica*, Venezia, Nella Stamperia di Carlo Palese, 1778, LXXp.
(17) FRESHFIELD (D.W.). *The Life of Horace Benedict de Saussure*. With the collaboration of Henry F. Montagnier, London, Edward Arnold, 1920, XII-479p.
(18) GOHAU (G.), «Du Système du monde à l'histoire de la terre», *Travaux du Comité Français d'Histoire de la Géologie*, n° 19 (1979) : pp. 1-8.
(19) GOHAU (G.), «Idées anciennes sur la formation des montagnes», *Cahiers d'histoire et de philosophie des sciences*, Nouvelle série, n° 7 (1983), 86p.
(20) GOHAU (G.), «La Naissance de la géologie historique : les "Archives de la Nature"», *Travaux du Comité Français d'Histoire de la Géologie*, Deuxième série, 4 (1986) : pp. 57-65.
(21) GOHAU (G.), *Histoire de la géologie*, Paris, Éditions La Découverte, 1987, 259p.
(22) HABER (F.C.), *The Age of the World : Moses to Darwin*. Baltimore, The Johns Hopkins Press, 1959, XI-303p.
(23)† HAIDINGER (K.), *Systematische Einteilung der Gebirgsarten*, Wien, Christian Friedrich Wappler, 1787, 82p.
(24)† LAMANON (R. de Paul de), «Description de divers fossiles trouvés dans les carrières de Montmartre près Paris, & vues générales sur la formation des pierres gypseuses», *Observations sur la physique*, 19 (1782) : pp. 173-194.

The Époques de la Nature and Geology during Buffon's later Years 385

(25)† LAUNAY (L. de), *Essai sur l'histoire naturelle des roches, précédé d'un exposé systématique des terres & des pierres. Ouvrage présenté à l'Académie Impériale des Sciences de St. Pétersbourg, en suite du programme qu'elle a publié en 1783*, St. Pétersbourg, De l'Imprimerie de l'Académie Impériale, 1786, 101 p. [Published in (4)] Also published separately : Bruxelles, Lemaire, and Paris, Cuchet, 1786, LXXVI-150 p.

(26)† LIMBOURG (R. de), «Mémoire pour servir à l'histoire naturelle des fossiles des Pays-Bas», *Mémoires de l'Académie Impériale et Royale des Sciences et Belles-Lettres de Bruxelles,* 1 (1777) : pp. 363-410. [Read 7 February 1774].

(27)† LIMBOURG (R. de), «Lettre à MM. les rédacteurs», *L'Esprit des journaux,* 9ème année (1780), 2 février : pp. 331-336.

(28)† MARMONTEL (J. F.), *Mémoires*, edited by Maurice Tourneux. Paris, Librairie des Bibliophiles, 1891, 3 vol.

(29)† NAIGEON (J.A.), *Encyclopédie méthodique. Philosophie ancienne et moderne,* Paris, Panckoucke-H. Agasse, 1791-an II, 3 vol.

(30) OLDROYD (D. R.), «Historicism and the Rise of Historical Geology», *History of Science,* 17 (1979) : pp. 191-213 and 227-257.

(31) RAPPAPORT (R.), «G.-F. Rouelle : An Eighteenth-Century Chemist and Teacher», *Chymia* 6 (1960) : pp. 68-101.

(32) RAPPAPORT (R.), «Problems and Sources in the History of Geology», *History of Science,* 3 (1964) : pp. 60-77.

(33) RAPPAPORT (R.), «Lavoisier's Theory of the Earth», *The British Journal for the History of Science,* 6 (1973) : pp. 247-260.

(34) RAPPAPORT (R.), «Borrowed Words : Problems of Vocabulary in Eighteenth-Century Geology», *The British Journal for the History of Science,* 15 (1982) : pp. 27-44.

(35) ROGER (J.), «Un Manuscrit inédit perdu et retrouvé : *Les anecdotes de la nature,* de Nicolas-Antoine Boulanger», *Revue des sciences humaines,* 71 (1953) : pp. 231-254.

(36) ROGER (J.), «La Théorie de la terre au XVIIe siècle», *Revue d'histoire des sciences,* 26 (1973) : pp. 23-48.

(37) ROGER (J.), «Le Feu et l'histoire : James Hutton et la naissance de la géologie», in *Approches des Lumières : Mélanges offerts à Jean Fabre,* Paris, Éditions Klincksieck, 1974 : pp. 415-429.

(38) ROGER (J.), «The Cartesian Model and Its Role in Eighteenth-Century 'Theory of the Earth'», in *Problems of Cartesianism,* T. M. Lennon, J. M. Nicholas, and J. W. Davis, eds., Kingston and Montreal, McGill-Queen's University Press, 1982 : pp. 95-125.

(39)† SOULAVIE (J.L. Giraud), «La Géographie de la nature, ou distribution naturelle des trois règnes sur la terre...», *Observations sur la physique,* 16 (1780) : pp. 63-73.

(40)† SOULAVIE (J.L.Giraud), *Histoire naturelle de la France méridionale,* Paris, 1780-1784, J.-Fr. Quillau, 8 vol.

(41)† SOULAVIE (J.L. Giraud), *Les Classes naturelles des minéraux et les époques de la nature correspondantes à chaque classe. Ouvrage qui a remporté le second accessit sur la question proposée par l'Académie Impériale des Sciences de St. Pétersbourg, pour le prix de 1785,* St. Pétersbourg, De l'Imprimerie de l'Académie Impériale des Sciences, 1786, 161p. [Published in (4)].

XV

Earth and Heaven, 1750–1800: Enlightenment Ideas About the Relevance to Geology of Extraterrestrial Operations and Events[1]

During the second half of the eighteenth century, what views were taken among geological scientists regarding the significance of extraterrestrial processes and events for the terrestrial effects with which their science was concerned? I hope that this brief answer to the question may have some interest for historical perspective on the rather confusing scene in geoscience during the period when geology was just beginning to emerge as a recognized discipline. My remarks are addressed mostly to how the Theory of the Earth traditions bore on this question. I also want to take into account the growing force, especially toward the end of the century, of an empirically-anchored methodology for geological investigation, which I believe had the effect of contributing to the marginalization of extraterrestrial considerations in the new geological science.

Systemic and Idiosyncratic Appeals to Extraterrestrial Agents

It is useful to recognize two different sorts of invocation of extraterrestrial processes and events in geological contexts during the later eighteenth century. One type of geological concern with extraterrestrial agents, which I shall refer to as *systemic*, concerns the outlook of scientific writers of wide perspective who presumed that a satisfactory comprehension of the Earth could not be complete without recognition of its existential condition as a planet, its constant and regular susceptibility to universal physical conditions. In such broadly situated scientific writings, for the period following mid-century, it would not do, for example, to ignore the still fresh perception (given recent developments in natural philosophy) that the Earth has an oblate figure not peculiar to our planet, a shape which is the natural result of circumstance for any rotating

---

[1] This essay is presented here in much the same form as delivered for the GSA Symposium at Seattle in 1994. Amplifications are added mainly in the notes.

spherical body not endowed with perfect rigidity.² A similar concern to take note of the Earth's regular reception of solar heat, as an essential element in the globe's natural economy, falls in the same category of awareness of regular or permanent extraterrestrial factors not uncommonly seen in Enlightenment science.³

This broadly systemic concern about extraterrestrial phenomena, based on maintaining awareness of the Earth's existential condition, is to be distinguished from what I shall call the *idiosyncratic* appeal to extraterrestrial agents. In this other category I have in mind the invocation of an ephemeral or momentary extraterrestrial event, a notable instance being the physical influence of a comet's near passage. I do not mean, of course, where we find an eighteenth-century writer discussing a comet sideswiping the Sun or Earth, that the author supposed such an event to defy what was known of the cosmic order or to fall outside the known laws of physics—to the contrary; but I do mean that the role assigned to such agents if identified with peculiar *moments* in the Earth's history was distinguishable from the role played by systemic kinds of considerations in Enlightenment science.⁴

An example of a somewhat skeptical regard for the consequences of idiosyncratic possibilities in the Earth's circumstances – its past history or its

---

² An example is John Whitehurst's *An Inquiry into the Original State and Formation of the Earth* (London: for the author by J. Cooper, 1778). Whitehurst declared (p. 2) that "Newton hath happily laid the foundation for a natural history of the terraqueous globe, by demonstrating its figure to be an oblate spheroid...." Notable among works situating the Earth systemically within the celestial system were geographies in the cosmographic tradition, such as the many editions (extending into the second half of the eighteenth century) of Bernhardus Varenius's *Geographia Generalis*. A well-known early Enlightenment specimen of 'planetary physics' by an author much interested in the Earth's shape is Pierre-Louis Moreau de Maupertuis, *Discours sur les différentes figures des astres* (Paris: Imprimerie Royale, 1732). See also John L. Greenberg, *The Problem of the Earth's Shape from Newton to Clairaut* (Cambridge, New York, Melbourne: Cambridge University Press, 1995).

³ See for example Jean-Jacques Dortous de Mairan, *Dissertation sur la glace* (Paris: Imprimerie Royale, 1749); Jean-Baptiste Romé de l'Isle, *L'Action du feu central bannie de la surface du globe, et le soleil rétabli dans ses droits* (Stockholm, Paris: P.-F. Didot, 1779) [and the second, enlarged edition, 1781]; Johan Gottschalk Wallerius, *De l'origine du monde, et de la terre en particulier*, transl. Jean-Baptiste Dubois de Jancigny (Warsaw, Paris: J. Fr. Bastien, 1780). See also the geological lectures of John Walker of Edinburgh: *Lectures on Geology*, edited with notes and introduction by Harold W. Scott (Chicago, London: University of Chicago Press, 1966), esp. pp. 60–71, 200–201. Useful discussion of the strong universalist tendencies in Enlightenment science is found in J. L. Heilbron, *Elements of Early Modern Physics* (Berkeley, Los Angeles, London: University of California Press, 1982), and Thomas L. Hankins, *Science and the Enlightenment* (Cambridge: Cambridge University Press, 1985).

⁴ Here I borrow phraseology from John C. Greene: a chapter section headed "A comet sideswipes the sun" in *The Death of Adam: Evolution and Its Impact on Western Thought* (Ames: Iowa State University Press, 1959), p. 24.

future destiny – is seen in the article on comets in J.-C. Valmont de Bomare's popular *Dictionnaire d'histoire naturelle* (1768), where the notion that a periodic comet might have been responsible for the universal deluge is treated as physically feasible, if unlikely:

> This opinion which can only be regarded as a conjecture of quite little weight, nonetheless holds nothing contrary to sound science, which informs us that the approach of such a Comet is capable either of an upheaval on the globe we inhabit, or of restoring [*de relever*] the Earth's axis, which according to M. de Maupertuis would gain for us a perpetual spring.[5]

Comets of course by now had long been taken to obey general laws, but in this view what made them possibly interesting to a physical science of the Earth was found in the unique events and irreversible changes they might conceivably have brought about, or could bring about in the future.

A rather less receptive attitude toward roles for idiosyncratic geological causes of any kind, terrestrial or extraterrestrial, was expressed by Peter Simon Pallas in *On the Nature of Mountains and on Changes which have occurred on the Globe*, read in 1777 before the Russian Academy of Sciences. Pallas took a stance opposed in general to mono-explanatory geological formulations: "... I dare to maintain that Nature uses obviously a variety of processes to create and destroy layers of rocks and change the surface of the Earth so that any hypothesis that is based only on one or a few of these processes cannot give a satisfactory explanation."[6] Such skepticism about the efficacy of single causes was common among geological writers during the later part of the eighteenth century; in this light a prevalent wariness or suspicion about explanations resting on idiosyncratic invocations of agents is hardly surprising.

It may deserve comment here that the distinction I propose between systemic and idiosyncratic appeals to extraterrestrial agents corresponds at best rather poorly to the more familiar opposition between uniformitarianism and catastrophism. Whatever may be the value of the uniformitarianism-catastrophism polarity in analyzing nineteenth-century geology (and its relevance there seems often more complicated than is generally represented in

---

[5] Jacques-Christophe Valmont de Bomare, *Dictionnaire raisonné universel d'histoire naturelle*, edition augmentée, 12 vols. (Yverdon: [no publ.], 1768–69), article "Comète," Vol. 3, pp. 263–67, on pp. 266–67.

[6] Albert V. Carozzi and Marguerite Carozzi, "Pallas' Theory of the Earth in German (1778). Translation and Reevaluation. Reaction by a Contemporary: H.B. de Saussure," extrait des *Archives des Sciences* (Société de Physique et d'Histoire Naturelle de Genève), 1991, Vol. 44, fasc. 1, pp. 1–105, on p. 36.

histories of geology), it is largely anachronistic for the eighteenth century, and its historical application there is often more hindrance than help.⁷

On the other hand there is a closer correspondence between this polarity differentiating the *systemic* from the *idiosyncratic* (which are inventions for this occasion) and a distinction – familiar in the geological vocabulary of the eighteenth century – between things, events or processes considered *regular* and *natural* on one hand, and *accidental* on the other. It was commonly presumed that consequences or products of the latter sort of process (idiosyncratic, accidental) would tend to lack order or regularity. Such disorder was in contrast to the ordered arrangement expected in geological features derived from regular or (as I refer to them here) systemic operations.⁸

Eighteenth-Century Theories of the Earth

There is no denying that in the half century before 1800 there was a continuing fascination, in the work of geological scientists, with the possibility of understanding the Earth's genesis and development in terms of cosmic processes. Just before 1750, two of the century's most talked-about theories of the Earth were published, each in its way seeking to give a comprehensive account of the Earth in a fashion that acknowledged the Earth's status as a planet. *Telliamed*, appearing in print in 1748 after circulating clandestinely in manuscript for a generation, tried to explain the Earth and its changes by reference to enormously long cycles of heating and cooling; *Telliamed* set the Earth presently at a point during a period of heating, slowly evaporating the

---

⁷   Comment on uniformitarianism and catastrophism as counter-productive labels for eighteenth-century geology is found in Rhoda Rappaport, *When Geologists Were Historians, 1665–1750* (Ithaca and London: Cornell University Press, 1997), p. 5. A classic discussion is Reijer Hooykaas, *The Principle of Uniformity in Geology, Biology and Theology*, 2nd impression, revised [original title: *Natural Law and Divine Miracle*] (Leiden: E.J. Brill, 1963). See also Hooykaas, *Catastrophism in Geology, Its Scientific Character in Relation to Actualism and Uniformitariansim*. Mededelingen der Koninklijke Nederlandse Akademie van Wetenschappen, Afd. Letterkunde, Nieuwe Reeks, Deel 33, No. 7 (Amsterdam, London: North-Holland Publishing Co., 1970).

⁸   On accidents in this context, see Rhoda Rappaport, "Borrowed Words: Problems of Vocabulary in Eighteenth-Century Geology," *British Journal for the History of Science*, 1982, 15: 27–44. The discussion is extended for one class of geological phenomena in Kenneth L. Taylor, "Volcanoes as Accidents: How 'Natural' Were Volcanoes to 18th-Century Naturalists?", in *Volcanoes and History*, Proceedings of the XXth Symposium of the International Commission on the History of Geological Sciences, edited by Nicoletta Morello (Genova: Brigati, 1998), pp. 595–618.

oceans and enlarging the extent of land.⁹ The *Theory of the Earth* published the next year by Georges Louis Leclerc, the comte de Buffon, took the Earth to have originated from a comet's close passage by the Sun. This idea was in fact less central in his 1749 *Theory* than it became for his *Epochs of Nature* three decades later, where he presented what was perhaps the eighteenth century's most influential – and most notorious – cosmogonical account of the Earth's history from hot disorganized matter to future decline as a congealed and lifeless globe.¹⁰

Both de Maillet and Buffon reflected a strongly favorable sensibility regarding the *systemic* type of extraterrestrial concern (de Maillet's perspective being Cartesian while Buffon's was Newtonian). But they were rather different in respect to their acceptance of certain *idiosyncratic* extraterrestrial influences. De Maillet's speculations embraced an ongoing sequence of terrestrial changes caused by cosmic circumstance. For Buffon, by contrast – with all due recognition that Buffon emphatically set the Earth within a cosmic system – the chief external event of consequence to the Earth came only at the very beginning, after which the Earth was, in a sense, on its own.

It is of course difficult to think of a self-respecting scientific writer of the Enlightenment who would refuse to acknowledge the pertinence in principle of systemic extraterrestrial considerations for a coherent understanding of the Earth. A key difference among such writers, however, might be the degree of confidence they chose to show that knowledge of the Earth could be derived directly from that which is systemic. An extreme expression of such confidence is found in the 1779 *Prospectus* written by É.-C. Marivetz and L.-J. Goussier for their *Physique du Monde*; here the authors declare that their work will demonstrate the general theory of physical geography as deduced from

---

[9] Benoît de Maillet, *Telliamed, ou entretiens d'un philosophe indien avec un missionnaire françois sur la diminution de la mer, la formation de la terre, l'origine de l'homme, &c.*, 2 vols. in 1 (Amsterdam: L'Honoré, 1748). This is accessible in a modern English translation by Albert V. Carozzi (Urbana, Chicago, London: University of Illinois Press, 1968).

[10] Buffon's *Histoire & Théorie de la Terre* appeared as the *Second discours* in the first volume of his *Histoire naturelle générale et particulière*, which eventually ran to 36 volumes (Paris: Imprimerie Royale, 1749–1789). The theory is elaborated upon in *Preuves de la théorie de la terre* in the same opening volume. *Des Époques de la Nature* appeared in the fifth volume (1778) of the *Supplément* to Histoire naturelle. The modern scholarly edition of *Époques de la Nature*, edited with introduction and notes by Jacques Roger, is in *Mémoires du Muséum National d'Histoire Naturelle*, Sér. C. Tome X (Paris: Éditions du Muséum, 1962). Recent historical studies include Gabriel Gohau, "La 'Théorie de la Terre' de 1749," in *Buffon 88: Actes du Colloque international pour le bicentenaire de la mort de Buffon*, edited by Jean Gayon (Paris: Vrin, 1992), pp. 343–352; and Kenneth L. Taylor, "The Époques de la Nature and Geology During Buffon's Later Years," in *Buffon 88*, pp. 371–385.

first principles of celestial physics.[11] It must be added that the way the authors then went about their business suggests that even they saw some extravagance in this assertion.[12] But in any case I think of Marivetz and Goussier's brave claim as far from typical, in fact as running against a strong tide of skeptical empiricism, geologically speaking, in the century's closing decades.

The examples of de Maillet and Buffon – the theories themselves and the fact that they continued to be discussed – illustrate the lasting force in customs of organizing much scientific discourse about the Earth around a cosmogonical theme, or what may seem to be the same thing, around a Theory of the Earth. One of the fruits of the acceptance of Copernicanism, the Theory of the Earth viewed the Earth now, through eyes trained in the New Philosophy, as a planetary entity, rather than in Aristotelian fashion as a set of homocentric regions or domains, each with its own qualitative characters. An object, as distinct from a region or a place, could very well have a history. In accord with this new scientific genre or way of conceiving the Earth, the objectives generally included giving a comprehensive account of how the Earth's surface features were generated and then transformed into the present state, and doing so in a manner consistent with accepted principles of natural philosophy.[13] For our purposes here, the key point is that a major ingredient of Theories of the Earth as ways of asserting the objectives of an earth science emphasized the Earth's status as a planet, thus as a member of a celestial family. In this perspective, one may well imagine that an author's *failure* to establish some direct links between the Earth and the larger celestial domain might be thought peculiar, even indicative of a scientifically defective treatment. This is reason enough to find it curiously impressive how conspicuously these links are in fact ignored, or given very little attention, in much that was published by geological investigators in the last three or four decades of the eighteenth century.

---

[11] Étienne-Claude de Marivetz and Louis-Jacques Goussier, *Discours préliminaire, et prospectus d'un traité général de géographie-physique* (Paris: Quillau, 1779), p. 1.

[12] Marivetz and Goussier, *Physique du Monde*, 5 Vols. in 7 (Paris: Quillau, 1780–1787). A deductive spirit does nonetheless inform much in this work.

[13] On general characteristics of theories of the earth, see especially Jacques Roger, "La Théorie de la terre au XVIIème siécle," *Revue d'histoire des sciences*, 1973, 26: 23–48; Jacques Roger, "The Cartesian Model and Its Role in Eighteenth-Century 'Theory of the Earth'," in *Problems of Cartesianism*, edited by Thomas M. Lennon, John M. Nicholas, and John W. Davis (Kingston and Montreal: McGill-Queen's University Press, 1982), pp. 95–125. See also Roy Porter, "Creation and Credence: The Career of Theories of the Earth in Britain, 1660–1820," in *Natural Order: Historical Studies of Scientific Culture*, edited by Barry Barnes and Steven Shapin (Beverly Hills & London: Sage, 1979), pp. 97–123; and Kenneth L. Taylor, "The Historical Rehabilitation of Theories of the Earth," *The Compass*, 1992, 69: 334–345.

Earth and Heaven, 1750–1800 7

Specific and Generic Senses of 'Theory of the Earth'

In weighing the influence of the traditions of Theories of the Earth during the eighteenth century, it is worthwhile noting an important if deceptively simple point to which François Ellenberger has called attention in his book on geology's development between 1660 and 1810. Ellenberger makes a case for distinguishing between two senses of the expression 'Theory of the Earth.' One sense refers to the *specific* and supposedly self-contained theories of the Earth offered by a good many others in addition to de Maillet and Buffon, ranging from Descartes through Burnet, Woodward, Whiston and Leibniz, through Scheuchzer and Bourguet to Werner and Hutton. The other sense is *generic* rather than specific, and refers to a widespread conviction that a coherent and systematic science of the Earth was in the process of being formed, but not yet really in hand, as the result of scientific work more empirically restrained and skeptical than was frequently visible in some of the specific theories of the Earth.[14]

This second, generic notion of the Theory of the Earth was clearly what certain geological investigators had in mind in saying that they hoped or assumed their work would contribute to the theory of the Earth. In a draft outline of a theory of the Earth written in 1796, Horace-Bénédict de Saussure said that "The Theory of the Earth is the science which describes and explains changes that the terrestrial globe has undergone from its beginnings until today, and which allows the prediction of those it shall undergo in the future." Such a theory, according to de Saussure, can only be built up by examination of the Earth as it is found: "the present state of the Earth is the only solid base on which the theory can rely."[15] The same conviction was registered by Barthélemy Faujas de Saint-Fond in his 1802 lectures at the Paris Museum of Natural

---

[14] François Ellenberger, *Histoire de la géologie*, 2 vols., Vol. II: *La grande éclosion et ses prémices, 1660–1810* (Paris, London, & New York: Technique et Documentation, 1994), pp. 13–16. My designation of different *specific* and *generic* senses of the Theory of the Earth corresponds to Ellenberger's distinction between *individual* theories (*les théories*, plural, each one different) and *the* theory of the Earth (*la théorie*, singular, an ideal objective held in common by geological thinkers).

[15] Quoted in translation from de Saussure's unpublished draft in Albert V. Carozzi, "Forty Years of Thinking in Front of the Alps: Saussure's (1796) Unpublished Theory of the Earth," *Earth Sciences History*, 1989, 8: 123–140, on p. 136. This fundamental motivation for geological observation–to generate satisfactory theory–is evident throughout de Saussure's published writings, including *Voyages dans les Alpes*, 4 vols. (Neuchâtel: Samuel Fauche, Louis Fauche-Borel: Geneva: Barde, Manget, 1779–1796). See for example the "Discours préliminaire," Vol. I (1779), pp. i–xx, esp. on pp. i–ii; the "Agenda," Vol. IV (1796), pp. 467–[539], on p. 467.

History, when he stated that geology is the science whose goal is to establish the theory of the Earth.[16]

De Saussure and Faujas were speaking for many of the generation of geological investigators during the last quarter of the century: to their knowledge no satisfactory theory of the Earth had yet been proposed, but it was widely believed that one must certainly be possible, and that it would come into existence through an inductive procedure by scrupulous attention to available phenomena. Many of them also assumed – as Cuvier evidently did in the decade after Faujas' 1802 lectures – that when this was accomplished the resultant theory could be counted on to have the general attributes of *the* Theory of the Earth – namely, global comprehensiveness, narrative coherence, and philosophical consistency – while of course lacking the patent flaws of the various specific but inadequate pretenders. It should not escape our notice that the rise of fieldwork towards its position as an indispensable procedure in late-eighteenth-century geology took place in the company of continued respect for the Theory of the Earth in its generic sense, even as the credibility of most specific theories declined.

Emerging Sense of Discipline: Proprietary View on Geology's Objects of Study

It is important also to acknowledge the fact that geology was an emerging new science, and that this was no secret. It is unnecessary, and probably fruitless, to try to fix with any precision a historical moment when geology became a recognized science. Broadly speaking, many will agree that the period we are now considering, between 1750 and 1800, was a critical in-between stage, in geology's emergence.[17] This era was rich with vital new developments in the work of individuals variously identified as natural philosophers, chemists, mineralogical naturalists, mining engineers, physical geographers, cosmographers, and so on. But regardless of the diverse disciplinary fields and occupational identifications of people cultivating scientific knowledge of the

---

[16] Barthélemy Faujas de Saint-Fond, *Essai de Géologie, ou Mémoires pour servir à l'histoire du globe*, 2 vols. in 3, Vol. I (Paris: Gabriel Dufour, 1809), p. 1.

[17] Roy Porter, *The Making of Geology: Earth Science in Britain, 1660–1815* (Cambridge: Cambridge University Press, 1977); Martin Guntau, *Die Genesis der Geologie als Wissenschaft*. Schriftenreihe für Geologische Wissenschaften, Heft 22 (Berlin: Akademie-Verlag, 1984); Rachel Laudan, *From Mineralogy to Geology: The Foundations of a Science, 1650–1830* (Chicago and London: University of Chicago Press, 1987); Gabriel Gohau, *Les Sciences de la terre aux XVIIe et XVIIIe siècles: Naissance de la géologie* (Paris: Albin Michel, 1990); David R. Oldroyd, *Thinking About the Earth: A History of Ideas in Geology* (London: Athlone, 1996), chap. 3.

Earth, the crucial reorganization and reconceptualizing of enterprises was just then in the process of yielding up an eventual consensus, to be realized early in the nineteenth century, about the legitimate standing of geological science. Speaking of conceptualization and institutions alike, until at the earliest just a short time before 1800, if then, geology can barely be said to have been by general assent a distinct scientific endeavor, by which particular investigators would hold recognized status within the sciences.[18]

However – and I think this is quite significant – in the century's closing decades there existed a growing cadre of observationally-focused investigators who sensed, and often articulated explicitly, their participation in the *establishment* of an emerging science. They manifested a proprietary attitude about the activity of doing geology, under whatever name (possibilities included variants of mineralogy, geognosy and physical geography, among others).[19] In thinking historically about their situation, we may need to overcome the complacency so easily bred by hindsight to realize that their success in persuading others of their enterprise's merit was not a foregone conclusion. In this set of circumstances, with its unavoidable dimension of scientific politics in discipline recognition, it should not surprise us to observe signs of proprietary attitudes toward both the act of geologizing, and the objects of geological study. Such signs are there, in the closing decades of the eighteenth century.[20]

---

[18] Kenneth L. Taylor, "The Beginnings of a French Geological Identity," *Histoire et nature*, 1981–82, No. 19–20: 65–82; Gordon L. Herries Davies, "A Science Receives Its Character," in G.L. Herries Davies and Antony R. Orme, *Two Centuries of Earth Science, 1650–1850* (Los Angeles: William Andrews Clark Memorial Library, 1989), pp. 1–28.

[19] Dennis R. Dean, "The Word 'Geology'," *Annals of Science*, 1979, 36: 35–43. See also Kenneth L. Taylor, "Geology in 1776: Some Notes on the Character of an Incipient Science," in *Two Hundred Years of Geology in America*, edited by Cecil J. Schneer (Hanover, NH: University Press of New England, 1979), pp. 75–90.

[20] It is in light of geology's still uncertain status during the later eighteenth century that we need to interpret, among other things, the doubts frequently expressed by geologists that the procedures of ordinary chemical investigation suffice to account for certain geological phenomena. One example: Faujas de Saint-Fond's insistence, in *Recherches sur les volcans éteints du Vivarais et du Velay* (Grenoble: Cuchet; Paris: Nyon aîné, Née & Masquelier, 1778), that "there is a Chemistry of Nature quite superior to that of Art" (p. iii). Denials that the 'laboratory of nature' can be equated with the comparatively puny, artificial resources of chemical laboratories were common in this period (and, it should be added, were frequently contested as well). Assertions of this sort–which by no means all denied the significance and use of chemical methods, but generally refused to accept them as sufficient to address all geological questions – can be understood as supporting a form of exclusivity of geological method, what could be called 'geological chauvinism': there are some problems with regard to terrestrial phenomena, according to this view, that belong only to geology and that will submit only to geological approaches.

What I am suggesting is that the historical effort to define and defend geological science might naturally have tended to accentuate those natural objects, most clearly in the domain of Earth-bound observers, and correspondingly to minimize emphasis on the importance of objects and processes in the domain of others, such as astronomers.[21]

Extra-Terrestrial Geological Factors Marginalized: Deluc and de Saussure

Let us look briefly at a pair of late-eighteenth-century geological figures, whose programmatic orientations illustrate a range of possibilities within a scientific climate tending to treat extraterrestrial factors as marginal to geology. The pair I have in mind – Jean-André Deluc and Horace-Bénédict de Saussure – may make an odd couple, even though they both came from the relatively small republic of Geneva. They represent two discernibly different temperaments in the geological science of their time: one rather adventurously given to comprehensive theorizing, the other more disposed to theoretical caution and tentative about how field observations point to a specific theory of the Earth.

Deluc was a merchant, diplomat, and naturalist who emigrated to England, where he became reader to Queen Charlotte. He wrote copiously about geological subjects in the closing decades of the eighteenth century, and into the following one. Often exasperatingly verbose, sometimes obscure, Deluc wrote at length about what he considered to be generally-acknowledged facts about configurations of strata, and their evident historical dislocations, and frequently juxtaposed these discussions with abstract doctrines of matter theory and chemical transformations affected by all the known imponderable

---

[21] Specimens of minerals, rocks, and fossils counted as belonging especially to geological scientists and as defining essential parts of their domain. But the geological objects which were perhaps the most strikingly claimed as geological inquiry's special possessions, during the closing decades of the eighteenth century, were mountains. Statements to this effect were made frequently by, among others, Déodat de Dolomieu, some of whose purplest prose designates mountains as high sanctuaries where brave geologists seek out the Earth's secrets. In one such passage Dolomieu named the likes of Pallas, Deluc, and de Saussure as heroic voyagers to remote summits, whose exploits set them apart as meriting the title of geologist. See "Rapport fait à l'Institut National, par Dolomieu, sur ses voyages de l'an cinquième et sixième," *Journal de physique, de chimie, d'histoire naturelle et des arts*, 1798/*an* 6, [new series] 3: 401–427, on pp. 403–4. This was reprinted for other audiences, not only in *Journal des mines*, 1798/*an* 6, 7: 385–402, 405–432, but also in *Magasin encyclopédique*, 1797/*an* 6, 3e année, Tome 5, pp. 148–156. On geology as a new science of specimens and of fieldwork (as well as of mineral distributions, rock formations, characteristic fossils, and a history of nature), see Martin J.S. Rudwick, "Minerals, Strata and Fossils," in *Cultures of Natural History*, edited by N. Jardine, J.A. Secord, and E.C. Spary (Cambridge, New York, Melbourne: Cambridge University Press, 1996), pp. 266–286.

fluids – caloric, phlogiston, light, electrical and magnetic fluids, and so on. The fact that Deluc's scientific reputation went into a decline in the nineteenth century from which it has never recovered should not prevent us from realizing that he was generally taken seriously by his contemporaries. Most of his geological discussions were resolutely terrestrial in scope, corresponding to contemporary efforts in mineralogy, geology, hydrology, and meteorology – all of which, Deluc doubtless hoped in common with many contemporaries, might enlarge understanding of the most interesting place we know, our immediate environment, the home of all known life.[22]

Consistently with views of such wide scope, Deluc was keenly attuned to a geological outlook according much respect to systemic sorts of extraterrestrial considerations, especially the regular role of the Sun as source of heat and light. Deluc's amenability to idiosyncratic extraterrestrial agents, however, was limited, perhaps surprisingly so. He was circumspect in giving credence to the geological relevance of supposed extraterrestrial agents. In a prolonged discussion, for example, of a contemporary's suggestion that Saturn's rings had been produced by impact of a bypasser celestial object, there arose a clear opportunity to connect this hypothesis by analogy with some comparable dislocating event on Earth. But Deluc swiftly turned aside from this prospect.[23] Why did he decline to engage this chance to connect geology with extraterrestrial events? I cannot be sure. But the decision situated him within the geological mainstream, and could not have harmed his reputation among contemporaries as a responsible geologist.

---

[22] On Deluc and his works, see the article by Robert P. Beckinsale in *Dictionary of Scientific Biography*, Vol. 4 (1971), pp. 27–29; Paul A. Tunbridge, "Jean André De Luc, F.R.S. (1727–1817)," *Notes and Records of the Royal Society of London*, 1971, 26: 15–33. Deluc's geological thinking is examined at some length in Albert V. Carozzi, "La Géologie: De l'histoire de la Terre selon le récit de Moïse aux premiers essais sur la structure des Alpes et à la géologie expérimentale, 1778–1878" in *Les Savants genevois dans l'Europe intellectuelle du XVIIe au milieu du XIXe siècle*, edited by Jacques Trembley (Geneva: Éditions du Journal du Genève, 1987), pp. 203–265, on pp. 207–219. The 'environmental-mindedness' of Deluc, encompassing the spheres of earth, water and air (geology, hydrology, meteorology), may be compared with that seen in others of the time, including J.-B. de Lamarck and James Hutton as well as de Saussure. A fundamental motivation for this breadth of interests, one supposes, must be the post-Buffonian precept that living things can be understood only in terms of their conditions of existence.

[23] "Dix-neuvième lettre de M. De Luc, à M. Delamétherie, Sur l'anneau de Saturne," *Observations sur la physique, sur l'histoire naturelle et sur les arts*, 1792, 40: 101–116. The author of the speculation about the origin of Saturn's rings was Ermenegildo Pini of Milan, author of a (specific) theory of the Earth. See "Seizième lettre de M. De Luc, à M. Delamétherie. Examen de la théorie de la terre du P. Pini, & premières remarques sur la Notice minéralogique de la Daourie, par M. Patrin," *Obs. sur la phys.*, 1791, 39: 215–230.

H.-B. de Saussure was a Genevan patrician, educator, and Alpine explorer. As a naturalist and natural philosopher, he might very properly be taken as the archetypal believer in the generic Theory of the Earth, while banishing all past and current specific pretender-theories as patently inadequate. In de Saussure's scrupulously related accounts of his journeys, one repeatedly meets his honest confession of perplexity about the meaning of what he has observed, his refusal to accept facile generalizations. His own experiences, particularly in the Alps, taught him that variety is the rule, constancy is elusive. Readers and colleagues knew de Saussure as passionately interested in geological theory, as keenly focused on observations thought to be particularly pertinent to success with theory, and as highly cautious if not skeptical about an easy path from observation to theory.[24]

De Saussure's writings display his pronounced interest especially in the Earth's surface features, while making plain his readiness to seize opportunities to examine any aspect of the physical environment, such as the qualities of the atmosphere at high altitude on Mont Blanc. However, while de Saussure's geological gaze was generally so fixed on what was exposed at the Earth's surface that one hardly expects to find him very much concerned about extraterrestrial considerations, it must be admitted that in his outline for geological observing he included questions about possible geological effects of the Earth's condition in the universe, including the possibility of its being hit by a comet.

By his own account, de Saussure's "Agenda" was an elaboration of his methodical habits in preparing for observation during his travels. As published, the Agenda became a kind of comprehensive list of things that might contribute in any way to solving the great problem of establishing a theory or system of geological science. Following an order starting at the macroscopic or cosmic end of things and moving to detailed matters more likely to fall in the way of the curious traveler, the Agenda begins with a short section entitled "Astronomical principles." Of the nine items in this section, four or five might be said to be basically 'geophysical': they are related to the Earth's shape, density, rate of rotation, and the like. The others are mainly addressed to extraterrestrial issues of the systemic variety, such as the Earth's general condition as a planet, and the dependence of climate on solar radiation. Similarly systemic in character, but

---

[24] The authority on de Saussure is Albert V. Carozzi. In Addition to "Forty Years of Thinking in Front of the Alps" (note 15 above), see Carozzi's "La Géologie" (note 22 above), pp. 219–250; Albert V. Carozzi and Gerda Bouvier, *The Scientific Library of Horace-Bénédict de Saussure (1797): Annotated Catalog of an 18th-Century Bibliographic and Historic Treasure*, Mémoires de la Société de Physique et d'Histoire Naturelle de Genève, Vol. 46, 1994; and Albert V. Carozzi and John K. Newman, *Horace-Bénédict de Saussure: Forerunner in Glaciology*, Mémoires de la Société de Physique et d'Histoire Naturelle de Genève, Vol. 48, 1995.

with implications for conceivable particular events of past or future, is a point about the intersection of cometary orbits with the Earth's celestial path. "6°. Paths of comets. If it is possible that they have intersected or do still intersect the [path of the] Earth in their orbits, and what would be the effects of such a collision?" The succeeding point, also formulated as a question, addresses in guarded terms the feasibility of the Earth's origins in a cosmic smashup: "7°. If it is, let us not say probable, but possible [*on ne dit pas probable, mais possible*], that a comet in plowing into [*en sillonnant*] the Sun, detached from it the Earth and the other planets?"[25]

Within the full panorama of geological possibilities, therefore, de Saussure would not exclude Buffon's hypothesis of terrestrial origins, however unlikely it seemed. But this qualified recognition of comets' imaginable role in the Earth's history belonged in the first of 23 chapters, nearly all of which are longer than that astronomical opening; and one senses that in de Saussure's eyes all of the others are fuller of geological promise. De Saussure's comprehensive geological enumeration covered all the bases, including extraterrestrial factors. Yet the Agenda leaves the reader in little doubt that de Saussure expected essential theoretical insight about the Earth's history and condition to be drawn from inspection of stones and strata, mountains and valleys.

Conclusions

For geological scientists in the second half of the eighteenth century, systemic extraterrestrial considerations were in a sense unexceptionable; who would rule out, for example, the significance of the laws of mechanics or the rule of law in the solar system? I have insisted on the continuing importance of the traditions of the Theory of the Earth throughout the eighteenth century, and those traditions directed attention to the Earth's membership in the family of planets. What one made of those systemic extraterrestrial considerations for geology was not a simple matter, but was I think subject to growing restriction. Idiosyncratic extraterrestrial effects suffered from an increasingly bad press as the century wore on. The geological response to cosmogonical histories like Buffon's was complimentary in the sense that his schema of organizing geological knowledge in terms of historical sequences gained favorable attention, but judgments about the reliability of Buffon's special version of those events or explanations for them were often harsh. Other theories of

---

[25] "Agenda, ou Tableau général des observations & des recherches dont les résultats doivent servir de base à la théorie de la terre," in *Voyages dans les Alpes*, Vol. 4 (1796), pp. 467–[539]. Chapter I, "Principes astronomiques," is on pp. 469–470.

the Earth tinged with an idiosyncratic aroma – identified with explanations referring to unique events rather than ongoing operations – tended to fare no better.

The generic ideal of the Theory of the Earth retained much strength. One of the outstanding features in the utterances of later-eighteenth-century geologists who maintained and pursued that ideal, is their fervent empiricism. Among those in quest of The Theory of the Earth, it was not only the utterances, but the actions that called attention to themselves. It was Buffon, after all, who opened his *Histoire Naturelle* in 1749 with about as fine an expression of critically sophisticated analytical empiricism as the century was to produce. But Buffon was mainly a literary naturalist, a reader and thinker who worked in the study; his scientific status had little to do with what he had actually seen.[26] By contrast, the traveling geological naturalists of the later part of the century made reputations as observers – reputations celebrated perhaps less widely in the larger culture, but very solid within the fellowship of naturalists. There were few in these circles with greater prestige than Pallas, de Saussure, or Dolomieu. These were observers with their eyes mainly on the ground.

In short, as regards this question of extraterrestrial causes, the main patterns of geological pursuits in the second half of the eighteenth century were caught in a state of dynamic tension. On one hand the fundamental traditions of the Theory of the Earth mandated respect for the Earth's planetary status and the integrative approach to earth science that implied. On the other, however, a decline in prestige of what were thought of as speculatively-founded specific theories of the Earth, the fruits of excessive confidence in reasoned imagination, was accompanied by a corresponding rise in the repute of geological fact-gathering, seen as the antidote to the errors of those theories. There was no weakening, but rather an enhancement, of geological investigators' commitment to the generic ideal of a Theory of the Earth, an objective to be gained through inductive discovery. Finally, it needs to be recognized that these champions of the incipient geology had a stake in being recognized as scientists with legitimate methods and interpretive techniques of their own, which allowed them to be distinguished from other kinds of natural philosophers or naturalists. Sometimes explicitly, but most

---

[26] Naturalists of the day might not uncommonly have basically bookish beginnings, and still grow into empirical geologists. In a recent essay I discuss the essentially 'literary' formative stages of the scientific career of Buffon's contemporary, Nicolas Desmarest, with emphasis on the usefulness of this mode of training for his emergence as an empirical naturalist: "La genèse d'un naturaliste: Desmarest, la lecture et la nature," in *De la géologie à son histoire: Ouvrage édité en hommage à François Ellenberger*, edited by Gabriel Gohau (Paris: Comité des Travaux Historiques et Scientifiques, 1997), pp. 61–74.

often by merely tacit exclusion of extraterrestrial considerations, geological investigation toward the end of the eighteenth century was identifying the integrity of the emerging science with the distinctively Earth-bound nature of the objects of study. If a distinct geological science was to have its place among the sciences, the objects of its study were best confined to the Earth. So the ambition of creating a specialized branch of science to understand the Earth's features and their development tended to be intertwined with a concept of terrestrial autonomy.[27]

---

[27] Some of the issues raised in this essay have been taken up in their nineteenth-century context by Philip Lawrence, "Heaven and Earth–The Relation of the Nebular Hypothesis to Geology," in *History, Theology, and Cosmology*, edited by Wolfgang Yourgrau and Allen D. Breck (New York and London: Plenum Press, 1977), pp. 253–281. Lawrence argued that a proper understanding of nineteenth-century geology requires historical recognition of prevailing relations of geological thinking with physics and astronomy; these relations included particularly geologists' acceptance of (1) a general conviction among astronomers that the Earth's origin is explained by the nebular hypothesis, and (2) "the geophysical processes most physicists believed had to follow from such an origin" (p. 253). Lawrence's discussion is altogether compatible, I think, with the proposition that nineteenth-century geologists' views had already been prepared, by attitudes taking shape within geological research during the closing decades of the eighteenth century, to suppose that their own science's business had to do principally with the phenomena available on the Earth itself.

INDEX

Académie Française: XII 79
Académie Royale des Inscriptions et
  Belles-Lettres (Paris): II 12–13
Académie Royale des Sciences (Paris):
  I 340, 346–8; II 15; III 1; IV 130;
  V 6; VI 2; VII 24 n1; XI 6, 25, 32;
  XII 82–3; XIII 70, 71
Accidents, in nature (*see also* Regularities):
  I 341, 350; IX 595–614;
  XIV 379–80; XV 4
Actualism (*see also* Uniformitarian principle): I 356; VI 1; VII 4
Adams, Frank Dawson: VIII 8
Agassi, Joseph: VIII 4
Age of the Earth: *see* Time, geological
Agriculture: I 344; II 16
Alembert, Jean Lerond d': I 340 n3
Alps (*see also* Mountains; Pyrenees): III 1,
  4–5; VI 1; VII 14; XI 2–3, 22, 23,
  26–8; XII 76; XIII 73–4; XV 12
Amiens, Académie des Sciences: I 341; II 2
Anderson, Johann: I 341; XIII 67
Annales des Mines: XIII 72
Antiquarianism; antiquarian scholarship:
  II 3, 11–14, 19–20; III 8; IV 130;
  V 2, 14; VII 10; XIII 68; XIV 377
Arduino, Giovanni: III 12, 13, 14; V 7 n3;
  VIII 6
Argenville: *see* Dezallier d'Argenville
Aristotle: X 372; XV 6
Auvergne: I 339, 340, 344–9, 354–6; II 3,
  18; III 3, 4, 6; IV 129–36; V 6, 8,
  9, 11, 17; VII 13; VIII 9; IX 606,
  608, 611–12; XI 26, 30–32; XIII 71;
  XIV 378

Bacon, Francis: V 2
Baillet du Belloy, Arsène-Nicolas: XIII 72
Bailly, Jean-Sylvain: I 355; XIV 377
Ballainvilliers, Simon-Charles-Sébastien
  Bernard de (*Intendant* of Auvergne):
  I 346; IV 133, 134
Bardonnanche, David Anselme de: IV 131
Barruel, Augustin: XIV 373

Basalt (*see also* Volcanoes): I 339, 340,
  346–8, 353–5; III 6, 8–9; IV 129–36;
  VII 13, 30–31 n 23; IX 606, 608–9,
  611, 612; XII 77
Beaumont, Élie de: *see* Élie de Beaumont
Beche, De la: *see* De la Beche
Bergman, Torbern: I 354; X 375
Berlin, Academy of: VI 1
Bertrand, Philippe: XIV 373
Besson, Alexandre-Charles: IX 603; XIII 72
Bible, geological use of: VIII 6, 13–14;
  XII 76; XIV 374
Boissier de Sauvages: *see* Sauvages
Boissieu, Jean-Jacques de: III 2, 7; IV 134
Bomare: *see* Valmont de Bomare
Bonamy, Pierre-Nicolas: II 7, 12
Bordeaux (and neighboring region): I 344,
  348; II 2 n3, 17; IV 134
Borel, Pierre: II 12
Born, Ignaz von: XI 19
Boué, Ami: XI 14–16
Bouguer, Pierre: I 340 n3, 341; XIII 67
Boulanger, Nicolas-Antoine: II 18; VI 4–5;
  VII 7–11; VIII 11; XIII 67;
  XIV 376–7
Bourguet, Louis: VI 1–7; VII 3, 6–7, 10,
  23, 25 n3, 26 n7; VIII 11; IX 603,
  609–10; X 373–5; XII 78; XIII 67;
  XV 7
Boutin, Charles-Robert: I 344; II 2 n3, 17;
  IV 134–5
Boutin, Simon-Charles: IV 134–5
Boyle, Robert: II 7; XII 78
Brémond, François de: II 7
Broc, Numa: II 13
Brochant de Villiers, André-Jean-Marie:
  XI 22, 27; XIII 72, 75
Brongniart, Alexandre: VII 15–16;
  XI 11–14; XIII 72, 75
Brydone, Patrick: IX 613–14
Buache, Philippe: I 341, 342; II 12;
  XI 6–8, 22, 25, 33, 34, 36; XII 82;
  XIII 68, 70
Buch, Leopold von: IX 599; XI 14

Buffon, Georges-Louis Leclerc, Comte de:
I 355; II 10–11; III 14; V 5–8,
10–12, 14–17; VI 4, 5, 6; VII 2–7,
10, 11, 12, 14, 20–25; VIII 5, 6, 8,
11; IX 602–3, 610–11; X 374, 376;
XI 33, 35, 36; XIII 67, 70, 75, 78;
XIV 371–83; XV 5–6, 7, 13–14
Burnet, Thomas: VI 1; VIII 5; XV 7
Burtt, Edwin Arthur: VIII 3–4
Butterfield, Herbert: VIII 3–4

Caesar, Julius: II 12
Camden, William: II 12
Canivet, Jacques (instrument maker):
IV 133
Canto, Jean-Baptiste: II 4
Carozzi, Albert V.: VIII 14
Carozzi, Marguerite: VIII 14
Cartesianism: *see* Descartes
Catcott, Alexander: VIII 14
Caylus, Anne-Claude-Philippe de Tubières
de: II 13
Central heat: *see* Heat internal to the Earth
Chambers, Ephraim: XII 79
Champagne (*see also* Soulaines): I 344,
348; II 14
Charpentier, Johann Friedrich Wilhelm:
XI 9, 10
Chemistry (*see also* Rouelle): II 5; VII 18,
32 n26; XII 77, 80–83; XIII 70, 71;
XV 10
Chronology: *see* Epochs; Time
Clerk of Eldin, John: XI 32
Collections: *see* Mineralogy
Collège de Troyes: *see* Oratory
Collège Royal, Collège de France:
XIII 72, 77
Collier, Katharine Brownell: VIII 4
Collini, Cosimo Alessandro: X 375
Collot, Mademoiselle: III 4
Comets: V 8; VIII 3; XIV 374, 375–6;
XV 2–3, 12–13
Condamine: *see* La Condamine
Condorcet, Marie-Jean-Antoine-Nicolas
Caritat, Marquis de: I 340 n3;
VII 24 n1
Continents (*see also* Oceans): I 342, 352;
II 11; VI 5; VII 3, 4, 7–10, 20;
X 376; XI 26, 29, 33, 35;
XIV 375
Copernicus, Nicolas; Copernicanism:
VIII 11; XV 6
Cordier, Louis: XIII 72, 75
Crystals, crystallization, crystallography:
I 351; III 9; XII 76; XIII 76, 78

Cuvier, Georges: I 355; VII 15–16;
XI 11–14; XIII 72, 77, 78; XV 8

Dailley (*ingénieur-géographe du Roi*):
IV 129, 131, 132, 133, 134
D'Alembert: *see* Alembert
Darcet, Jean: XII 81; XIII 71, 72
Daubenton, Louis-Jean-Marie: VI 4;
XIII 70, 72
De la Beche, Henry: XI 14, 15, 17
Della Torre, Giovanni Maria: *see* Torre,
Giovanni Maria della
Deluc, Jean-André: I 353; VII 15, 16;
XII 78; XIII 71; XV 10–11
Deluge(s): VI 4–5; VII 10; VIII 3; XII 84;
XIV 376; XV 3
Denudation: I 339, 344, 345, 348–50, 352;
II 12; III 10, 11, 14, 16; V 11; VI 4,
6; VII 3, 10–11, 23, 29 n22; XI 30,
31; XII 76, 77
Descartes, René; Cartesianism: II 9; VIII 6,
7; IX 600; XI 36; XV 5, 7
Desmarest, Anselme-Gaëtan: II 5
Desmarest, Nicolas (*see also* Géographie
physique): I 339–356; II 1–20;
III 1–16; IV 129–36; V 3–9, 11–12,
14, 17; VI 5–6; VII 6, 11–13, 16;
VIII 9–11; IX 605, 611–13; X 373,
375, 376; XI 26, 30–32; XII 81;
XIII 69, 71, 77; XIV 373, 374 n10,
378–9
Dezallier d'Argenville, Antoine-Joseph:
XII 79
Dictionary of Scientific Biography: VIII 9
Dietrich, Philippe-Frédéric de: XIII 70, 71
Dijon, Académie de: V 7
Dolomieu, Déodat de: I 354; VII 15; VIII 8;
X 375; XI 26; XIII 71, 72, 73, 74–7;
XV 10 n21, 14
Dortous de Mairan, Jean-Jacques: I 355;
XIV 376
Drury, Susannah: I 346
Dufay, Charles-François de Cisternay: VI 2
Dufrénoy, Armand: XIII 72
Duhamel, Jean-Baptiste: XIII 72
Dupain (artist): IV 133

Earthquakes: I 341–3; II 8; IX 610; XII 80,
84; XIII 68; XIV 375, 381
École des Mines: XIII 72, 75
École des Ponts et Chaussées: XIII 71
Edwards, George: XII 79
Electricity: II 7–8; XIV 381
Élie de Beaumont, Léonce: X 377; XI 15,
18, 36, 39, 40; XIII 72

Ellenberger, François: V 12; VII 1, 15; XIII 69; XV 7
Encyclopédie: I 343–4, 347; II 6, 10; VI 4, 5; VII 10, 11
Engineers: *see* Mining
English Channel: I 340–41; II 2, 5, 11–12; XI 7–8, 22, 25
Enville, Marie-Louise de La Rochefoucauld, duchesse d': I 340 n3; III 1
Epochs; *Époques* (*see also* Historical Geology): V 1–17; VII 13; XIV 371–83
Erosion: *see* Denudation
Erratic rocks: III 11
Etna: I 343; IX 606, 608, 613

Falconet, Camille: II 13
Farey, John: XI 14
Faujas de Saint-Fond, Barthélemy: XIII 72, 73, 75–7; XV 7–8
Feller, François-Xavier de: XIV 373
Ferber, Johann Jacob: III 13; XI 19; XIII 75
Fieldwork: II 3, 14–20; IV 129–36; VII 13, 18; VIII 13–14; XIV 379, 383; XV 8
Fleuriau de Bellevue, Louis-Benjamin: XIII 75
Flood: *see* Deluge
Foncemagne, Étienne Lauréault de: II 12–13
Fontenelle, Bernard le Bovier de: V 14; XIII 69
Fortis, Alberto: III 8, 12, 15; V 7 n3
Fossils: I 351; III 11; VI 2; VII 15, 16, 22; XI 11, 14; XII 76, 79, 83, 84; XIII 67, 68, 69, 70
Foucault, Michel: XIII 68
Franklin, Benjamin: II 7
Freiberg (Saxony): X 370; XI 9; XII 75
Füchsel, Georg Christian: VIII 6
Fusion: *see* Crystals

Geikie, Archibald: VIII 5–10
Genssane, Antoine de: XIII 71
Geognosy: XI 9, 12, 14; XV 9
Géographie physique (Desmarest): I 350–52; II 11; III 2–3, 12–16; VII 13; X 373; XIV 373
Geography, physical (*see also* Denudation; Mountains; Oceans; Physiography): I 342; IX 600; XII 77; XIII 68, 77, 82–3
Giant's Causeway: I 346; IX 606
Gilbert, William: II 7
Gillet-Laumont, François-Pierre-Nicolas: XIII 75
Gillispie, Charles Coulston: I 353; VIII 8–9

Giraud Soulavie: *see* Soulavie
Gironde: *see* Bordeaux
Gmelin, Johann Georg: XIII 67
Gohau, Gabriel: VII 15, 18; XIV 371, 379
Gonthier: I 345 n20; II 15
Goujet, Charles-Pierre: II 12–13
Goussier, Louis-Jacques: XIV 373; XV 5–6
Granite: I 351; III 9
Greene, John C.: VIII 9
Greenough, George: XI 14
Grosley, Pierre-Jean: I 342 n8, 345 n20; II 12–13, 15, 16, 17; III 4–5
Gua de Malves, Jean-Paul de: XIII 71
Guettard, Jean-Étienne: I 339, 340, 353; II 16, 18; IV 130; VIII 8, 9; IX 611; XI 4–9, 33; XIII 67, 69, 70
Guillot-Duhamel, Jean-Pierre-François: XIII 71, 72
Guyenne: *see* Bordeaux

Haidinger, Karl: X 375
Haller, Albrecht von: VI 5
Halley, Edmond: II 12; XI 36
Hamilton, William: III 14; IX 613; X 375; XII 76
Hassenfratz, Jean-Henri: XIII 72
Hauksbee, Francis: II 7, 9, 10; VII 11
Haüy, René-Just: XIII 71, 72
Heat internal to the Earth: I 355; XII 76–7; XIV 373, 374–6, 381; XV 4–5
Hellot, Jean: XIII 71
Hénault, Charles-Jean-François: II 12–13
Histoire naturelle: *see* Buffon
Historical geology (*see also* Epochs): I 347–52; III 11–12, 15–16; IV 130; V 1–17; VI 3; IX 613; X 371–2, 377; XI 1, 9, 11–15; XII 84; XIV 371–83
Historical scholarship: *see* Antiquarianism
Holbach, Paul Henri Thiry, Baron d': XII 78–9; XIII 70
Hooke, Robert: VIII 6; IX 608
Hooykaas, Reijer: I 349
Humanist scholarship: *see* Antiquarianism
Humboldt, Alexander von: X 377; XI 36
Hutton, James: I 352, 355; VII 15, 17; VIII 2, 3, 8; IX 596–8, 609, 613, 614; XI 32; XII 75–6, 77; XIII 77; XV 7

Iceland: I 341; V 12
Industry (*see also* Manufactures; Mining): I 340, 345; II 16–18, 20; IV 135; V 6; XI 4; XIII 68, 71
Institut de France: V 6; XIII 71
Internal heat: *see* Heat internal to the Earth

Italy: I 347–8; II 16; III 1–16; IV 131;
    IX 605–9, 613–14

Jameson, Robert: XI 9
Jardin du Roi (*see also* Muséum): V 6;
    XIII 70, 72
Jars, Antoine-Gabriel: XI 19–21; XIII 71
Joly de Fleury, Guillaume-François: I 342
Journal de physique: XIII 72, 77
Journal de Verdun: II 7, 12
Journal des Mines: XIII 72
Jussieu, Antoine de: XIII 68
Jussieu, Bernard de: II 15; VII 11

Kircher, Athanasius: I 342; XI 2–4
Klaproth, Martin Heinrich: I 354
Koyré, Alexandre: VIII 3
Kuhn, Thomas S.: VIII 4

La Condamine, Charles-Marie de: III 15;
    XIII 67
Lacurne de Sainte-Palaye, Jean-Baptiste de:
    II 12–13
La Grange, N. de: IX 605
Lalande, Joseph-Jérôme Le Français de:
    IX 607
Lamanon, Robert de Paul de: VII 16;
    X 375; XIII 71
Lamarck, Jean-Baptiste de: VII 17;
    XIII 72, 76–7
Lamétherie, Jean-Claude de: I 353; VI 4, 5;
    X 375; XIII 72, 73, 76–7
Landforms: *see* Denudation; Geography,
    physical; Mountains; Oceans;
    Physiography
La Rochefoucauld d'Enville, Louis-Alexandre de: I 347; III 1–2, 6–7, 13, 14;
    IV 131
Laudan, Rachel: VII 7, 18; VIII 14
Launay, Louis de: X 375
Laurent, Goulven: XIV 378 n19
Lavoisier, Antoine-Laurent: VII 13–14, 24,
    31–2 n24–6; VIII 12, 13; XII 81
Laws, natural (*see also* Regularities): II 9;
    V 15, 17; VI 1–7; VII 1–36; VIII 3,
    11; IX 600, 612; X 372–3; XI 33;
    XII 84; XIII 67, 79; XIV 379; XV 3
Lefèbvre d'Hellancourt, Antoine-Marie:
    XIII 72
Lehmann, Johann Gottlob: VIII 6; XIV 377
Leibniz, Gottfried Wilhelm von: V 14; VI 1;
    VIII 6; XV 7
Lelièvre, Claude-Hugues: XIII 71, 72, 75
Lhuyd, Edward: XII 79
Limoges: I 344 n16; IV 132

Lipari Islands: I 343
Loménie de Brienne, Étienne-Charles de:
    I 343 n3
Longuerue, Louis du Four de: I 343 n11
Ludot, Jean-Baptiste: I 340 n3
Lyell, Charles: VIII 1–6, 10, 11, 14; XI 15;
    XII 84–5; XIII 78

Maclure, William: XI 15
Maillet, Benoît de: VIII 6; XIII 67; XV 4–6
Mairan: *see* Dortous de Mairan
Malesherbes, Chrétien-Guillaume de Lamoignon de: I 340
Manufactures, inspection of: I 344 n16;
    II 17–18, 20; IV 132, 134
Maps, mapping: I 346–7; IV 129–36;
    VII 7–9; IX 612; XI 1–40; XII 77;
    XIII 68, 71
Marivetz, Étienne-Claude: XIV 373;
    XV 5–6
Marsigli, Luigi Ferdinando: III 12–13; XI 4,
    22, 24, 26, 29
Mayer, Jean: XII 81
Mendes da Costa, Emanuel: XII 79
Meteorology: VII 3, 4, 23; XI 36; XV 11
Michell, John: VIII 8
Mineralogy, mineralogists, mineral collections: I 354; II 6, 14; III 7; IV 132,
    134; VII 2, 3, 22; XI 4–11, 15–16;
    XII 77, 78–80, 82, 84; XIII 67, 70,
    72, 74; XV 9, 11
Mining, mining engineers: XI 2, 9, 15,
    19–21; XIII 68, 71, 72
Monnet, Antoine-Grimoald: XII 81
Montet, Jacques: I 347
Moro, Anton Lazzaro: I 352; III 12; VIII 5;
    IX 608
Mountains, mountain ranges (*see also* Alps;
    Pyrenees): I 342, 343, 352; II 8, 11;
    III 4–5, 12–13; IV 130; VI 2, 3, 5;
    VII 3, 7, 10, 13, 14, 20–22, 24 n1;
    IX 603, 613; X 374, 376–7; XI 3,
    19, 22, 23, 25, 26, 28, 32–34, 36–40;
    XII 79, 82, 84; XIII 67, 74–6;
    XIV 380, 381; XV 10 n21
Muséum National d'Histoire Naturelle (*see
    also* Jardin du Roi): XIII 72, 76
Muthuon, Jacques-Marie: XIII 72

Neptunism: I 339, 346, 352–6; IV 130;
    VIII 10; IX 610; XII 76–7; XIV 376
Newton, Isaac; Newtonian science; Newtonianism: II 9; VII 11, 17; IX 600;
    X 375; XI 36; XV 5
Nollet, Jean-Antoine: II 5, 7

Observations sur la physique (see also *Journal de physique*): XIII 72
Oceans, phenomena of (*see also* Continents; English Channel; Tides): I 341, 342, 346; II 2; III 10, 11, 13, 14; VI 2, 5; VII 3–4, 7, 10–11, 14, 20–23, 24 n1; VIII 12; IX 610; XI 22, 25, 26, 29; XII 84; XIV 373, 375, 378, 381; XV 5
Omalius d'Halloy, Jean-Baptiste-Julien d': XI 14
Oratory, Oratorian Collège de Troyes (*see also* Troyes): II 4, 12, 13; IV 130
Ospovat, Alexander M.: X 369–70, 373

Palassou, Pierre-Bernard: X 375; XI 32, 36, 37; XIII 71
Pallas, Peter Simon: VI 5; VII 24 n1; VIII 3, 6, 8, 14; X 375; XI 22, 26; XIII 75; XV 3, 14
Paracelsianism: XII 81–2
Paris: *see* Académie Royale des Inscriptions et Belles-Lettres; Académie Royale des Sciences
Paris Basin: VII 16; XI 11–12;
Passeri, Giambattista: V 7 n3
Pasumot, François: I 345 n20, 346–7; IV 129–36
Pennant, Thomas: XII 79
Periodization of Earth's history: *see* Epochs
Peru: I 341
Pezet (*ingénieur-géographe du Roi*): IV 131, 133, 134
Physiography (*see also* Denudation; English Channel; Geography, physical; Mountains; Oceans): I 339, 348, 350; III 10–11, 13; IV 130, 135; V 11; VI 6; VII 4, 20–22; XII 77, 78, 82–3; XIV 373, 374–6, 380
Picot de Lapeyrouse, Philippe: XIII 71, 75
Pictet, Marc-Auguste: XIII 75
Plutonism (*see also* Heat; Volcanoes): I 354–6
Pott, Johann Heinrich: XII 79
Prévost d'Exiles, Antoine-François: XIII 68
Primary rocks, formations (*see also* Secondary rocks): I 344, 351; IX 602; XI 36; XIV 374, 377
Pyrenees (*see also* Alps; Mountains): III 4, 5; XI 32, 36, 37, 38; XIV 379

Ramond de Carbonnières, Louis: XI 36, 38
Rappaport, Rhoda: I 353; II 14; VIII 9, 13; IX 598–600
Raspe, Rudolf Erich: I 347; IX 608

Ray, John: IX 608
Réaumur, René-Antoine Ferchault de: VI 1; XIII 67, 69
Regularities in terrestrial phenomena (*see also* Laws, natural): I 350; II 7, 9–11; III 10; V 13, 15, 17; VI 1–7; VII 1–36; VIII 11; IX 595–614; X 370–78; XI 36; XII 380; XV 4
Religion: *see* Bible
Revolutions (*see also* Uniformitarian principle): I 345; V 13; IX 598–605
Rivers, phenomena of (*see also* Denudation; Springs): I 342, 344–5; II 12; III 4, 10–11, 13, 14; VI 6; VII 3, 10, 11–13, 21, 23, 29 n22; X 377; XI 3, 19, 22, 31, 32
Roger, Jacques: V 6, 15; VII 7; VIII 11, 13; XIV 371, 376
Romé de l'Isle, Jean-Baptiste Louis: I 354; XIV 373
Rouelle, Guillaume-François: I 351; II 5, 15; III 12; VII 11, 14; VIII 11; XII 80–82; XIII 69, 70; XIV 377
Rouen, Academy of: II 12
Royal Academy of Sciences, Paris: *see* Académie Royale des Sciences, Paris
Royou, Thomas-Marie: XIV 373
Rudwick, Martin J.S.: V 14; XIII 78

Sage, Balthasar-Georges: XIII 70, 71, 72
Sainte-Palaye: *see* Lacurne de Sainte-Palaye
San Sebastián: I 345
Santorini: IX 606
Saussure, Horace-Bénédict de: I 354; VI 5; VII 14–15; VIII 3, 6, 8, 14; IX 603; X 375; XI 26, 28; XII 76; XIII 73–6; XV 7–8, 10, 12–14
Sauvages, Pierre-Augustin Boissier de: XIII 67
Saxony (*see also* Freiberg): I 346; XI 9–10
Scheuchzer, Johann Jacob: XI 19, 23, 26; XII 79; XV 7
Scripture, sacred: *see* Bible, geological use of
Seas: *see* Oceans
Secondary rocks, formations (*see also* Primary rocks; Tertiary rocks): I 344, 351; IX 602; XI 36; XIV 377
Seneca, Lucius Annaeus: II 12; III 13 n33; IX 597, 605–6
Shorelines: *see* Oceans
Smith, William: XI 11, 13, 14
Société Géologique de France: XIII 72

Somner, William: II 12
Soulaines (Champagne): II 8, 15
Soulavie, Jean-Louis Giraud: IX 603, 612–13; X 375
Spallanzani, Lazzaro: III 12
Springs (*see also* Rivers): I 343; II 8, 11; III 11; XI 2–4; XII 80; XIII 68
Steno, Nicolaus: III 5, 13
Strange, John: IX 605–9, 612, 613
Stratigraphy, strata (*see also* Geognosy): II 7; III 8, 11, 13–14; VI 3; VII 3, 7, 13–16, 18, 22; IX 602; X 377; XI 1, 6, 9, 11–17; XII 76, 78, 84; XV 10
Suite de la Clef: *see Journal de Verdun*
Switzerland: I 346; III 1; VI 1; XI 3
Systems: *see* Theory of the Earth

Tacitus: II 12
Targioni Tozzetti, Giovanni: III 5, 12, 13, 14, 15
Technology: *see* Industry; Manufactures
Telliamed: *see* Maillet
Tertiary rocks, formations: IX 602; XI 15, 18
Theory (or Theories) of the Earth: II 6; V 3, 4, 11–17; VI 1–3, 5–7; VII 2–7, 10, 13, 16–18; VIII 1–14; IX 600, 609–10; X 375, 378; XI 33–40; XII 77, 81–2; XIII 67, 73–4, 77; XIV 371–83; XV 1, 4–8, 10–14
Tides (*see also* Oceans): VII 4, 22–3; VIII 12; XIV 375
Tillet, Mathieu; II 2 n3
Time, geological (*see also* Epochs; Historical geology): I 351; V 1–17; XII 76, 78; XIV 377–8
Torre, Giovanni Maria della: III 14; IX 607
Troyes: I 340 n3, 342 n8, 345 n20; II 4, 12, 15, 16; IV 131
Trudaine, Daniel-Charles: I 340, 345; II 17; IV 134
Turgot, Anne-Robert-Jacques: I 340 n3, 344 n16, 345 n20; II 16; IV 132

Twyne, John: II 12

Uniformitarian principle; uniformitarianism (*see also* Actualism; Revolutions): I 339, 349–50, 356; V 11, 13; IX 610; XV 3

Valmont de Bomare, Jacques-Christophe: V 12–15; VI 5; XII 79; XIII 70; XV 3
Varenius, Bernhardus: II 12
Vauquelin, Nicolas-Louis: XIII 72
Venel, Gabriel-François: XII 82
Vesuvius: I 343; III 7, 14; IX 606–8, 613
Volcanoes, volcanic phenomena (*see also* Auvergne; Basalt; Etna; Vesuvius): I 339–52; III 3, 6–9, 11, 14–15; IV 129–36; V 6, 11, 17; VII 21; VIII 9, 10; IX 595–614; XI 26, 30–32; XII 77, 78, 80, 84; XIII 74, 75–6; XIV 375, 378, 380–82
Volvic: I 340
Vosges: II 16; III 4
Vulcanism (*see also* Neptunism; Volcanoes): I 339, 352–6

Walker, John: IX 603; X 375
Wallerius, Johan Gottskalk: I 354; IX 603–4
Watelet, Claude-Henri: II 13, 16, 17
Webster, Thomas: XI 14
Werner, Abraham Gottlob; Wernerian theory (*see also* Neptunism, Geognosy): I 339, 353–5; VIII 2, 3, 9, 11, 14; X 370; XII 75; XIII 77; XV 7
Whiston, William: VI 1; VII 22; VIII 3; XV 7
Winckelmann, Johann Joachim: III 8
Winds: *see* Meteorology
Woodward, John: VI 1; VII 7, 15, 22; VIII 5; XV 7

Zittel, Karl Alfred von: VIII 5, 8